Scientific Culture and the Making of the Industrial West

SCIENTIFIC CULTURE AND THE MAKING OF THE INDUSTRIAL WEST

Margaret C. Jacob

New York Oxford
OXFORD UNIVERSITY PRESS
1997

Oxford University Press

Oxford New York
Athens Auckland Bangkok Bogotá Bombay Buenos Aires
Calcutta Cape Town Dar es Salaam Delhi Florence Hong Kong
Istanbul Karachi Kuala Lumpur Madras Madrid Melbourne
Mexico City Nairobi Paris Singapore Taipei Tokyo Toronto

and associated companies in

Berlin Ibadan

This book is a substantial revision of *The Cultural Meaning of the Scientific Revolution*, Copyright © 1988 by Alfred A. Knopf, Inc.

Published by Oxford University Press, Inc.
198 Madison Avenue, New York, New York 10016

Oxford is a registered trademark of Oxford University Press

Library of Congress Cataloging-in-Publication Data

Jacob, Margaret C., 1943–
Scientific culture and the making of the industrial West / Margaret C. Jacob.
p. cm.
Substantial rev. ed. of: The cultural meaning of the scientific revolution. 1988.
Includes bibliographical references and index.
ISBN-13 978-0-19-508220-3

1. Science—Social aspects—History. 2. Science—History.
3. Science and industry—History. I. Jacob, Margaret C., 1943–
Cultural meaning of the scientific revolution. II. Title.
Q175.5.J3 1996
306.4'5'094—dc20 96-26493

In memory of
Margaret O'Reiley Candee
(1906–1996)

Contents

Acknowledgments

When the press asked me to do a new edition of *The Cultural Meaning of the Scientific Revolution* (New York, Knopf-McGraw Hill, 1988) I had no idea that it would turn into a major rewrite. Nancy Lane deserves credit for prodding a reluctant author to think again. Research was aided by National Science Foundation grant no. 9310699, which gave me access to the labor of a number of talented graduate students: Mirjam van Tiel and Willeke Los in Dutch history; Jeff Horn and David Smith in French archives; and most recently, Dale Bowling in French history, Axel Utz in German, and Lizabeth Zack on the French colonies. Professor Jeff Horn, now at Stetson University, and I have continued the collaboration. David Smith worked in the Montpellier archives especially. My research from Delaware to Vizille was underwritten by the grant and the facilitation provided by Ronald Overman of the foundation. The grant continues to fund an ongoing project on the cultural factors in early industrialization in both British and Continental archives. Librarians everywhere helped, and my special thanks go to the Birmingham City Library; to Mme. M. Deschamps at L'Ecole des Ponts et Chaussées in Paris; state archivists in Rotterdam, Liège, Antwerp, Troyes, and Lyon; and also the Bakken Library in Minneapolis. The librarians at the Rare Book Room at Van Pelt Library, University of Pennsylvania, have made every chapter in this book easier to write.

Professional and personal debts need acknowledgment; repayment is beyond my means. Advice came from Joel Mokyr and Alice Amsden, J. R. Harris especially, Eric Robinson, Joyce Appleby as always, and Wijnand Mijnhardt on matters Dutch. Betty Jo Dobbs who died in 1994 talked about many of these issues with me—a missed friend. Lynn Hunt makes life a joy and deserves a lifetime subscription to Michelin's red guide, without which no archive anywhere should be visited. Zekini merits longer afternoon walks in the park and more treats for having waited so patiently under so many different desks in several countries.

My mother, Margaret O'Reiley Candee, died on the day that the proofs for this book arrived. Although she had never finished secondary school in rural Ireland, she read and commented upon all that I published. No voice will ever replace hers.

M. C. J.

Le Bourget-du-Lac
Summer 1995

Abbreviations

AD	Archives départementales, France
AN	Archives nationales, Paris
BCL	Birmingham City Library, Birmingham, U.K.
ECAM	L'Ecole centrale des arts et metiers, Paris
ENPC	L'Ecole nationale des ponts et chaussées, rue des Saints-Pères, Paris
JWP	The James Watt Papers, acquired in 1994, Birmingham City Library

Scientific Culture and the Making of the Industrial West

Introduction

This book travels back and forth across the English Channel and ranges through many countries. It seeks to understand the cultural origins of an international phenomenon that by 1800 had begun in northern and western Europe: the industrialization of manufacturing, mining, and transportation. A new scientific understanding of nature preceded mechanized industry and most important, assisted in its development. So Part I of this book looks at seventeenth-century science in the wake of Copernicus, not from the scientists' perspective, but mostly as educated people would have understood it. Part II then explores what science had to do with making power technology (like the steam engine) more thinkable, even more doable. Wherever possible the method employed is comparative. This method permits an examination of how science fitted into different cultural contexts from seventeenth-century England to late eighteenth-century revolutionary France.

Comparison also allows the distinctive features of a cultural system to stand out. Put another way, the law of universal gravitation is the same in every culture, but its discovery, its use, and its application are the work of human beings encoded with values, entwined within social networks, decorated with symbols of status, people whose ideas are glorified or forbidden by religious beliefs and practices. Thus the story to be told about eighteenth-century people only begins with the invention of modern science, with the conceptual or intellectual foundations of science from Copernicus and Galileo to Newton. It then proceeds to explore the complex differences in the way the mechanization of nature was absorbed within the major language groups to be found in Europe and the American colonies. The English, French, and Dutch—with a quick look at the Germans and the Italians—occupy center stage in this account of the deployment, the "placing" of science within formal and informal settings. The social and the cultural are inextricably linked.

In *The Cultural Meaning of the Scientific Revolution* (1988), I explored many of the same themes found in this book. This one should be seen as its intellectual heir. But when asked to do a new edition of *The Cultural Meaning*, I wanted to write an expanded, more nuanced version, reflecting my continuing thought and research. Both books describe the new science in some detail; this book gives greater attention to the eighteenth century within a comparative framework. The new and internationally circulated science that begins with Copernicus (1543) and culminates in Newton's *Principia* (1687) performed many intellectual, ideological, and utilitarian tasks. They differed depending on cultural context and national circumstances. In Catholic Europe the new science could only be used selectively because the Church objected to various aspects of its teachings. In Protestant Europe the new science could be fashioned into arguments that supported authority in both politics and religion. In all places new knowledge about nature arose within matrices of other knowledge systems: within mental universes that also held theological, philosophical, social and political, artistic, and increasingly global and ethnographical information. So, for example, the Renaissance style of art that privileged realism contributed profoundly to Galileo's ability to imagine valleys and mountains on the moon when in fact all he could see in his telescope was shadows.[1]

By 1700 scientific knowledge could offer uniform and universal knowledge about nature, and it could be made widely available because it was published in languages accessible to well-educated Euro-Americans, either Latin or French. Yet precisely at that moment when these extraordinary conceptual tools came "on the market," the marketplace had dramatically expanded. After the mid-seventeenth century as Europeans traded, explored, conquered, or enslaved in new places amid new peoples, their mental universe became more complex. Non-Western peoples challenged ingrained assumptions about human nature or the universality of belief in the monotheistic deity. For the same elites who consumed science commerce also brought unprecedented supplies of capital and encouraged applied science, which in turn promoted inventiveness. By 1780, first in Britain, capital and ingenuity promoted industrial development, the application of machine technology to mining, transportation, and manufacturing. After 1700 and largely before 1800 all Western peoples discovered more worlds, natural, geographic, technological, cultural, or simply human than had ever been the case before or since.

Industrial development occurred first in Britain for reasons that had to do with science and culture, not simply or exclusively with raw materials, capital development, cheap labor, or technological innovation. How scientific culture was used or understood in European settings differed, however, with local contexts and circumstances. In some places and not in others mental shifts led to entrepreneurial activity and to the more rapid application of power technology. Discovering the shifts requires a more comparative approach to the immediate origins of Western industrialization than I had been able to offer in *The Cultural*

Meaning. The approach also means standing back somewhat, trying to be a visitor from far away and not judging so much as asking about how those cultures worked. What made the British absorb and use science—invent a culture of practical science—that was different from what can be seen in France? What did the progressive Dutch do with Newtonian science in their educational system? These are the kinds of questions that permit a comparative method to flourish.

The perspective found in this book also treats culture as a structure in its own right, found in the mind, but also encoded in the objects available to people or invented by human ingenuity as displayed in this period largely in mechanisms and mathematical formulae.[2] Then and now a person's cultural makeup engages in a dialectic with the world, a complex shifting and altering depending on new as well as habitual experiences. For example, James Watt, who invented the modern steam engine, brought to his workshop his religiously inspired habits of disciplined work and the profit motive. These were designated as the natural impulses of free-born Englishmen. He also brought mechanical and mathematical knowledge and manual skill to what was about two years of intense activity spent modifying an old engine and then building a new one. Once built, however, the engine changed him and his family. It is not sufficient to say that the engine passively carried mechanical culture embedded in the motions of its noisy beam and the creaking of its valves. Anyone who worked with it also became its servant; all who did sought in their way to keep abreast of it or to master it. The business side of the engine made Watt into a fretful capitalist; it also turned him into a gentleman of science who wrote, spoke, dressed, and lived differently from the young man we encounter in his letters from the 1750s. Imagine the transformative effects the machine could have on the coal men who hour by hour fed it or their overseers who watched every movement of men and machine. Then when applied to cotton manufacturing the machine changed the work habits, discipline, salary, family life, leisure time, and expectations—the cultural universe—of women as well as men.[3] The kind of people who walked around, touched, or understood engines and mechanical devices from the time of Galileo to the age of steam receive the most attention in this book.

Like its predecessor, this new book starts with the conceptual roots of mechanical science, going back to the new scientific learning largely of the seventeenth century and briefly examining its integration into Western consciousness. In the first four chapters of Part I, readers of *The Cultural Meaning* will recognize many arguments, restated, reprinted, or compressed. Then the focus and argument shift. The elements of the natural world encoded in science were not peripheral to industrialization and Western hegemony; rather they were central to it. Neither, I am suggesting, would have occurred in the circumscribed mental universe that postulated a finite universe, an earth-centered cosmos, a natural order animated by spirits, observable but not mathematicized and mechanized. Once the book puts Newtonian mechanics and its diffusion in the Enlightenment in place, the discussion turns to the absorption of applied mechanics in particular social settings. Two rather unexpected figures emerge as fascinating and central: the entrepreneur and the engineer, the key figures in the eighteenth-century development of mechanized industry.

Because the science of mechanics in all its branches became so central to generating industrially usable knowledge, the book looks in some detail at the post-*Principia* deployment of mechanics by entrepreneurs and engineers. Wherever possible I try to make porous the boundaries between what people then called natural philosophy and the useful arts, science and technology to us. Pick up any English language textbook in Newtonian science from after 1700—books that were critically important in making Newton's science intelligible—and just try to separate "pure" from "applied." Our modern categories do not work. What we call technology was part and parcel of what most Newtonian practitioners called mechanics.[4] Newton may not have thought in such applied terms, but his followers did.

So, there is no point spinning our wheels seeking anachronistic distinctions. The more interesting and larger issue should be: Did the new science integrate differently in the British social and cultural landscape from what happened in western Europe during the eighteenth century? The answer is yes; more important, the differences attributable to the various regional or national settings help explain relative industrial progress or retardation.[5]

The comparative approach to the industrial applications of scientific learning throws new light on the single most important application derived from it. The late eighteenth-century application of scientific knowledge and experimental forms of inquiry to the making of goods, the moving of heavy objects whether coal or water, and the creation of new power technologies dramatically transformed human productivity in the West. But the transformation happened more rapidly or belatedly in certain regions and countries than in others. In Britain the rate of productivity grew between 1800 and 1860 at about three times the rate for 1700 to 1760; from 1760 to 1801 the rate was about twice what it had been in the preceding period.[6] Put another way, there is a remarkable cultural story to be told, a mental landscape to be painted, which may help explain why the British were the first to industrialize. By 1750 roughly the same scientific knowledge was available in all the major languages, but it was taken up differently, hence assigned different meanings, in different countries and regions. It was "packaged" differently. In other words, the *context* of discovery or dissemination is very important in the history of science and its role in early industrialization.

Note that I am referring to the history of science, and not simply or separately about the history of technology. One of the strategies running through this book seeks to pry science out of the pristine sanctuary reserved for it by a previous generation of historians. Once cloistered, "pure science"—they said—had nothing to do with industrialization. This book adds its voice to a growing chorus that integrates science with worldly concerns. In this case the worldly are entrepreneurs and engineers possessed of practical scientific acumen and eager to profit from the advantages it offered them. Indeed I shall be arguing that, from at least as early as the mid-seventeenth century, British science came wrapped in an ideology that encouraged material prosperity.

Isaac Newton (1642–1727), the founder of modern mechanics. (Portrait from a private collection.)

Getting at the cultural, rather than dwelling solely on the economic side of modern industrial society, requires that science be seen socially. The eighteenth century was the critical moment for both science and economic development. In that century literate Westerners embraced science as never before. First in Britain, then gradually throughout Western Europe, they learned it in schools and lecture halls; they picked up its contents from general textbooks; they read about scientists and their exploits in newspapers and journals; they came to *believe* in science and its power. Governments sent spies—generally to Britain—to find out the latest technological breakthroughs in manufacturing or mining. By the early nineteenth century ministers of state encouraged the teaching of science and mathematics in grammar and secondary schools for both boys and girls. An international competition for technological development had begun and continues to this day.

By the final decades of the eighteenth century, the British also burst through the labor barrier, particularly in textiles. Entrepreneurs applied water, machine, and steam technologies, rather than simply the age-old methods of hand or horse labor, to manufacturing, mining, and transportation.[7] The resulting profits suddenly gave a new British elite a place in international power politics that ended only with World War II. By the 1780s French ministers of commerce and industry believed that one element in British success lay precisely in the ability to be inventive in the sciences.[8] The Belgians (in mining and some textiles) and the Swiss (in watchmaking) were not far behind the British in industrial development, but the French only began to industrialize in earnest after 1800; the Dutch, Germans, and Americans on a national scale were later still.

These two related transformations, the unprecedented assimilation of science and the beginnings of the Industrial Revolution, need to have their relationship sketched, however briefly. We need to recapture what contemporaries living with both phenomena understood implicitly. When an early nineteenth-century scientific lecturer compared the achievements of French and British science in the previous century, he compared the "pure" French scientists, Laplace and d'Alembert, to the practical engineers, James Watt and John Smeaton, who are "no less distinguished for their success in improving the practice of the useful arts and manufactures." [9] He did not even use the terms "pure" and "applied." When a young English day laborer and gardener rose in the world after the Napoleonic wars he did so by attending lectures in natural philosophy and chemistry, becoming knowledgeable in engineering, and then traveling the world only to wind up in Boston where he could make instruments to illustrate the various sciences. There he put in motion the first Mechanics Institute so that other young men could travel the same path to upward mobility.[10]

Looked at globally and comparatively, and then put in human terms, the most important cultural meaning to be extracted from the Scientific Revolution—by no means the only meaning—lay in the creation first in Britain by 1750 of a new person, generally but not exclusively a male entrepreneur, who approached the productive process mechanically, literally by seeing it as something to be mastered by machines, or on a more abstract level to be conceptualized in terms of weight, motion, and the principles of force and inertia. Work and workers could also be seen in these terms, and the brutal human costs of early factory life are not unrelated to the employers' and engineers' ability to think mechanically. In so doing they reduced manufacturing costs by using machinery in place of people. Little wonder that by 1800, rather than adopting a cynical and defeatist detachment from science, women taught it in their academies for elite girls while radical reformers, seeing that machines would only grow in importance, allied with workers who understood the potential of applied mechanics and sought to make the subject their own. The novelist, Elizabeth Cleghorn Gaskell, presented in 1848 a fictional portrait of scientifically minded weavers in Manchester who worked with Newton's *Principia* open at the loom.[11]

But before ordinary workers and literate women could imagine science as a knowable body of information with many applications, a vast conceptual transformation taking two centuries to complete first had to occur. When Polish astronomer Copernicus argued in 1543 that the sun lay at the center of the universe; when Italian artist, courtier, and experimenter Galileo was placed under house arrest in 1633 for saying that Copernicus had been right all along; when French aristocrat and philosopher Descartes presented his new method for thinking scientifically in his *Discourse on Method* (1637); and when western European gentlemen and aristocrats founded scientific academies generally from the 1660s onward, none of them could have foreseen the industrial transformations that lay ahead. They did not think about mechanizing cotton, or using steam engines in mines, or applying the laws of motion to the movement of water in rivers and canals.[12] Yet by the last quarter of the eighteenth century that was precisely what was starting to happen. The scientific legacy of Copernicus, Galileo, Descartes, and especially Boyle and Newton—reworked in textbooks and lectures—had helped to make the concrete applications of power possible.

This book addresses the question "Why Britain first?" partly by asking, Why not France or The Netherlands (north or south) first? The French governments before 1789 wanted the most advanced technologies; the Dutch had capital in abundance and their southern provinces were near the Belgian coal fields. The Austrian Netherlands (Belgium) was run by a progressive government with domestic coal readily available. Yet in none of those places did a town like Birmingham arise, which by 1785 was a hub of industrial activity, marveled at by French and Dutch visitors alike.[13] They said that in the town people even walked with alacrity, their faces displaying "a pleasing alertness." This exceptionally skilled workforce was weaving textiles, making guns and "toys"—buttons, pots, pans, watch chains, any small metal object—at an unprecedented volume. By 1800 the population in the town exploded and so too did poverty as new workers flooded in seeking employment by early industrial capitalists. In the history I have chosen to write here I dwell on the mindset of employers, not on the evils of the employment, because to understand the latter you must appreciate how human agents were able to create conditions so favorable to themselves.

Many historical forces created Birmingham and its most famous partnership, that of the steam engine manufacturers, James Watt and Matthew Boulton. Here we will discuss them, the engineer and the entrepreneur, as paradigmatic. When we delve into the letters and diaries of the Watts, all the economic forces well known to historians turn up: profit taking, the cost of labor, the demand for new sources of power, and the general market for consumption in everything from steam to coins themselves (which Boulton actually mass produced in his metal works).[14] But read on. The Watts display a vast cultural universe of religious values, political beliefs, self-identities, psychologies, and for our purposes most important, applied scientific learning and scientifically informed attitudes. Of all the new sciences to come from the seventeenth century and rest in the new industrial sites, Newtonian mechanics and the new chemistry were the most important and useful industrially. In this short book we will focus almost entirely on mechanics; chemistry would require another book.

As Watt well knew, mechanics concerned the motion of fluids and solids, the weight and pressure of different substances, and mechanical devices, pumps, levers, weights, and pulleys, as well as electricity and light. In the eighteenth century, thanks primarily to Newton's work, mechanics became an organized body of readily accessible knowledge. More important, it became something of a rage. People like Matthew Boulton and his friends paid good money to go to lectures, or to see electrical demonstrations, or to watch mechanical toys dance and play instruments. And in Britain, later on the Continent, new cultural figures emerged: itinerant lecturers, civil as distinct from military engineers, scientifically or mechanically literate entrepreneurs like Josiah Wedgwood (of blue porcelain fame) and Boulton himself, later French scientific reformers like Jean Chaptal, and technicians like the Periers brothers.

Most important for this book, mechanical science permitted engineers and entrepreneurs to have a conversation while standing at a coal mine that kept flooding, or when figuring out the best size steam engine to attach to a previously horse-driven battery of textile machines, or when dredging a harbor or laying a canal over hilly terrain. Scientific savvy also gave them an advantage over the semiskilled workers. Industrial entrepreneurs with mechanical training could see how the parts of the whole factory fit together, how in precise terms the division of labor between human and machine enhanced profit, and how much extra labor could be extracted from human beings using levers as well as brute strength. A French spy said of English work patterns: "There is no country where labour is so divided as here. No worker can explain to you the chain of operations, being perpetually occupied with one small part: listen to him on anything outside that and you will be burdened with error. This division is purposeful, thus resulting in inexpensive handwork, the perfection of the work and the security of the property of the manufacturer."[15] We may want to side with the worker in this drama, but we should also realize that one of the few ways he or she escaped the tedium of mechanized labor was either to stay at home or to become mechanically skilled, an overseer of machines, a skilled operator, or a small-time entrepreneur. Unskilled or artisanal hand labor was being paid less; by 1820 in Britain it was on its way out.

Hydraulics, hydrostatics, pneumatics—all branches of mechanics—may not grip innovative, ambitious Euro-American minds today, but in contemporary Korea mechanical skills are highly valued, and international competitions for inventions are eagerly entered and often won. From a cultural perspective Koreans of the late twentieth century look a little like late eighteenth-century English or Scottish industrialists: appliable learning captures the imagination. It is generated as well as used; by 1800 it became a tool for survival or success in an industrializing world.

The challenge for the historian is to figure out how and why mechanical knowledge and ways of thinking were taken up, or generated by, eighteenth-century Westerners with entrepreneurial interests. Rather than looking for the Newtons, or later the Laplaces, this book focuses less on scientific genius and more on the nature of the cultural values and matrices that fostered application and disciplined curiosity. The main characters are first seventeenth-century visionaries and natural philosophers, then eighteenth-century profit seekers, promoters of scientific learning, coffee house lecturers, civil engineers, chemists turned industrialists, liberal clergymen, and not least political revolutionaries—in England in the 1640s and 1650s, in France in the 1790s.

You may wonder why a book that sets its trajectory toward understanding the cultural foundations of the industrial West looks more at scientists and less at technologists. But the latter and not the former have already enjoyed considerable attention in books about early industrialization. To redress the imbalance the approach taken here emphasizes science, yet sees its culture as intimately related to technology. Think of fraternal twins, born into a family particularly eager for profits and improvement: they have different *personae*, different looks, but are still profoundly related. Here the focus is on the scientific sibling. She has generally been presumed to have been so abstract and learned, university trained and groomed, as to be above any involvement with the scruffy business of machine making or profit taking. It was once thought that self-educated tinkerers, unversed in science, associated only with the more practical down-to-earth technological sibling. With his assistance the tinkerer fulfilled the economic aspirations placed on both children. He made machines by trial and error, bleached fabrics or spun them with greater ingenuity and more cheaply, or slowly perfected the raising of beams or the action of steam condensers. All that happened, to be sure. But look a little more closely at industrial moments and you can also find the scientific twin at play, generally in the form of rational, textbook-learned mechanics. The history of early industrialization has, by and large, missed the twin who was an offspring of Newton's *Principia*.[16] Indeed the older historiography about science and technology presumed that their kinship was remote; they were second cousins at best. To put the genealogy unveiled in this book another way: the Scientific Revolution had more to do with the Industrial Revolution than has generally been presumed.

The Scientific Revolution, a term invented only in the mid-eighteenth century, describes earlier and specific intellectual innovations. In 1543 Copernicus argued both mathematically and rhetorically in *De revolutionibus orbium coelestium* (*On the Revolution of the Heavenly Orbs*) that the sun lay in the center of the universe. In the next generation Kepler established the orbits of planetary motion; his contemporary Galileo discovered the keys to the local motion of earth-bound bodies. In the 1660s Robert Boyle in England perfected an air pump that displayed the vacuum convincingly, discovered the laws of gases and laid out the basic experimental methods of verification through replication. His contemporary, Issac Newton (1642–1727), demonstrated the law of universal gravitation in his *Philosophiae Naturalis Principia Mathematica.* Therein he proved the importance of Kepler's planetary laws and elaborated on Galileo's mechanics. As a result in one paragraph a sparse outline of a very complex story can be laid out.

But the outline is nowhere near as interesting as the history it raises. For reasons we still do not fully understand, the new Western science that depended on a sun-centered picture of the universe and the motion of our earth, was a very different science from what could be found in other cultures at the same time. Mathematics applied to the heavens permitted a picture of the sun-centered universe to defy "common sense," what human beings observe day in and day out. The new science also took up the telescopic observation of visible bodies in motion in the heavens and the eye's rigorous examination of local motion here on earth.[17] The new scientific protocols and philosophies required that the results of observation be described largely according to mechanical principles, that is, actual contact between bodies, their push-pull in motion or when being put into motion. The new science also elevated mathematical analysis to unprecedented heights of importance. Deeply influenced by the search for universals associated with Renaissance neo-Platonism, the leaders of the Scientific Revolution sought universal natural laws. Refined into a method, experimentalism provided through replication a way to confirm or reject any claim about a law being universally true.

The questions raised by the Scientific Revolution persist to this day: Why did Western elites find this science so appealing? How did they reconcile it with religious belief? And not least, how did aspects of the new science get picked up and given mechanical application to specific industrial and technological needs, to the achievement of an unprecedented impact and control over the natural environment? The answers offered to such big questions will always be controversial.

Some historians now take issue with the word "revolution" even being applied to a transformation in scientific thinking that had fits and starts, often accompanied by magical and mystical elements coexisting along with laboratory work and systematic collecting.[18] They object that before 1800 the scientific way of thinking and being in the world had little impact on the lives of the vast majority of Euro-Americans or their colonies. But sometimes small elites effect mental revolutions, not to mention political ones. For those who consumed the new scientific learning, joined the hundreds of new academies, made original contributions however small, examined local terrains, then became "gentlemen of science" or engineers, a conviction grew: something truly extraordinary was happening. Boyle saw the scientific way as a revolution as early as the 1650s. A century later the industrialist, Josiah Wedgwood, said that "a revolution" in manufacturing was at hand and he urged his friends to profit from it.[19] By Wedgwood's time a consensus about science had formed among the educated. They saw the revolutionary nature of the mental shift that began with Copernicus and was consolidated so brilliantly by Newton. By the 1820s the conviction became commonplace that now too in industry, particularly in cotton, another kind of revolution was being effected.[20]

A process of cultural assimilation had given science its revolutionary attributes. The Enlightenment of the eighteenth century completed the assimilation and made scientific progress into an article of Western faith. Until quite recently Westerners believed that science and technology offered not only universally true systems of knowledge, but also inevitable progress, material and cultural. Once it became clear to everyone what extraordinary changes and wealth the British were achieving through manufacturing and transportation, the story told in this book ends. *From a cultural perspective*, by 1815 the Industrial Revolution was over. By that date governments and educated elites throughout the West came to realize that basic science had to be taught to as many people as possible and scientific education was critical to winning the industrial race, to national wealth and power.

One short book can only do so much. Here I am not trying to explain why or how Copernicus, Newton, or a host of now less famous natural philosophers (to use the term they would have understood) conducted the experiments they chose to pursue, or solved the mathematical or technical problems that consumed their interest. As the Bibliography demonstrates, many excellent historical accounts of the major achievements of the Scientific Revolution now exist. Instead I want to know how within a specific and evolving cultural framework the transition from seventeenth-century science to late eighteenth-century industrialization changed the values and perspectives of Westerners forever.

M. C. J.

University of Pennsylvania
July 1996

I

INTELLECTUAL FOUNDATIONS

1

The New Science and Its New Audience

As all Copernicus's contemporaries knew, it is not self-evident that the earth moves. Indeed for centuries most people believed that they stood on an earth that was fixed and stationary; the sun and all the planets revolved around it. The Alexandrian astronomer and mathematician, Ptolemy, who died around 178 A.D., put the geocentric, earth-centered wisdom into one great book, the *Almagest*, and in the sixteenth century his arguments still seemed to make eminently good sense. Surely it appears obvious that, "If the earth had a single motion in common with other heavy objects, . . . living things and individual heavy objects would be left behind riding on the air, and the earth itself would very soon have fallen completely out of the heavens."[1] Ptolemy wrote common sense when he said that if the earth moved people would be left riding in the air. Part I of this book will outline the replacement of common sense about the natural world with an uncommon conception of nature as uniform and mechanized.

That a moving earth could fall right out of the heavens was plausible in 1600, but not in 1700. By then many educated Westerners, particularly in northern and western Europe and in the English-speaking colonies, had abandoned Ptolemy's system for Copernicus's heliocentric system. The gradual move to a Copernican universe also set in motion what became a revolution in the ways educated Westerners viewed nature, both physical and human. Common sense based solely on what the eye could see stopped being adequate and so too did simple fear and awe at nature's power. In their place nature became conceptually domesticated; we might say it was "naturalized." Motions and forces of bodies joined the realm of knowable, if abstract, items in a crowded universe that by 1700 had expanded to include images of new continents and peoples, the controlling effects of state bureaucracies, and vast commercial networks spread across the Atlantic. All these discoveries about nature and people accumulated into an unprecedented challenge to orthodox Christian beliefs and the clergy, both Catholic and Protestant, who supported them. With the new science initiated by Copernicus and augmented by Galileo, Descartes, Newton, and Boyle, came exceptional confidence but also arrogance. Westerners knew more about nature; some presumed that this made them superior to other peoples and cultures.[2]

A scholarly priest from Poland and not a seafaring explorer set the slow transformation in the Western understanding of nature on course. Nicholas Copernicus (1473–1543) benefited from being educated in Renaissance Italian universities where he almost certainly learned neo-Platonic thought and, of course, Aristotelian physics. After he went back to Poland, where he spent his life (about which we know very little), Copernicus became an ecclesiastical administrator, a lawyer, and a part-time astronomer. He seems an unlikely candidate for the honor of starting what in retrospect and many centuries later would come to be known as the Scientific Revolution.

The new Italian cultural revival fueled by the art and philosophy of the ancient Greeks and Romans, as well as a fascination with Arabic science, had enhanced the status of mathematical learning. The humanism to which Copernicus had been exposed had many faces: civic and aimed at public service; philosophical, neo-Platonic and aimed at recapturing the principles of symmetry; and harmony, the divine perfection implanted by the infinite power of the Creator. Copernicus's humanistically inspired education had taught him that truth about nature lay in abstract, mathematical elegance. His astronomy grew directly out of Renaissance culture as well as his study of the ancients, Aristotle, Plato, Ptolemy, and Euclid.

Betting on the greater truthfulness of mathematical rather than sensory evidence, as early as the 1520s Copernicus placed the sun in the center of the universe. In a single imaginary leap he achieved a greater mathematical elegance and simplicity than Ptolemy had offered nearly 1,500 years earlier. Other than simplicity and elegance Copernicus's leap had little meaning, in and of itself. It opened the heavens to greater numerical scrutiny because heliocentricity eliminated many of the circles within circles, or epicycles, that Ptolemy had to assign the moving planets to explain their heavenly positions closer or farther from the earth. But Copernicus had no certain proof for a sun-centered universe. When he finally published his ideas in 1543 he turned instead to rhetoric: "Why then do we hesitate to grant [the Earth] the motion which accords naturally with its form that of a sphere, rather than attribute a movement to the entire universe whose limit we do not and cannot know? And why should we not admit, with regard to the daily rotation, that the appearance belongs to the heavens, but the reality [of it] is in the Earth?"

In 1543 Copernicus asked a good question about the hesitation of his contemporaries to accept heliocentricity. His question even relied on the very philosophy that they used to support Ptolemy. Aristotelian notions decreed that round, spherical bodies move naturally in circles according to their "form," that of the sphere. Copernicus retained the circular motion of Aristotle; he just wanted the planets alone to engage in it. Only in one important area did Copernicus break with Aristotle. The Copernican style expressed a willingness to move away from the appearances of the heavens and to search for an abstract reality beneath the appearances. An increasingly sophisticated group of astronomers and natural philosophers right up to Isaac Newton (b. 1642) would engage in the same search. Invariably it meant that Aristotle, whose philosophy would always save the appearances and work from them, had to be discarded.

Challenging any aspect of Aristotle as interpreted by medieval theologians raised many complex issues. The philosophy of Aristotle, vastly modified by scholasticism, stood as the commonplace wisdom of university teachers and preaching clergy alike. It was one of the intellectual foundations on which rested the theology of the Church. Of course, a scholastic style of argumentation did not always work outside the confines of university disputations and logical syllogisms. People generally do not talk in theorems. Yet theorems were an essential component of intellectually rigorous Christianity explicated in the scholastic mode. Aristotelianism informed the style and content of what the Church's leading intellectuals taught, and their power was significant.

As well as practicing a distinctive logic, the clerical scholastics used Aristotle to enshrine the basic Christian division between body and soul, securing it by the doctrine of immaterial forms. They alone gave shape and meaning to inert matter while permitting the substance of a body, for example bread, to be transformed by the priest into the body of Christ. At the Mass the Eucharist retains the shape of bread but its essence, or form, has become divine. Form imparted meaning to nature; the motion of a body was directed by a purposefulness; heavy bodies fall to earth, for example, because it is in the nature of heaviness, imparted through form, to seek that which is heavier. Water rushes to fill a space because nature abhors a vacuum. Armed with Aristotle, it was relatively easy for Christian theologians to argue that God endowed nature with its teleology, its purpose.

Given the importance accorded to Aristotle and Ptolemy, it made perfectly good sense, particularly in the absence of an overarching theory to explain the earth's motion, to resist adamantly the motion of the earth which, to be sure, human beings, then or now, could not sense. Despite his rhetorical appeal, Copernicus understood perfectly well why educated contemporaries led by their clergy resisted, why they placed their faith in centuries of learning that located the earth in the center of a closed universe, surrounded by luminous bodies, planets visible because of their light but not materially real.

The Ptolemaic or geocentric system also worked reasonably well if, for example, you were trying to navigate a ship or calculate the date for Easter. Astronomers had been making calendars for centuries on the basis of a geocentric model. Its mathematics were excessively complex, but the model could predict the positions of the planets. Most important, the geocentric universe fitted neatly into the Christian drama of creation, which placed humankind and hence the earth at the center of a divine plan. To dislodge geocentricity would require what Copernicus could not have imagined. An enormous transformation in mentality would be necessary, taking 200 years to complete, before heliocentricity, and all that it implied, gained widespread acceptance among the educated of Europe. The displacement of Ptolemy and Aristotle would also require a new audience for science, one far broader than what for centuries had been largely clerical.

Inventing Bridges to a New Audience

Among the educated elite in sixteenth- and seventeenth-century Europe we can identify certain key groups whose acceptance or rejection of the new science would set its fortune. Put somewhat abstractly and in the words of an historian of Copernicanism, "the flowering of new world views must be considered within the context of complex sociocultural systems."[3] Renaissance princes and their courts were one such system. They offered the possibility of patronage or, more important, the protection and promotion of new ideas as compatible with their power. The papal court and the pope in Rome formed one nexus of princely patronage. Just before his death in 1543, Copernicus dedicated his treatise to the pope. It is not clear that he ever read it but that is not the point: access to papal authority was highly desired by many late Renaissance philosophers and naturalists.

Within the half century after Copernicus's death, however, even more new men (and some women) joined the select company of the highly literate, constituting a new and expanded audience for science. As trade increased throughout Europe wealthy merchants of the early seventeenth century applied simple mathematics to day-to-day transactions; they weighed goods and kept accounts. To their commercial interests, Galileo—who became the seventeenth century's most famous Copernican—appealed when he argued that his geometrical approach to physics took into account the real world of everyday material objects. When errors crept into his physics, he explained to his readers that it was because the experimenter was like "a calculator who does not know how to keep proper accounts."[4] Galileo presumed that it would be largely unthinkable for his abstractions not to bear relation to reality, just as it would be bizarre for "computations and ratios . . . not to correspond to concrete gold and silver and merchandise."[5] The increasingly common commercial application of mathematics to everyday bodies and motions encouraged the growth of a mathematical and mechanical science.

By 1600, noticeably in the Low Countries, Italy, and southern England, merchants or commercially oriented aristocrats bought and read more books; in some cities they also controlled local government. Scientists and philosophers courted their favor. Galileo Galilei (1564–1642) wooed the mercantile aristocracy and courts of the Italian city-states and when threatened by his boldness the group that had most to lose, namely the clergy, took to their pulpits to denounce him to the larger population. Galileo figured out, perhaps earlier than anyone, that it was critically important to capture the attention of the new public.

In any bid to win a following, the clergy always had the edge. They were key players in European intellectual life. As Galileo discovered, both merchants and princes were less skilled in philosophical abstraction than were their clerical teachers and preachers. The clergy, whether Catholic or Protestant, were the purveyors of the written and spoken word. They controlled all the universities, the pulpits, and in many cases access to the right to publish; learned discourse, and hence the very language of natural philosophy, had been their province for centuries. When the well-educated clergyman spoke from the pulpit he translated complex metaphysical assumptions about the universe and its relationship to the deity into the daily language of religious piety. If the clergy could not—or would not—make that translation, then the language of natural philosophy, in short the language of what became the new science, stood separately from that of commonplace religiosity. Mathematical language would always be out of the ordinary, belonging generally to those gifted at it. Philosophical, descriptive, and rhetorical language about nature could be made comprehensible, at least to the educated. But someone had to make the translation.

Without the assistance of the clergy little about the new science could be learned by the illiterate or semiliterate; even the literate looked for clerical guidance, or at the very least, worried if the clergy attacked ideas as impious or ungodly. In those countries where the clergy embraced the new science, or were at least neutral toward it, science flourished. Where science remained suspect or was persecuted, as occurred in parts of Catholic Europe dominated by the Inquisition, relative intellectual stagnation in science was the price to be paid.

Decade by decade after 1600 confidence in heliocentricity had grown. New observational data and new mathematical formulations of planetary motion encouraged the selected use of Copernicus's mathematics without necessarily the acceptance of the revolutionary hypothesis that lay at its heart.[6] Slowly, fitfully, but in retrospect inevitably, the new learning translated out of Latin into all the major languages became the province of the educated, the literate consumers of books and commodities. As scientific learning spread it had the effect of increasing the distance between what the uneducated believed and the educated, the makers of higher culture, now assumed about nature and the heavens.

Under the impact of science the division between the cultural universes of the poor and the prosperous grew in early modernity. Indeed science and heliocentricity were key elements in effecting the division. Yet in the immediate decades after the death of Copernicus, actually advocating a separate understanding of nature for ordinary folk and one for the elite was greeted with suspicion, especially by the Catholic clergy. They had battled to bring the mass of Europeans affected by Protestantism back to the church. One of the principal strategies of this so-called Counter-Reformation was to champion popular piety, to make religion emotionally accessible to all. The moment the scientists *tried to make an appeal to* an increasingly literate audience they risked conflict with church teaching aimed at the general populace. Precisely this tension between learning appropriate for the educated and the beliefs of the masses undermined Galileo's appeal to the literate.

During the sixteenth century three developments widened the popular audience for both religion and science. The printing press and the Protestant Reformation promoted literacy, while the commercial revolution gave previously unknown men (and a few women) access to the press and publication because what they had to say, whether about religion, medicine, or mechanics, would sell in a world of increasing literacy, prosperity, and economic opportunity. The sixteenth century witnessed an expansion in the marketplace of Europe that also coincided with constant inflationary pressures. Put quite simply, the creation of an articulate popular culture, one that was occasionally heretical and hostile to the traditional magistrates, occurred simultaneously with a widening gap between rich and poor. All social and economic evidence we have from the period immediately prior to Galileo's confrontation with the church points to an increase in poverty for the majority in most parts of Europe. When combined with an increasing prosperity for many aristocratic and mercantile elites, particularly for those who could use their land or capital to take advantage of the new market pressures, inflation worked to widen cultural as well as economic gaps.[7]

In addition, the Protestant Reformation, quite apart from its obvious appeal to the magistrates of the cities, and indeed to the heads of the new nation-states, offered ordinary people a vision, often millenarian in nature, of a better future order here on earth. This popular millenarianism, when combined with distinctively Protestant doctrines such as predestination and the priesthood of all believers, gave the uneducated a disciplined path by which they might achieve that better future order without the Roman Catholic Church controlling access to grace and salvation.

We will find the millenarian vision an important rationale for the acceptance of the new science in Protestant countries. The English philosopher Francis Bacon (1561–1626), a precise contemporary of Galileo, offered the new science as one of the avenues by which that millenarian reformation might be achieved. But he did so in language that specifically repudiated any association between millenarianism and the culture of the people, or between science and the contemporary opponents of church and state.

In consequence of all these developments a gradual and widening separation in the traditional relationship between the culture of the few and the many developed, with regard to science in the course of the seventeenth century. The elite, far from wishing any longer to foster popular culture, seek now to control and redirect it. The new audience for science and the courtship it required presented Galileo with an unprecedented opportunity for a large and educated following. But such a courtship, as we are about to see, was perilous given the Church's interest in piety and learning.

Galileo's Confrontation with the Church

In 1616 the theologians of the Congregatio Sanctae Inquisitionis condemned the proposition: "Sol est centrum mundi . . ." (The sun is the center of the universe . . .). They promulgated the decree to all the offices of the Inquisition in the world and also placed Copernicus's book on the revolution of the heavenly orbs on the Index of Forbidden Books. They acted in response to a letter they had received from a Florentine Dominican who complained that the "Galileisti," that is, a group of Galileo's more aggressive followers, openly taught that the earth was moving. Prior to the letter, Galileo's clerical and Aristotelian enemies had attacked him from the pulpits of the city; indeed these enemies had formed a secret group that had as its expressed purpose the discrediting of Galileo and his doctrines.

The politics of clerical versus lay power formed the social setting of Galileo's confrontation. There was also a precise philosophical setting for the drama. Galileo embraced the new astronomy not as a hypothesis, but as a truth about nature. He was a philosophical realist. Galileo believed that "the true constitution of the world was worth investigating" and that Copernicus had actually discovered knowledge about "the true arrangement of the parts of the world."[8] In other words Galileo believed that verbal and mathematical representations of what is seen can also capture the way nature actually is. His realism extended to problems in mechanics, and his confidence increased because he was a skilled experimenter who, as early as the 1580s, worked on problems in local motion, on pendula and projectiles.

In 1609 Galileo managed to increase magnification of his telescope by a factor of ten and with it he surveyed the heavens. There he observed for the first time what he came to identify as satellites around the planet Jupiter and earthlike formations, valleys and hills, on the surface of the moon. The planets and moon began to resemble the earth and all were seen to possess motion. The abstract mechanical vision of nature—all physical bodies being composed of matter and motion, possessing only shape and size—lies in embryo within the conclusions Galileo drew from his experimentalism. To Galileo the Aristotelian and Ptolemaic universe started to seem less credible.

With the philosophical armor of realism, the confidence instilled by experimentation, and his powerful court patrons in Florence and Rome, Galileo, humanist, artist, experimenter, courtier, and scientist, took on the clergy regarding the subject of Copernicanism and its implications. Arguing for hills and valleys on the moon and spots on the sun, Galileo openly endorsed the Copernican system. He meant it not as a hypothesis for debates in logic, but as the way the heavens work. The ensuing confrontation between Galileo and the Church became a symbol in its age, and well beyond, of the supposedly inevitable conflict between the new science and traditional Christianity.

By the time of the 1616 decree, Galileo had received a degree of international recognition. As early as 1604 his lectures in Padua had attracted over a thousand listeners, and in 1610 he published a highly successful and easily read account of a new body, the so-called *supernova*, that had appeared in the heavens a few years earlier. When he became court mathematician to the Grand Duke of Tuscany, who resided in Florence, Galileo continued his aggressive search for converts to the new science, and assiduously wooed the grand duke. In his public role of popularizer and in his court role as mandarin, Galileo challenged the monopoly on scientific education enjoyed by the clerical teachers of the local universities. Many of them may have learned nothing about astronomy since mastering Aristotle's work on the heavens, *De caelo*.

The clergy's competence was directly challenged by the new science, and not surprisingly they spearheaded the attack on Galileo and used Scripture as their immediate weapon. As one Florentine Aristotelian put it when he used Aristotle's physics against Galileo and tied it to the preservation of a literal reading of the Scriptures: "All theologians without a single exception say that when Scripture can be understood according to the literal sense it must never be interpreted in any other way." A similar warning was issued to Galileo by a cardinal of the church who stated in 1612 that Copernican doctrine could be maintained only if it was assumed that the Bible naively speaks about the earth's immovability "in accordance with the language of ordinary people." But that was a risky assumption because it drove a wedge between learned and general culture, a wedge that the church, already traumatized by the Reformation, wished to avoid. But Galileo would not back off from the assumption about biblical naiveté.[9]

In 1615, while defending his own and Copernican ideas of the universe, Galileo insisted that "the mobility of the earth" is "a proposition far beyond the comprehension of the common people." Galileo knew perfectly well, indeed was eventually told personally by the pope, that his scientific learning was still esteemed despite the decree of 1616, that he could always hold the Copernican idea as a hypothesis. But as a realist, he believed in more than that.

Throughout his career, both before and after the condemnation of 1616, Galileo believed himself in possession of a special knowledge. He insisted that the new science was a discourse separate from the language of ordinary people and that the mechanical philosophy—the notion of particles possessed of weight and measure and in constant push and pull—described the natural world better than any alternative explanation. With these assumptions he could hold to the truth of Copernican doctrine as well as to the laws of the new mechanics. The notion of "forms" was irrelevant. He could also have the confidence and audacity to present both to the educated elite for their support. He may even have imagined that he possessed the power to influence the church at the very highest level of its hierarchy, in a circle to which he had frequent and long-standing access. Galileo, courtier, was also a man of the church. With friends in such high places how could he go wrong?

As a result of his self-confidence Galileo imagined as late as 1632 that it would be possible to have the decree of 1616 overturned. He thought that there could be one understanding of nature for the people and another for the educated. He castigated those clergy "who would preach the damnability and heresy of the new doctrine, that is, of the Copernican doctrine, from their very pulpits with unwonted confidence, thus doing impious and inconsiderate injury not only to that doctrine and its followers but to all mathematics and mathematicians in general."[10] The new science was unsuited to pulpit discussion—that is, to what Galileo and some of his closest friends and supporters had come to understand as the proper content of popular religiosity.

One of Galileo's friends, Giovanni Ciampoli, explained to him the gulf he saw between their mutual scientific learning and what was now fit for popular dissemination: "I have spoken to no one yet who did not judge it a great irrelevance for preachers to want to enter their pulpits and discuss such lofty and professional subjects among women and ordinary folk, where there exists such a small number of well-informed people."[11] It must be remembered that at this time a Florentine Dominican had publicly attacked Galileo and his followers, indeed all "mathematicians," as being identical with astrologers. He had therefore attempted to slur the new science by associating it with magic and naturalism, beliefs still commonly found in the larger culture. Throughout the seventeenth century the new science would seek to distance itself from those beliefs, indeed in some instances to wage war against them. In Galileo's defense of himself and his science we can see the first stage of the century-long struggle against the common sense notions of nature found widely in the general population.

Galileo openly displayed his distance from "women and ordinary folk" in his various published defenses of the new astronomy against the strictures of the church. He argued that there were now two professional elites, mathematicians and theologians, and both had the obligation to take great care in what they said to the people. The theologians, Galileo said, have long held that the Bible is full of passages "set down . . . by the sacred scribes in order to accommodate them to the capacities of the common people, who are rude and unlearned." These passages were capable of a deeper meaning, one that has always been the responsibility of theologians to discover. Galileo allied the new science with the exegetical tradition that maintained an esoteric learning separate from, and unsuited for, the masses. He argued, "Hence even if the stability of heaven and the motion of the earth should be more than certain in the minds of the wise, it would still be necessary to assert the contrary for the preservation of belief among the all-too-numerous vulgar." The issue here, as Galileo presents it, is the danger of popular heresy: "The shallow minds of the common people" must be protected from the truth about the universe lest they "should become confused, obstinate, and contumacious in yielding assent to the principle articles that are absolutely matters of faith."[12]

In arguing as he did that "wise expositors" must look beyond the vulgar, literal meaning of Scripture, Galileo unwisely put himself in disagreement with the decree of the Council of Trent (1546), which prohibited any attempt "to twist the sense of Holy Scripture against the meaning which has been and is being held by our Holy Mother Church." That decree had been issued in direct response to the Protestant Reformation and the multitude of new biblical interpretations offered by learned Protestant theologians and also by the numerous unlearned Protestant sects that had sprung up all over Europe.

The Aristotelian opponents of Galileo, led by the scientifically learned Jesuits, got the church's hierarchy to endorse the older astronomy. They were able to embrace the counter-reform teachings of Trent at a time when the church feared any new voices, even those that attempted to restrict their learning to the discourse of the educated.[13] In 1632 Galileo was put on trial by the Inquisition, and the following year he was condemned to house arrest. From that moment, all that he published had to be smuggled out of Italy to the freer presses of the Dutch cities. He had lost his struggle with his clerical opponents, and other Christian supporters of scientific learning, such as the French natural philosophers, Mersenne and Descartes, saw his defeat precisely in those terms. The English poet, John Milton, visited him in Italy in 1638 and wrote, "There it was that I found and visited the famous Galileo, grown old, a prisoner to the Inquisition for thinking in Astronomy otherwise than the Franciscan and Dominican licensers of thought."[14]

The conflict between Galileo and the church had not been inevitable, the result of the unceasing warfare between science and religion. Rather it occurred largely because of historical circumstances. The Protestant Reformation of the sixteenth century had placed the leadership of the church at the center of doctrinal confrontation with Protestant theologians, as well as with "heretical" intellectuals of the period. Many of them were desperately seeking a way out of the impasse created by the irreconcilable split between Protestants and Catholics. By 1600 the church saw enemies on every side: Protestants, strong particularly in northern and western Europe, in possession of their own universities, and even dominant in certain cities and states; skeptics, hostile to doctrinal orthodoxy of any kind and found most commonly among the lay elite, particularly in France; and not least, new and heretical philosophers, frequently from a clerical background, who sought to revive the religiosity of the ancient pagans as the foundation on which some sort of new universal religion might be constructed. One such prophetic philosopher, the Italian Dominican Giordano Bruno, traveled to the major courts of Europe proclaiming this pagan naturalism with all its magical associations as the alternative to the doctrinal orthodoxies of both Protestants and Catholics. Bruno was also a spirited promoter of heliocentrism. In 1600, by order of the Inquisition, Bruno was burned at the stake in Rome; but his ideas did not die with him and they may have caused the Italian church to spy in Galileo a latter-day Brunian.[15]

As a result of all these challenges to its authority the Catholic church took the arbitration of theological matters out of the community of scholars and conferred it on a bureaucratic institution in Rome—namely, into the hands of the clerical administrators of the Inquisition. This shift away from the community of scholars as the final arbiters in doctrinal matters—a process well under way by the early seventeenth century—provides the context within which Galileo's condemnation in 1633 occurred.

Without this context of the Reformation and the Counter-Reformation, the condemnation of Galileo and its effects become extremely difficult to understand. Similarly, without the new audience for science in place, Galileo would not have gotten as far as he got in attempting to forge in Florence and the other Italian city-states a link between the educated laity and the new science. Operating in the tradition of Renaissance humanism, concerned about the power of the Inquisition and its clerical supporters, Galileo sought allies and patrons among court aristocrats as well as merchants. He argued that science was uniquely suited to the interests of this new audience.

Given what historians now know about the formation of distinctively urban and elite cultures in the early modern period, we would have to conclude that Galileo was the first scientist to appeal to a newly empowered literate culture, and, more important, that he had a reasonably accurate grasp of its values and assumptions.[16] What he did not reckon on was the immense power of the Roman Inquisition. He may also have failed to realize how out of touch with and unsympathetic the local bureaucracy of the Inquisition had become to an intellectual tradition of free scientific inquiry that had once flourished in the late medieval universities and that had never intended, nor in fact constituted, a threat to the doctrinal foundations of Christianity.

Galileo's friends in the scientific academy of Florence also believed it would be possible to win the church over to heliocentricity, to place the new science at the center of Catholic learning. Galileo may have shared their preoccupation and, as it turned out, their miscalculations. Certainly Galileo attempted to argue the Copernican case not only on scientific grounds but also on theological ones. In so doing he embarked on a very risky course, that of appealing to his elite audience in Italian while at the same time arguing as a layman on matters in which theologians held strongly and professionally sanctioned opinions. More precisely, theologians spied in Galileo's atomic theory of matter—small particles in collision creating the changes we see around us—a threat to the doctrine of transubstantiation. The doctrine states that the priest has the power to turn bread and wine into the body and blood of Christ. To do that, immaterial forms must be detachable from inert matter; otherwise how would the bread and wine retain its visible shape and taste but undergo a transformation in its essence? The matter theory inherent in the mechanical philosophy rendered "forms" irrelevant, thus it undercut the established explanation for transubstantiation.[17] The doctrine of the Eucharist would present problems for every Catholic mechanist of the seventeenth century. Yet first and foremost in the list of Galileo's heresies, heliocentricity contradicted certain passages in Scripture that make reference, in passing, to the sun's motion.

These tensions and misconceptions within the church, and between Galileo and the church, suddenly brought the new science out of the domain of the universities and the learned disputations of the natural philosophers. Science made its way onto the intellectual agenda of all educated Europeans. What might have remained matters for debate among experts—for example, the relative merits of the Copernican system in relation to the geocentric system of Ptolemy, or the possibility of reconciling heliocentricity with the teachings of Aristotle—now became topics of widespread intellectual interest. Added to that purely philosophical and theological disagreement came Galileo's proclamation that the new mechanical philosophy forms the basis of a privileged learning suitable only for the educated few. To express this appeal to elite culture in the language of Galileo, science is fit only for "the minds of the wise" and not for "the shallow minds of the common people."

The Elements of the Mechanical Philosophy

Galileo went further in his departure from Aristotle and Ptolemy than simply accepting the motion of the earth. He conceptualized heliocentricity as the necessary corollary of new mechanical assumptions about bodies that were antithetical to those taught in all the schools and universities of the time. Quite simply he assumed that not only the earth, but also all bodies in the heavens, were real, physical entities, hence subject, at least in principle, to all the pressures and forces exerted on earthly matter. At the root of his matter theory Galileo assumed the existence of hard, impenetrable particles as the building blocks of nature; he was an atomist. Throughout the century most mechanists would also be atomists or corpuscularians partly because Galileo's telescope had provided compelling proof of the physicality of the heavens and hence the uniformity of matter. Atomism of ancient lineage offered an elegant philosophical assumption that could explain uniformity.

In 1609 through his own (in hindsight, primitive) telescope, Galileo saw the heavenly bodies more clearly than anyone before him. He dimly perceived the shadows on the surface of the moon and proclaimed it rough and mountainous. Previously he had seen a "new star" among the supposedly fixed stars. The Aristotelians had assumed the heavens to be perfect, therefore immaterial and unchangeable; clearly, according to what Galileo was seeing, they were wrong. But more than the assumption of perfection would have to go if Galileo's observations and mechanical experiments were correct. For example, Aristotle would have weight inhere in bodies and their speeds as freely falling bodies be proportional to their weight. Galileo argued that in motion, bodies fall at speeds determined not by their weight (or their shape), but by the resistance they encounter in the air—in short, that speed actually increases "the moment and force of the weight." And speed, as well as resistance, are measurable. In his words, "a material or corporal substance . . . has boundaries and shape, . . . relative to others it is great or small, . . . it is in this place or that, . . . it is moving or still, . . . it touches or does not touch another body"; and despite making every effort of imagination, Galileo claimed that he could not divorce a body from such primary qualities. But weight, taste, color, and smell Galileo called secondary qualities, and they can be imagined away: "I hold that there exists nothing in external bodies . . . but size, shape, quantity, and motion" (*Il saggiatore*, 1624). In the universe constructed by Galileo and the other mechanical philosophers who were his contemporaries, heliocentricity was only one part of a larger conceptual whole. At its center lay the assumption that bodies and motion—both of which are capable of mathematical application and observation—are the objects on which the new science must now focus its attention. In that fundamental sense and assisted by the mechanical philosophy Copernicus led to the steam engine.

But before Westerners reached the point of application, the whole of Galileo's vision would have to be assimilated. Briefly summarized, Galileo's science demanded the acceptance of a mélange of fundamental and new assumptions: The world around us consists of bodies subject to mechanical laws; these can be uncovered by the senses, by observation, and by experiment. The larger universe partakes in mechanical operations because the earth is a body, like the planets, that moves in boundless space; and not least, if there are scriptural texts that state or imply otherwise, these must be understood as simply the use of commonplace metaphorical language intended for the benefit of "the vulgar," not as inherent contradictions between Scripture and natural philosophy. Galileo maintained that God's word could not ultimately contradict God's work. But Aristotle could be contradicted because, in Galileo's words, "reasons persuade me—and Aristotle himself taught me to find peace of mind in that which I am persuaded by reason and not solely by the authority of the master . . . Philosophizing must be free." Once free, Galileo assured his readers, the science of mechanics and its branches may be quite useful "when it becomes necessary to build bridges or other structures over water, something occurring mainly in affairs of great importance." When appealing to his audience Galileo also sought to persuade them of the usefulness of the new science. As we shall see in Part II when we look at the eighteenth century, Galileo put mechanics on a rational footing without ever imagining the extraordinary feats that could be accomplished with it.

The Effect of Galileo's Condemnation

The writings of Galileo, his subsequent trial and condemnation, moved the new science into the forefront of learned discourse throughout Europe. Anyone attracted by the ideas of Copernicus, if living in Catholic, as opposed to Protestant, Europe, now had to think very carefully about how to announce that support. In France, for instance, the clerical opponents of papal intervention in the affairs of the French church saw in Copernicanism a new weapon in their struggle; the Jesuits, with their strongly ultramontane (propapal) conception of religious authority, sided with the Inquisition's condemnation. In Protestant countries, on the other hand, support for Copernicanism could now be construed as antipapal and hostile to the power of the Catholic clergy. What a splendid incentive for its adoption. This ideological linkage was to prove critical in creating the alliance between Protestantism and the new science.

Empirical science continued to be practiced in Italy after the public condemnation of Galileo. Yet the major philosophical innovations were now to occur elsewhere. After his condemnation, science in the seventeenth century became an increasingly Protestant and therefore northern and western European phenomenon. Much controversy has been generated among historians trying to explain that linkage, but the relationship can be sustained if we emphasize two points. The first is the ideological link, so attractive to Protestants, between opposition to the powers of the Roman church and its clergy and support for Copernicanism—and it should be borne in mind that those powers were frequently justified philosophically by the use of Aristotelian arguments. The second point concerns the dissemination of scientific knowledge. It is a truism that the enterprise of science depends on the communication of new knowledge. In early modern Europe that meant the printing of scientific books.

After the condemnation of Galileo, books at the vanguard of the new science—that is, those advocating the mechanical philosophy and heliocentricity—had to be published where the Inquisition had no authority. In practice that meant in Protestant Europe: in the German cities, in England, and most especially in the Dutch Republic, which had only just won its independence from Spain and the Inquisition.

In the 1690s when Anglican clergy in London were preaching Newton's science complete with atomism, Italian followers of the new science were under trial in Naples.[18] Among the charges brought against them by the Inquisition had been their supposed belief "that there had been men before Adam composed of atoms equal to those of other animals, that all was taken care of by nature, that there was no God . . . that the sacraments should not be recognized." The accused were simply followers of the new mechanical philosophy. By 1700 English science in its Newtonian form was first accepted on the Continent in the Dutch Republic and disseminated there by Dutch scientists as well as by the French language press based in the Dutch cities. Many historical consequences can be traced back, in some ultimate sense, to the "victory" over Galileo of a few Aristotelian professors, some Florentine and Jesuit clergymen, and the bureaucracy of the Roman Inquisition.

The Social Utility of Science

The promoters of the new science from the followers of Galileo onward believed in its social utility. Scientific learning and disciplined practice would channel the energies of the great, promote discipline among the lower orders, and if wisely pursued, result in unprecedented benefits for everyone. It is right, Italian as well as English reformers argued, that ordinary people—artisans and peasants—should be kept out of politics, which is tricky and unpredictable. Politics presupposes a secret wisdom known only to princes. But, as distinct from Galileo's argument that science was fit only for the elite, other theorists argued that science can be made available to everyone. Nature is everywhere the same: its operations, unlike those of politics, are regular and predictable. Thus scientific inquiry can be safely encouraged among the populace, and to the extent that ordinary people commit themselves to the study of nature, politics can become what it ought to be, a monopoly of the elite, with a consequent reduction of disorder produced by popular rebellion.[19]

This argument for the widespread practice of scientific inquiry rests on its supposed social utility in maintaining traditional authority. It is first enunciated, although never put into practice, in Counter-Reformation Italy at least partly in response to the church's condemnation of Galileo and the new astronomy. It was to become a powerful argument, one that we shall later hear from English apologists for the Royal Society; indeed a version of it appears in Thomas Sprat's official *History* (1667) of the society. In that Protestant context, where clergy and scientists could be allied in a common enterprise intended to support the Church of England, the argument gained widespread acceptance. Eventually, in the eighteenth century, the power promised by science became one of the most important justifications for the promotion of scientific inquiry. Yet before it could be accepted science had to be rendered safe.

The English version of the social utility argument was probably developed quite independently from the Italian ones. What is important about the argument is that in early modernity it appears in a variety of contexts, but always with the same intention. Science can increase the wealth and power (both social and military) of existing elites. It can be a force for social stability, generally not for social reform, and its purpose is to increase the prosperity and wealth of the state. In every period where we find the argument put forward with particular force we will also find radical thinkers who oppose it, who would have science directly serve the people, to the benefit of all humankind.

The genesis of the English vision of social utility came from a servant of the state, the Lord Chancellor of England, Francis Bacon (1561–1626). In the first instance Bacon sought to render monarchical government increasingly effective, to rationalize its operations and to harness science to the service of state building. As a lawyer and politician he turned to natural philosophy as part of his statesmanship.[20] Indeed he envisaged a vast program for the collection of information on every aspect of human mores, law, and nature; his empiricism knew no bounds. Yet Bacon possessed a very precise sense of the utility of knowledge and saw, with remarkable perspicuity, that the mechanical arts could make an unprecedented contribution "to the endowment and benefit of man's life." In this respect Bacon saw more clearly than any of his contemporaries the extraordinary advances that had already been made by mechanical artisans in shipbuilding, navigation, ballistics, printing, and water engineering. He also knew the disdain with which the educated and titled held such unlettered men. He chided, "It is esteemed a kind of dishonor unto learning to descend to inquiry or meditation upon matters mechanical, except they be such as may be thought secrets, rarities, and special subtilties." He attacked the "supercilious arrogancy," so much a part of the aristocratic culture of his time. In place of the hunt and war-making, the perceived activities of truly masculine leaders, Bacon offered a new vision of the truly cultured and educated man.[21]

Such a man (Bacon was very precise in giving a gender identity to scientific activities) should pursue science because it is truly "masculine." The study of nature, rather than hunting and killing or theological squabbling, repudiates the "degenerate learning that did chiefly reign amongst the schoolmen." In *The Advancement of Learning* (1605) Bacon directly assaulted the old clergy of the Roman church, "their wits being shut up in the cells of a few authors (chiefly Aristotle their dictator) as their persons were shut up in the cells of monasteries and colleges." In so doing he allied scientific learning with Protestant culture of the kind that had been institutionalized after the Henrican Reformation: Erastian, in that it favored the domination of bishops and ministers by kings and local aristocracy, and national, in that it eschewed the sectarian divisions that were commonplace in Continental Protestantism and opted instead for a single national Anglican church. Such Protestant gentlemen as Bacon would create should cultivate the sciences in imitation of the way that solicitors interrogate the common law, and they should observe the activities of the mechanical artisans so as to achieve a natural philosophy that would be practical and progressive, one capable of "perpetual renovation." What better way to preserve and enhance governments, he asked, than "to reduce them . . . to first principles, to a rule in religion and nature, as well as in civil administration?" The Protestant state would flourish, Bacon believed, under the aegis of a strong central administration led by an intelligent king and a unified church. The state should pursue common enterprises of legal and scientific endeavor, everything from shipbuilding, navigation, ballistics, printing, and water engineering.[22]

Not all the learned men of science would do the same thing in Bacon's vision. He left ample room for a variety of scientific activities, "some to be pioneers and some smiths, some to dig and some to refine and hammer," some to be "speculative, and others operative." For Bacon, as for all his early modern followers, the division of labor between theory and application was two sides of the same coin. The modern great divide between "pure" and "applied" science is a nineteenth-century invention; it was simply not understood in this earlier period. The Baconian vision created a space for the engineers and entrepreneurs of its world; artisans also could contribute to fostering the strong Protestant state. Theorists needed to understand what they did and convert their practices into general principles.

We will follow Bacon's language in our discussion of seventeenth- and eighteenth-century science. The domain of science cannot be divided up into amateurs on one side and professionals on the other. Instead, as Bacon explained in his posthumously published and immensely popular utopian tract, *The New Atlantis* (1627), there would be many different workers in the vineyard of science. Most accessible to ordinary men would be the model of "the merchants of light," those who would use "the books, and abstracts, and patterns of experiments" to propagate science in the marketplace. Contemporaries inspired by this utopian vision interpreted it as the call for a vast, European-wide program for the dissemination of science, one unprecedented as an ideal before or since. In his utopian paradise Bacon would also have "Lamps," those who would take care "to direct new experiments, of a higher light, more penetrating into nature than the former," that is, those who, like learned judges, would draw out of science "things of use and practice for man's life." When calling for the application of science to industry, French revolutionaries of the 1790s would publicly invoke the name of Francis Bacon. Long before them, English readers of Bacon would miss his immediate concern to foster the ambitions for state building of his king, James I. Instead they would find in Bacon's message guidance for scientific pursuits of every kind, from collecting, to observing, experimenting, and inventing.

In his advocacy of the validity and usefulness of the new science, Bacon repudiated the secretiveness and exclusivity of the magician; he also urged the rejection "of fables and popular errors." His science enshrined the work ethic, "the laborious and sober inquiry of truth," as the correct method of inquiry in opposition to "the high and vaporous imaginations" found in natural magic, superstition, and in Aristotelian emphasis on "sympathies and antipathies, and

[the] hidden properties" of things. Bacon eloquently allied this sober, mechanically oriented science with the Protestant Reformation:

> We see before our eyes, that in the age of ourselves and our fathers, when it pleased God to call the Church of Rome to account for their degenerate manners and ceremonies, and sundry doctrines obnoxious and framed to uphold the same abuses; at one and the same time it was ordained by the Divine Providence, that there should attend withal a renovation and new spring of all other knowledge.[23]

Bacon saw the renovation of the sciences as the work of divine providence, and, as a true English Protestant, he possessed a very precise sense of the role of providence in history. The renovation of science and philosophy prepared the way for a larger scheme, a vast unfolding, a "great instauration" of learning that would precede the end of the world. Only then would human beings be liberated from the effects of their original fall from grace.[24]

The millenarian impulse found in Bacon turns up over and over again in seventeenth-century English Protestantism. Particularly after Bacon's death, English Puritans and then their heirs, the Dissenters, took up Bacon's call for restoration and renovation. It must be reckoned as one of the main motivations for the cultivation of scientific inquiry in seventeenth-century England. In the hands of Puritan reformers who saw the Anglican church as corrupt, Baconianism became part of a revolutionary vision. The Baconian impulse as it developed became utopian, even millenarian. The millenarianism of Isaac Newton and some of his followers later in the century can be related to this Baconian background. Similarly English scientists' militant promotion of science as one of the foundations, both practical and ideological, of state power owed much to their unique and compelling search for the millenarian transcendence of historical time. It should be emphasized, however, that the millenarianism of Bacon and his followers always located control and leadership in the new paradise firmly in elite hands.[25] Bacon himself had no use for the Puritan reformers of his day, people who in the next generation became open opponents of court and king.

An emphasis on Bacon's millenarianism inevitably entails a recognition that there were profoundly mystical elements in his thought. He would steal the vision of the magicians by unlocking nature's secrets, yet he would discard their secretive methods. Bacon believed that the ancient myths and fables contained a hidden wisdom, and in his search to recover and vastly augment that wisdom he resembles ill-fated Bruno as well as various hermetic thinkers of the sixteenth century, not least the Swiss medical reformer Paracelsus (1493–1541). Paracelsus used the magical and Platonic tradition, which emphasized the correspondences between the human body and the heavens, to legitimate an empirical and experimental approach to the study of disease, and his program was blatantly anti-Aristotelian. [26]

Bacon condemned the magical elements in the thought of Paracelsus, but he had to admit that the latter's natural history was extremely useful. The magical tradition in early modern Europe at moments promoted scientific inquiry.

Magic, long associated with heresy, inspired the search for alternative natural philosophies to that of Aristotle and the scholastics; its literature, typically alchemical or astrological, could also promise a dramatic unfolding of nature's secrets, and its systematic exploration could promote empiricism. At their core the magical arts promised to reveal a single, unifying philosophy of nature. For that reason Bruno could be a Copernican of sorts; Bacon could attempt to steal the zeal of the magician while excluding him from the company of the empirical gatherers; Newton could practice alchemy throughout his career; and the German scientist and mathematician Gottfried Wilhelm Leibniz (1646–1716) could dabble in astrology. Yet in its adoption of the mechanical philosophy and in its repudiation of popular beliefs, the new science of the early modern period ultimately rendered magic irrelevant to the needs and interests of the educated elite.[27]

In a self-conscious bid for the widest possible readership among the literate in England, Bacon published most of his important works in English and not in Latin. Yet they were quickly translated into Latin and published in Continental editions generally emanating from Amsterdam. As early as the 1620s Bacon's ideas were known on the Continent, particularly in select philosophical circles in Paris.[28] There his emphasis on the systematic collection of data, on an easily applied empiricism, appealed to botanists and collectors at the newly established Jardin des Plantes. Similarly his vision that science promised to lighten the human burden, to provide domination over nature, appealed to reforming German Protestants of the same period who wished, like Bacon, to include natural knowledge in the millenarian reforms promised in the coming age. Finally Baconianism played an inspiring role in the founding of the French Academie des Sciences in the 1660s,[29] and as we shall see in chapter 8, the academy played a complex, if often inhibiting, role in French industry right up to the Revolution of 1789.

As early as 1620 Bacon's writings were known to the important Dutch mechanical experimenter, Isaac Beeckman, and he in turn had a major influence on the French natural philosopher, Descartes. Then during the 1640s Bacon's ideas were discussed at the University of Leiden. The Netherlands in general, and Leiden in particular, were the most important centers for innovative natural philosophical education in seventeenth-century Continental Europe. At the end of the century the most advanced medical teacher and practitioner of his time, Herman Boerhaave (1668–1738), a professor at Leiden, could barely contain his rhetorical enthusiasm for the promise of medical advancement offered to those who would heed Bacon's call to experience nature for themselves. Perhaps even more fascinating in terms of the European spread of the Baconian vision, we can find his utopian and humanistic writings about science, among other topics, translated into Dutch in the 1640s and 1650s.[30]

Bacon's utopian tale, *The New Atlantis* (1627), is a good, short fantasy that had widespread appeal. It presents an island paradise dedicated to peace and scientific progress, and it was nowhere more popular than in England. But its appeal to Continental Protestants can also be seen in a Dutch translation of 1656, pocket sized, in clear and simple prose. Bacon appealed to the literate in one of the most heavily mercantile and urbanized areas of Europe.[31] In the spirit of Bacon's local and lifetime search for powerful patrons to promote the work of science, the posthumous Dutch translation was dedicated to a prince, in this case Frederic Henry, Prince of Orange. Given the diffusion of millenarian ideas among Continental Protestants, we may postulate that Baconian ideas about a future paradise based on science also evoked interest born of millenarian fervor.

By the eighteenth century, reformers were no longer convinced that human progress required any disruption of historical time. They easily ignored Bacon's millenarian side and concentrated on his call for a scientific empiricism intended for the future relief of the human condition. In the midst of the European Enlightenment the French encyclopedists, led by Diderot and d'Alembert, invoked the memory and ideals of Francis Bacon in the first great *Encyclopèdie* (1751). The most impressive assemblage of knowledge then known, it sought to make all branches of learning unified and accessible, and it emphasized the applied arts. In the 1790s Bacon's *Novum Organum* (1620) was translated into German at a time when German scientific societies were proliferating and the old professorate was being challenged by newer men profoundly interested in the practical application of the sciences to the problems of society and industry.[32] In the 1830s one branch of the leadership of the British Association for the Advancement of Science, an organization committed to the practical and industrial application of science, also invoked the memory and vision of Bacon. By that time, however, Bacon's vision of a science based on the collection of facts could be opposed to a more theoretical and professionalized vision of science, one whose supporters wanted the enterprise of science to be dominated by "great minds" and by the search for general laws.[33] This experimental and heroic alternative to the Baconian vision they, and we, could describe as Newtonian, and even by the early eighteenth century in England it had come to replace Baconianism as the dominant scientific ideology. But in the emphasis on utility found among so many Newtonian lecturers and experimenters, or in the scientific interests of artisans like the Watt family, the Baconian vision lived on, eventually subsumed under the larger rubric of Newtonianism. Together both ideologies gave science an increasingly practical and mechanical focus. In an age dominated by sectarian strife, censorship, revolution in England and rebellion in France, the Baconian vision in the hands of state builders or reformers urged Westerners to turn to science and its application. All the following chapters describe the turn toward nature mechanized and the accompanying search for applications of its mechanical powers. At every turn the Baconian legacy inspired visionaries as well as industrialists.

2

The Cultural Meaning of Cartesianism: From the Self to Nature (and Back to the State)

When on hearing that the word K-I-N-G signifies supreme power, I commit this to my memory[.] It must be the intellectual memory that makes this possible. For there is certainly no relationship between the four letters K-I-N-G and their meaning, which would enable me to derive the meaning from the letters.

Descartes's Conversation with Burman—J. Cottingham

In the 1630s Galileo's confrontation with the church and its Aristotelian leadership became only one symptom of a wider, more universal crisis over which philosophy commanded absolute authority in intellectual matters, as well as over who should establish the criteria by which that authority might be asserted. At stake lay centuries of received wisdom, the synthesis of Aristotle and Christianity known as scholasticism. Armed with scholastic explanations all church doctrines (like transubstantiation) were buttressed and explicated. Armed with the notion of a kingly form impressed on the majestic body, people did not have to memorize words to describe his *gloire*, as if it and the words could be detached. A king's essence resided within him.

The slow destruction of scholasticism, such a powerful and culturally unifying way of thinking, threatened to unleash socially dangerous beliefs and values. Some were of ancient origin, others grew out of folk beliefs and practices. All were heresies long condemned by all orthodox theologians and philoso-

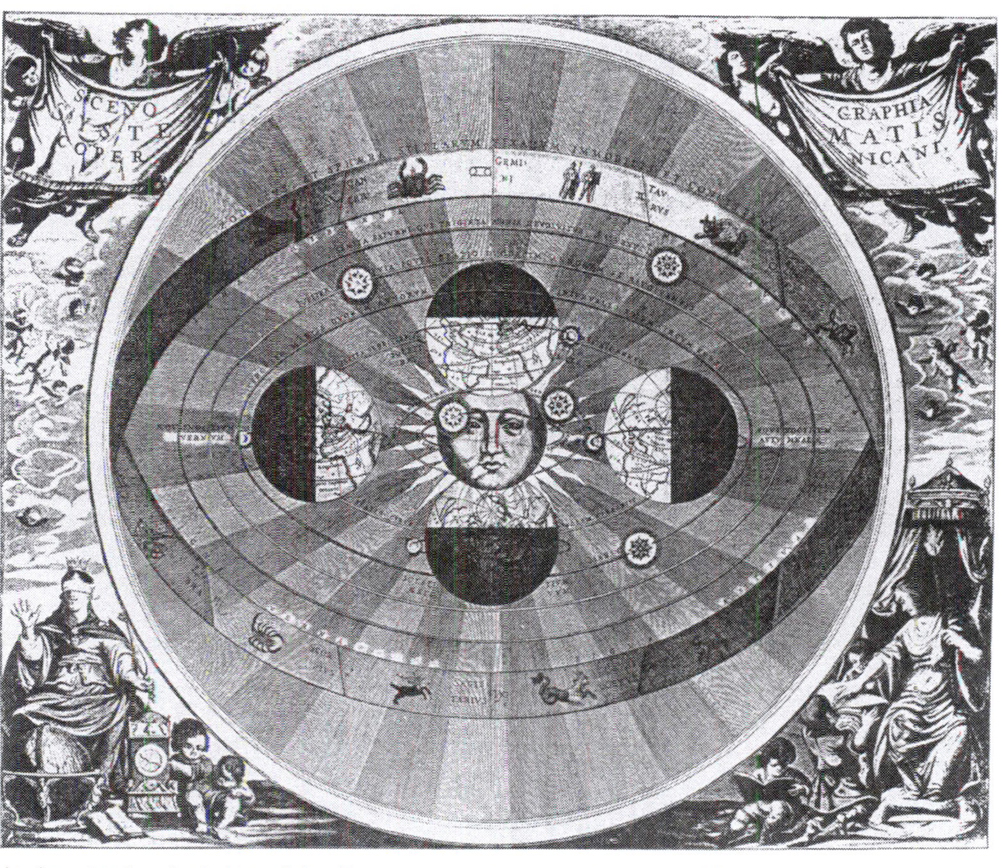

An imaginative depiction of the Copernican universe as the young Newton might have seen it had he consulted Andreae Cellarii's contemporary *Harmonia Macrocosmica*, 1660. (From the Mansell Collection.)

phers. Any synthesis of the new mechanical philosophy with those heresies, either with naturalism of popular and ancient origin or with materialism, would provoke the authorities in either Protestant or Catholic Europe to attempt its repression. Within this context the post-Galileo proponents of the new science, who in most cases shared the values and assumptions of ecclesiastical and state authorities, searched for an acceptable way to understand nature mechanically. The mechanical philosophy required a formulation that would be ideologically advantageous, yet another guarantor of religious orthodoxy, social order, and political stability.

By the early seventeenth century, however, any ideological construction intended to secure and unify the polity seemed destined to fail, yet another victim of the doctrinal struggle between the Reformation and the Counter-Reformation. A century of religious controversy and open warfare between Catholics and Protestants had left many a civilized observer convinced that the only alternative to brutal intolerance was skepticism, a refusal to believe anything doctrinal with absolute certainty. Such skepticism among the educated elite was profoundly dangerous to the maintenance of order and stability in society as a whole. Nonbelief, when rigorous, systematic, and searching, was considered a threat to all orthodoxy; no institution is safe if people simply stop believing that certitude is possible when it comes to the doctrines justifying its existence.

Somewhat ironically and dangerously, skepticism as a mode of thought and argumentation was used in the sixteenth century by Protestants and Catholics alike—always in refutation of the other. It also gained currency as the result of the reintroduction of ancient skeptical writers, in particular Sextus Empiricus, whose writings appeared in print in Latin (1562), in English (around 1590), and in a privately circulated French translation of 1630. Giordano Bruno, for instance, makes mention of the skepticism he observed among academicians in the course of his various European travels. Skepticism had become fashionable in the highest social circles.

In the 1630s, as René Descartes (1596–1650) watched the condemnation of Galileo, he turned to skepticism. But his motives were not those of its promoters. Instead he sought to refute the more extreme version that led to relativism, and also to steal the fire of the skeptics, to announce their method as a way the individual can subject all ideas to examination to arrive at new truths about nature. Skepticism justified a personal, intellectual odyssey.[1] In Descartes's version of skepticism as method, science became an unprecedented source of individual expression.

The Threat Posed by Skepticism

By far the most sophisticated version of skepticism was put forward by a late sixteenth-century French layman and adviser to the king, Michel de Montaigne. In the midst of the French wars of religion of the 1570s Montaigne lost his faith in human reason, in its ability to know anything with certainty. With devastating pessimism he labeled reason a "puny weapon," and his purpose was "to crush and trample underfoot human arrogance and pride."[2] He cited the controversy surrounding heliocentricity and identified the conflicting opinions found in the new science as one more reason to acknowledge the human predicament, the futility of searching for actual truth. There was also a social element in skepticism. Montaigne spoke for the lesser French nobility or gentry and the upper bourgeoisie who had seen the great aristocrats mutilate the state and dominate the court, using religion as their pretext and always to their own advantage.[3] By contrast the gentry would practice a rigorous self-control "that marks them out as the class most appropriate to take political control and order society in a new and humane way."[4] They would distance themselves from *les grands*, retreat from state affairs if need be, and bide their time, waiting for a new kind of order in the state, one that brought peace while giving them greater purpose and place. Descartes belonged to the same social layer to which Montaigne's ideas appealed.

In purely intellectual terms Montaigne gave voice to a profound intellectual crisis provoked largely by the Reformation, but also by the new science. But his answer provoked a reaction that in turn resolved the crisis. In response to skepticism Descartes offered the first intellectual synthesis of modern thought to rest entirely on the individual's ability to know nature mathematically and experimentally. With knowledge, Descartes promised, came mastery. You can do things because you think. His discovery of the uses to which science could be put laid a new foundation of orthodoxy, both political and religious. Scientific enquiry became a viable alternative, a way of rejecting both the fashionable skepticism of Montaigne's generation and the scholasticism of the clergy and the schools. Descartes's reliance on the self, disciplined by his method, could establish an entirely new metaphysical foundation for doing science, or for practicing one's religion, or for giving allegiance to authority of church and state. It was also consonant with the ethos and restlessness of the French gentry from which Descartes, *un gentilhomme*, a military man and a private scholar, came.

We can set the stakes as Descartes may have imagined them as he anxiously watched the church's condemnation of Galileo in 1633. Should followers of the new science retreat into skepticism and deny the possibility of having a true picture of the world? Or why not fideism, just accepting on faith whatever the clerical authorities in their wisdom designate as true? We can paint mental portraits of such late sixteenth- and early seventeenth-century skeptics as Descartes may have personally encountered: someone capable of converting from one religion to another at will; or a cynic about claims of divine sanction, or indeed any sanction for kingly authority, some one who kept forgetting what K-I-N-G meant; or a practitioner of no particular ethical creed associated with doctrinal Christianity. Such a person could live solely according to the moral dictates imagined to exist in nature. In short, skepticism could endorse a practical as well as theoretical naturalism, a way of being in the world that contemporaries called *libertinage*. It in turn could sanction ruthlessness, aggression, or power for its own sake pursued by cynical but dangerous men.

Naturalism was commonplace in early seventeenth-century France, just as it had been in late sixteenth-century Italy. In Toulouse, Guilio Cesare Vanini (d. 1619) deified nature by referring to it as God.[5] He was burned at the stake for so doing; so too was another pagan naturalist, Fontanier, burned to death in Paris in 1622. In that same decade alchemists were banned from the city and a prominent *libertin* was put on trial. The chaos of the sixteenth century had unleashed a rich and dangerous diversity in French intellectual life. There was an unprecedented challenge to Catholic orthodoxy as buttressed by Aristotelianism. Heterodoxy in the hands of *les grands* could also challenge the absolute authority of the monarchy and thereby threaten the sovereignty of the state.

Among the lesser French aristocracy and bourgeoisie, and even among the clergy, progressive elements were convinced that only strong monarchy could check the excesses of the great nobility, *les grands*, many of them ultramontane, pro-papal or even pro-Spanish. Other seekers for a way out of the morass created by skepticism wanted to enhance the Catholic Church's power. They wanted a common set of values, aside from blind faith, which could rally elites against the independent Protestant communities still powerful in certain districts and cities. For such monarchical Catholics as Father Mersenne, one of the earliest Parisian advocates of mechanical science, a new philosophical foundation for orthodoxy seemed absolutely essential. For Mersenne religion ensured the well-being of the state; indeed he believed, as he put it, that there should be *gendarmes temporels et spirituels* to enforce order and orthodoxy.[6] He and his mechanistically inclined friends, among them Descartes, repudiated the naturalists as well as the Aristotelians and sought in the new science the foundation for a new and scientifically progressive Christian orthodoxy.

The quest of Mersenne and his friends, undertaken amid the fear of continuing political instability, gave a ferocious intensity to French natural philosophical debates of the early seventeenth century. The debates must also be seen as waged within terms set by the skeptics. They had praised the new science insofar as it challenged Aristotle and then went on somewhat perversely to use the challenge presented by Galileo's achievements to argue that nothing in science could lay claim to certainty.

The skeptics attacked science precisely at the time when scientific learning aroused unprecedented interest among the educated laity. By 1632 a curious Parisian institution had developed to cater to their interest. This center, or Bureau d'addresse as it was called, was founded by one Théophraste Renaudot (d. 1653), an intellectual in the employ of the chief minister of state, Richelieu, thus a minor bureaucrat in the centralizing state that he struggled to create. Renaudot was also a publisher, a man-about-town, and devotee of the new science. At his bureau minor noblemen and *gentilhommes*, merchants, bankers, and lawyers jostled at weekly meetings with tradesmen, even artisans, and shared their common enthusiasm for all new knowledge of a practical sort, and for science in particular.

To our great benefit Renaudot published accounts of the gatherings that reveal the extraordinary eclecticism—for its detractors the extraordinary confusion—that reigned in the minds of the cultured classes over ways and means by which nature might be understood.[7] They sought some sort of "correct" method by which the natural world might be explored. Still dominant were the Aristotelians; but they were joined in debate by cabalists, believers in number mysticism, as well as by Paracelsians. These critics of the medical profession were at the Parisian meetings in great numbers, and they had taken up the ideas of the sixteenth-century German medical reformer Paracelsus. In a naturalistic vein he had argued for natural remedies, for consultation with the stars, for a return to the healing power of nature, in opposition to the blood-letting and high fees practiced by the official doctors. At the bureau some interest was even shown in applied mechanics, but it was the sciences useful for health or for trade that interested the bourgeois, as well as noble, seekers of truth. Before news of the condemnation of Galileo reached Paris, visitors to the bureau even argued for Copernican heliocentricity.

The weekly meetings reveal the existence of a market for science in early seventeenth-century Paris—comparable to what existed in London or Amsterdam—and they also remind us that natural philosophical language had now entered common parlance. Within educated discourse, Aristotelianism, although still commonplace, aroused profound dissatisfaction; but no coherent alternative—except naturalism with its heretical associations or skepticism—existed to replace it. For those who wanted a learned, scientific discourse that ensured orthodoxy in religion and order in the state, this impasse was a very dangerous state of affairs.

The systematic Aristotelianism of the scholastics was simply irreconcilable to the new science, either to the heliocentric doctrines of Copernicus or to the mechanics of Galileo. Without the philosophical revolution initiated by Descartes the extraordinary intellectual potential of the new science would have remained anathema to orthodox Christianity, especially to Catholicism. To that extent science could not have been assimilated into the mainstream of European high culture outside of Protestant Europe. Descartes transformed mechanical and mathematical science into the foundation of an entirely new understanding of nature with direct implications for human institutions.

We should ask: why did that philosophical revolution occur first in France and not in either Italy or The Netherlands, the only two other Continental places where the new mechanical science had reached such a degree of experimental maturity? In other words, why was it Descartes, and not the great mechanical scientist of the Low Countries Isaac Beeckman (1588–1637)—a profound influence on him—who achieved the philosophical synthesis that made Descartes the greatest natural philosopher and advocate of the new science to be found in his generation? Before we can appreciate Descartes's achievement, which rendered mechanical science but one aspect of an entirely new foundation for all human inquiry as well as the source of a new cosmic order, we should examine briefly the thought of Beeckman, the only other mechanical philosopher outside of Counter-Reformation Europe capable of producing such a grand natural philosophical synthesis based on mechanical principles.

Beeckman and the Mechanical Philosophy in the Netherlands

Even before Galileo, Isaac Beeckman must be recognized as the first mechanical philosopher of the Scientific Revolution. There were other mechanists before and contemporary with him, but none of them developed a systematic philosophical approach to mechanical problems, one that speculated as to the atomic construction of matter and designated this mechanical philosophy of contact between bodies as the key to all natural forces, to every aspect of reality from watermills to musical sound. When Descartes first met Beeckman in the Dutch town of Breda in 1618 the French philosopher quickly came to recognize him as "his master." They discussed every aspect of motion. Beeckman taught him to think systematically of clusters of atoms and empty spaces as the building blocks of the phenomena we see around us.

Yet the rather humble Beeckman never developed his mechanical philosophy into an entire philosophical way of thinking, indeed of living, as Descartes was to do. We might speculate that this sort of grand cosmology was temperamentally alien to Beeckman, a Dutch burgher, son of a manufacturer, and leave the matter at that. But we should not dismiss the very real social, religious, and political differences—that is, the contextual factors—that separated the Dutch cities of Beeckman's time from the Paris of Gassendi, Mersenne, and Descartes. In the first instance Beeckman was a Protestant in a republic now dominated, after its rebellion against Spain, by a Calvinist clergy and, more important, by a Calvinist lay magistracy that exercised considerable authority over the clergy. There were fierce struggles between types of Dutch Calvinism and all involved political stakes, personal reputation, even fines, possibly imprisonment. But the clergy distinct from the laity exercised no hegemonic power comparable to what they enjoyed in France.

Beeckman's own religiosity was intensely pietistic and individualistic, close indeed to the spirituality found among the more extreme English Puritans of the early seventeenth century for whom Bacon had had so little regard. Beeckman's extreme Protestantism gave him absolute confidence that "God had so constructed the whole of nature that our understanding . . . may thoroughly penetrate all the things on this earth."[8] Protestants like Bacon routinely used the metaphor that God reveals himself in his word, the Bible, and his work, nature. Beeckman seems never to have struggled with the fear of atheism when he approached either atomism or the mechanical philosophy; his Calvinism saved him from the struggle that gave birth to Descartes' complex synthesis. Beeckman as a Protestant Copernican did not have to cope with Jesuits or with the Inquisition's condemnation. Equally important, Beeckman encountered Aristotelianism in the Dutch schools and universities, but never did the Calvinist clergy entrenched in the universities enjoy within them the monopolistic power possessed by their counterparts at the Sorbonne. In a Dutch context one did not have to construct an entirely new foundation for learning to salvage Christian Orthodoxy from the pretensions of the clergy, nor did Beeckman have to fear that intellectual dissent would literally destroy the Dutch polity or himself. Whatever profound disagreements separated Dutch Calvinists from one another in the early seventeenth century—and there were many—neither clergy nor magistracy sought to shore up a monolithic state power as the only alternative to internal chaos. Indeed quite the reverse is true. The stability of the Republic rested very much on local and urban power in the hands of merchants and noblemen who had recently managed to free themselves from Spanish imperial power.

There is one other particularly relevant aspect of the Dutch situation that permitted Beeckman to develop his mechanical interests to the fullest. The Dutch cities were the most mercantile in the European world; hand-powered industry flourished. Beeckman himself made candles but also laid water conduits. As his mechanical interests developed, he easily mixed with merchants, navigators, and doctors. With his friends he met in his own private mechanical "society"—*het collegium mechanicum*—where both practical and learned men could apply their mechanical interests to watermills or the problems of navigation, and this at a time when Dutch commerce was everywhere expanding. The very practical men whom Descartes would praise in his famous *Discourse on Method* (1637) were already well established and secure in both an economic and political sense in the early Dutch Republic. Indeed Descartes tells us that the energy, access to printers, and relative toleration of the Dutch Republic were among the reasons he chose in 1628 to live there for what became nearly twenty years. Not surprisingly, his system of natural philosophy would take hold in the Dutch Republic by the 1660s. The rigid Calvinist clergy opposed it but they could not stop it.[9]

The Social Meaning of Cartesianism

Certain unique conditions found in early seventeenth-century French society fostered the Cartesian synthesis. The conditions permitted, indeed demanded, an intellectual revolution of the sort Descartes provided if they were to find resolution. High on the list of conditions was the all-pervasive concern to restore certainty in learning without encouraging the monopoly enjoyed by the scholastic clergy. There was also the need to provide a new foundation for ethical and political conduct supportive of the central government. Without a new foundation no long-term stability would be possible. In addition the condemnation of Copernicanism put intellectual reformers into a profound quandary: conform and lose the chance to reform or dramatically reorientate philosophy away from Aristotle and the schools.

Although personally close to many clergy and educated by the Jesuits, Descartes spoke primarily to secular elites. He wrote in French to "a lay audience . . . open to new ideas."[10] He offered science as an ally of their interests and passions, and he made clear his desire to bend both this new learning and the conflicting interests of disciplined elites into the service of strong central government. The only alternative to royal absolutism in early seventeenth-century France appeared to be the chaos of religious hatred and civil war led by conspiratorial grandées. In the face of these possibilities many French philosophers and contemporaries of Descartes, clergymen such as Mersenne and Gassendi, searched for a new intellectual order. It was Descartes who achieved the synthesis that set the new mechanical science into a framework wherein it could be embraced not as a naturalist heresy but as a profound truth. Amid the maze of skeptics, libertines, naturalists, and hermeticists, Descartes cut a wedge that allied the new science with the individual's ability to will the attainment of his or her own knowledge. In his letters to women Descartes made clear his conviction that his message was a universal one. His proclamation, "I think, therefore I am," when allied to the support he and his followers self-consciously offered to legitimate authority—in the French context to absolute monarchy and in the Dutch context to the urban magistrates—allied science to the social goals of order and stability. In the absolutist state that governed France until 1789 the Cartesian ideal of science in the service of order imposed by the state's representatives remained a consistent goal.

Deeply convinced that God alone guaranteed the possibility of truth, Descartes believed himself to have a sacred mission to revise completely the accepted methods of learning and to establish the method of mathematical reasoning as the key to all learning.[10] He tells us that the mission came to him in a dream in the year 1619, long before 1637 when he published the famous *Discourse on Method*.[11] By 1619 Descartes had come to define matter as would the mathematician. It is simply extension, that which occupies space, and all qualities such as color and weight are simply accidents, as it were, that result from the size or relative motion of matter. Imagine in your mind a three-dimensional triangle with length, breadth, and depth and then project it into space. What lies inside its borders (supply no color; imagine no particular substance like wood, etc.) is filled with a conglomerate of minuscule particles; that is matter. The mind must grasp the abstract configurations of matter clearly and distinctly, just as the mathematician conceives of simple numbers, lines, and curves, or as the practitioner of the mechanical arts approaches a problem in simple, local motion, abstractly reducing it to lines and points. If you want to imagine how the triangle moves imagine a larger triangle in collision with it. Then make a few such triangles out of various substances and start colliding them, measuring distance traveled, etc.; try, like Galileo, to uncover the laws that govern bodies in motion. Remember that air too has particles in constant contact; Descartes (unlike Newton) permitted no void to exist in nature. Such an approach to nature, mathematical in origin and deductive, abstract, and incremental in application, was irreconcilable with the formal, often memorized education offered in the schools. Nor did it work well with the scholastic definition of matter, which held that qualities like color or texture were inherent in bodies. In the scholastic conception a form would configure a shapeless matter into the triangle; it would be textured and colored, weighted by the nature of the substance that made it—earth is heavy and tends to fall, fire is light and volatile, rising, and so on—not because of its size or the pressures exerted on it.[12] The philosophical revolution that stands as central to the creation of modern science—namely the mechanical conception of nature—had already occurred in Descartes's mind before he sought to convert his audience in the *Discourse.*

If we ask ourselves whom Descartes sought to address when he published the *Discourse* (first in Holland in 1637; then, when approved by the censors, in Paris), we find one answer in the scientific treatises to which that essay was intended as a preface. The *Dioptrics*, for example, aroused the hostility of later seventeenth-century scientists, because they believed that Descartes had not in fact demonstrated in that treatise the laws of reflection and refraction of light. And he had used the tract to proclaim his discovery. Such objections obscure Descartes's intention. Operating with the assumption that light is instantly transmitted, and having explored its trajectories, Descartes simply proclaimed and illustrated the reflection and refraction of light, reasoning by mechanical analogies with bouncing balls and other moving bodies. The power that mechanical concepts gave permitted proclamations. Among this first generation of Continental mechanists the intellectual exhilaration did not necessarily foster empirical work.

The treatise on optics contained much more than the laws for which it eventually became famous. It dealt with light, the eye, the senses, the way the retina forms images, telescopes, and not least the best method for cutting lenses. In short it was aimed initially at practicing, intelligent lens grinders.[13] As Descartes put it: "The execution of the things that I shall say must depend on the work of artisans, who ordinarily have not studied at all. I shall attempt to make myself intelligible to everyone, and to omit nothing, nor to assume anything that one might have learned from the other "sciences."[14] Descartes's egalitarian voice should not blind us to his own genteel origins, but it should alert us to his purposefulness. He meant to speak beyond the inner circle of natural philosophers, courtiers, and clergymen who practiced science. He wrote for any man—indeed for any woman—possessed of good reading skills and some formal or artisanal education. Late in the twentieth century it is fashionable to attack Descartes for thinking abstractly about "man." But knowing what he actually said and to whom reduces that view to a caricature.

With an engraving of a simple peasant whose labors are illuminated by the light of the deity adorning its title page, the *Discourse* addressed an audience that had not been trained at the scholastic colleges or that, had it been, was also quite possibly disaffected from the old methods of learning. In the *Discourse*, actually, as noted, a preface to three scientific treatises, Descartes sought to convert practical but educated men of business and trade, among others, to the new mechanical philosophy, indeed to the new method of thinking that would be illustrated in the treatises that followed. While Descartes as a gentleman moved largely in the circles of the minor nobility, among the polite and erudite of Paris or Amsterdam, he sought also to capture the attention of just the sort of men who could have frequented the Bureau d'addresse. These were precisely the same groups to whom the policies of the absolutist kings of this era, Louis XIII and later Louis XIV, sought to appeal without at the same time dangerously alienating the old feudal elites.[15] Descartes's appeal sought out the same audience, one that could appreciate the benefits to be derived from stability and expanding commercial activity. The preface to the three treatises became the most famous text that Descartes ever wrote, probably the most famous and widely read document of the Scientific Revolution.

Descartes in his study writing his *Discourse* and in the process stomping his foot on Aristotle. (With permission from the Bibliothèque Nationale, Paris.)

The Cartesian message focused on the self and self-discipline. It proclaimed the self, its interests and passions, as the first arbitrator of knowledge. Descartes appealed to his readers with all the rhetorical brilliance of a new prophet intent on proclaiming a brilliant novelty.[16] He begins the *Discourse on Method* by proclaiming the relative equality of common sense, the most universal of human attributes. His call is directly to people of common sense; indeed he assures them that his own mind is really quite "ordinary."[17] Taking a swipe at the scholastics, he notes that even they had to admit this basic human equality in "forms" or "natures," but not in "accidents." Descartes as a mechanist will now proceed to eliminate even "accidents" from nature and philosophy.

Despite being of ordinary mind Descartes informs his readers that he has found a new method "to increase my knowledge," that he has made actual progress in the search for truth. He will not presume to teach it, only "to demonstrate how I have tried to conduct my own" reason (p. 28). Descartes eschews the pedantic role of the clerical schoolman, so that by "my frankness

I will be well received by all." He then proceeds to demolish the learning of his youth, even though he was at "one of the most famous schools in Europe." All that he learned in letters and in "occult and rare sciences" did not in fact give him "clear and positive knowledge of everything useful in life." Of course, he performed some valuable intellectual exercises, but in the end he "was assailed by so many doubts and errors." Descartes appeals directly to the skeptics; indeed he even identifies with their plight.

The way out of the morass of pedantry and skepticism lies in latching onto that which will satisfy "the curious" (for example, those who enjoy intellectual life for its own sake) and also "lessen man's work" (p. 30). Mathematics demonstrates "the certainty and self-evidence of its reasonings," not least, it is useful for the mechanical arts. It can discipline the common sense we all possess; it can teach us to make our thoughts "clear and intelligible." It alone is perfectly suited to one who must think cautiously about matters "that affect him closely," and its reasonings are far superior to those taught by the men alone in their studies who "produce no concrete effect" and are out of touch with common sense (p. 33). Theology, by the by, can be left to those who have some special grace from heaven and not to "a mere man" such as Descartes. Likewise traditional philosophy is far too uncertain (p. 32). As for alchemy, astrology, and magic, with their naturalist associations, they are best left to "those who profess to know more than they do."

Descartes's rhetorical strategy is to lay himself bare, to confess details of his own life. He did not just stumble on his method; first he had to explore "the great book of the world," the courts, the armies, where as a *gentilhomme* by rank and education he met "people of different humors and ranks" (p. 33). Out of lived experience he came to rely only on himself and his mind, disciplined by mathematics. In one of the most powerful metaphors in the *Discourse*, Descartes repudiates the wisdom of the ages, comparing it to "old cities" built on the foundations of ancient and medieval ruins. With a vision one imagines as shaped by the ordered and relatively new cities of the Dutch Republic, with their geometrical and planned regularity, Descartes would have us build cities designed as those cities might have appeared to him, by "a single architect . . . by the human will operating according to reason" (p. 35). The Cartesian man or woman—for Descartes spoke with an egalitarian voice to his many female correspondents and, he hoped, readers—wills to use her reason, to create what she wants for herself and "to lessen man's work."

By appealing to selves and wills Descartes is now on dangerous ground in relation to the needs of public order. Such a self-willed person might not bend his will or reason to the state. Descartes admonishes his readers that only the laws of God and the state have made Europeans civilized. Consequently it is "unreasonable for an individual to conceive the plan of reforming a State by changing everything from the foundations"—how senseless it would be to demolish "all the houses in a town for the sole purpose of rebuilding them." Rather the Cartesian method aims solely at ordering the individual's life (pp. 36–37). However difficult that personal quest may be, it is far easier than the difficulties that arise "in the reformation of the least things affecting the State"

(p. 37). Only "meddling and restless spirits"—we may think here of the great aristocracy or upstart lesser bourgeois—in possession of "neither the birth nor the fortune to administer public affairs" (p. 38) are forever plotting the reform of the state, and Descartes makes it quite plain that none "could suspect me of this madness" (p. 38). Breaking with the egalitarian voice he has employed so far, Descartes says that very few individuals are capable of the disciplined thinking that he has willed for himself—the vast majority are either confused or simply follow "the opinions of others" (pp. 38–39).

Descartes is calling on all restless spirits who might be tempted to meddle in the state to join him in an enterprise of quite a different sort, to rebuild solely the foundations of their own minds, "to try one's hand at architecture" (p. 45) of a most radical sort. To reorder the mind, to bend the will stoically to conquer one's self, "to change my desires rather than the order of the world" (p. 47)—this is the task at hand.[18] The reward Descartes promises for those who follow his scientific method is nothing less than mastery over nature. By comparison, he would have us believe, altering the imperfections of the state is a paltry matter, and a dangerous one. Descartes saw, perhaps earlier than any other European except those clustered around Galileo in Italy, that science in the right hands promised order and progress in the material realm without threatening to unleash the disorder that the early modern states dreaded above all else. He even gave a list of rules to be followed: "Never accept anything as true that I did not know to be evidently so," avoid prejudice, include in your reasonings only that which presents "itself so clearly and so distinctly" that it cannot be doubted; that is, focus on real objects or on rules that explain their workings, order your priorities, begin with the simple and go to the complex, impose an apparent order even if none actually exists, and keep complete records and lists of what you are doing.

Descartes's method is both scientific and rational, although not rigorously experimental by post-Newtonian standards, and it stands as the first clear articulation of the new scientific methodology to be found in modern Western thought. This model for intellectual clarity depends heavily on Descartes's experience as a mathematician, who "keeps the right order for one thing to be deduced from that which precedes it" (p. 41). Hence deduction, rather than induction based on experience or experiment, became the hallmark of seventeenth-century Cartesianism. We need not think that Cartesians could not be experimental, for they were in late seventeenth-century Italy.[19] But by and large French Cartesians remained exclusively theoretical. The combined legacy of Cartesian and scholastic teaching may account for the fact that by the 1790s French colleges were noticeably deficient in teaching devices needed for mechanical applications (pp. 180–81).

The most stunning aspect of Descartes's way of scientific thinking was the radical charter it gave to the individual. While seeking always "to obey the laws and customs of my country," Descartes, and presumably those who would follow him, must doubt all other intellectual authority. Only the self, more precisely, the thinking mind—"I think, therefore I am" (the famous *cogito, ergo sum)*—can be taken as given. The first obligation of the scientific person is to embark on an intellectual odyssey that begins in doubt and ends with an affirmation of self. Interestingly, Descartes chooses to undertake that odyssey, as he tells his readers, while living in The Netherlands, where the society is already highly disciplined (p. 51), and where "busy people more concerned with their own business than curious about that of others" permit the philosopher to live in peace. When Descartes did embroil himself in ideological brawls with Dutch anti-Cartesians and when trying to foster the teaching of his philosophy at Utrecht or Leiden he consistently sided with the lay magistrates against the strict and theocratic clergy. To imagine that Descartes had no political interests, even in his adopted homeland, is to ignore the evidence that we now possess.

Although Descartes deeply valued order and stability, ironically his major legacy entailed a radical individualism. One of the most startling aspects of his message is its insistence that even the idea of God must be perceived in the human mind before the being it describes can be acknowledged as real. Descartes's personal theism is unquestionable and manifest throughout his writings, but his method for affirming God's existence left little necessity for taking the preaching of ecclesiastical authorities as the main source of the individual's religiosity. Equally disturbing to those authorities was the tendency within Descartes's own thought and that of some of his followers to treat all of nature, including the operations of the human body, as explainable solely by reference to mechanical laws (p. 62). Without the "forms" of the scholastics, nature is solely matter in motion; from the circulation of the blood to the movement of light, "the rules of mechanics . . . are the same as the rules of nature" (p. 72). For some of Descartes's followers his philosophy implied human equality. Only the existence of the soul renders human beings separate from the material order. Descartes's radical separation of mind from body could open the door to a scientifically based materialism. In addition Cartesian thinking about the deity left little room for the nonrational, purely emotional experience of the divine so commonplace in early modern enthusiasm as practiced by male and female prophets and visionaries.

The Cartesian Legacy

Descartes's immediate French followers, such as Rouhault (1618–1672), later Malebranche and Regis, used his system in support of orthodox Christianity and rationally justified royal absolutism. Rouhault did the first Cartesian textbook, *Traité de physique* (1671), which went through twelve editions and was translated into English in 1723—by then with Newtonian footnotes that contradicted the text and undoubtedly confused many a Cambridge undergraduate. As late as 1740 Madame du Châtelet, one of the earliest French Newtonians, was trying to undermine its influence. But when it first appeared in French, Cartesians were battling for recognition, having been excluded from the new and royally sponsored, Académie des Sciences.[20] In 1663 Descartes's writings had been put on the Index of Forbidden Books, and there was a growing clerical opposition to his legacy. But Louis XIV's chief minister, Colbert, was open to Cartesian and Baconian teachings and he was intent on fostering commercial and scientific development, thus he believed enhancing the influx of bullion into the kingdom.

Rouhault focused his discussion of liquidity and the porousness of hard bodies on improving methods for separating gold from silver; his treatment of salt emphasized its commercial uses, and air pressure is discussed in relation to military weapons and fountains. Rouhault initiates a Cartesian style of application that emphasized the commercial and the military uses of science, but did not pay particular attention to local motion or mechanical devices needed to teach the laity how to harness mechanical principles in the service of industry. The Cartesian version of application remains commonplace in French science into the 1750s.

Cartesianism, as explicated by the first generation of Descartes's orthodox followers, gave ideological support to the linkage between science and state power. Competition against the growing power of English science, also institutionalized in 1662 in the Royal Society of London, further increased the importance accorded to science in the regime of the Sun King. Cartesian science, it was argued by Rouhault, discredited the naturalistic errors of "the vulgar,"[21] as well as of the Aristotelians, and it deserved a unique place within the culture of the educated elite. Knowledge of cosmology assists the study of geography and hence navigation and trade, while it was also essential that people understand the nature of metals, minerals, and salts, as well as medicine. The understanding, the ability to reason, belongs to peasants as well as philosophers, and all need to reason for themselves.

The first generation of Cartesians bent mechanical science to the needs of commercial capitalism and to the policies of Colbert. That Cartesianism might undermine Catholic doctrines, such as transubstantiation, as critics argued,[22] came to be seen as irrelevant to the consensus about science promoted by Colbert and eventually the Academy. In 1661 the right to censor all but religious books was removed from the doctors of theology at the Sorbonne and placed in the hands of the chancellor, who had the right to appoint royal censors. Slowly the tide would flow away from the scholastics. The arguments of Rohault and the other Cartesians were intended to assist its passage.

The most ideological and popular explication of Cartesian science came from a famous academician of the Academy of Science, Bernard de Fontenelle (d. 1757). His *Conversations on the Plurality of Worlds* (1686) went through five editions within four years of its publication, and some twenty-five editions in various languages by the mid-eighteenth century.[23] It presented a very simplified Cartesianism for the edification of the *noblesse* and *philosophes* of the salons, but it also specifically charged the Cartesian universe with the task of providing a model for the absolutist state. The scientific universe was to be the province now of the same elite that frequented the court and the salons. Knowledge of it permitted them to achieve a remote and superior knowledge of all other "worlds": those of "the people" and of all foreigners. Most important, Descartes's philosophy revealed an underlying order amid the apparent perturbations and dangerous motions that sometimes characterize the material world. Fontenelle would have similar laws work in society "which would fix people in those spheres of life that are natural to them." Just as the little planets obey the force of the larger planetary motions, so too the smaller bodies in all places will render homage and respect to the larger preponderance of the state. It was an ideal that fitted well with the goals of the Academy as prescribed by Colbert. The scientist must be a disinterested servant of the state[24]; the great nobility should ponder closely his exemplary role.

It is worth noting that Fontenelle's conversations were with an aristocratic lady. At the Parisian salons fascinated by the new science, we find the first evidence for a significant interest in it among women. The Cartesian message of thinking systematically for oneself had the potential to enlist any educated human being as its follower. Although this interest in the new science among early modern women never provided them with access to institutional science and hence to power, their interest can be shown to have existed as early as the 1650s in both England and France and to have increased in the course of the eighteenth century.[25] Science, even when in the service of absolutism, could promote an intellectual freedom that ultimately boded ill for those who would monopolize its power. One of the first systematic French feminists was the Cartesian, François Poullain de la Barre, and his egalitarian arguments published in the 1670s were cited by women for many decades to come.[26]

With the textbook consolidation of Cartesianism as a wholly new and comprehensive system for explaining the physical universe and for thinking scientifically, its earliest proponents sought through the printed word and public lecture to capture a new and even larger audience for science. This phenomenon of propagation, linked as it was to the printing press and to the expanding world of secular fraternizing—the salons, coffee houses, and cafés—was perhaps the single most important factor in rendering the new science into a unique and vital aspect of Western culture. We need to acknowledge that it was Cartesian propagandists and lecturers in the period prior to 1700, operating necessarily outside the traditional French schools and universities, who first sought the largest audience for science ever assembled in seventeenth-century Europe. They operated even more effectively in the Dutch Republic where by the 1660s Cartesianism was firmly implanted in the universities.

A quick survey of Cartesian literature for the laity reveals the intentions of its authors and the general tendency of Cartesian science to promote order in the state as well as commercial development. Following in the footsteps of Jacques Rouhault, Pierre Sylvain Regis lectured in Paris and the provinces. He spread Descartes's message while tying it to the need for absolute authority and order in the state. Reason would guide the self-interested subject in this ordered probity, just as the laws of Cartesian physics guide the universe. For Regis, physics and morality were intimately connected, and both must serve the cause of reason and order in society and government. His lectures moved effortlessly from an explication of the Copernican universe based on Cartesian principles, to the earth, the nature of air, water, and salt, the properties of metals, fermentation, vegetation, simple chemistry in general, wind, the motion of the sea, the role of God in creation, anatomy, the principles of civil society, the nature of the passions, and only last, mechanical devices intended largely to illustrate Cartesian mechanics.[27]

In many respects the lectures of Regis, first published in 1690, anticipate the popular format adopted within a decade by the English and Dutch followers of Isaac Newton. But in one critical respect Cartesian lectures differ from their Newtonian successors: relatively little attention was paid by Cartesians to mechanical devices intended for industrial application. The Cartesian natural philosophers of seventeenth-century Europe possessed in many cases a vision of the role of science far in advance even of that imagined by Colbert. But when they sought an audience, both philosophers and listeners responded to the economic order everywhere around them. To that extent Cartesian science would remain, in its mainstream, ideologically absolutist in politics and practically commercial in application. In contrast, after 1700 the science of the Newtonians with its debt to Baconian empiricism became the science of constitutional monarchy and of early industrialization.

Yet the Cartesians were the true pioneers of modern science. They often labored in a hostile climate. The Sorbonne had little use for them; eventually the archbishop of Paris closed down Regis's lectures "in deference to the old philosophy." Despite this opposition, Cartesian science spread in France, although not as rapidly as it did in Protestant Europe. Its French growth, however slow, eventually established Descartes's science at the Academy of Sciences and at about 170 liberal arts colleges (out of 400) that actually taught natural philosophy. Once ensconced, Cartesianism proved hard to dislodge. As we shall see in chapter 5, by the 1740s when Newtonian mechanics had captured the educational system in England, in Scotland, and at the major Dutch universities, the French colleges continued to teach Descartes relentlessly. As a result educated Frenchmen of the generation prior to the 1750s missed any formal education in practical Newtonian mechanics as well as in the entire Newtonian philosophical outlook.

3

Science in the Crucible of the English Revolution

No single event in the history of early modern Europe altered the fortunes of the new science more profoundly than the English Revolution. At its outbreak in 1640 Descartes's *Discourse* was being read and appreciated in select circles while the earlier writings of the utilitarian prophet of science, Francis Bacon, were also undergoing a marked revival. Just as in France, the issue of the acceptability of the new science remained hotly contested. Any dramatic alteration in the concept of the natural world required the integration of socially relevant beliefs and the needs of authority—the integration, in short, of religion and an ideology of political order.

In mid-seventeenth–century England the stakes raised by the need for order were dramatically high: the intellectual fermentation provoked by natural philosophy occurred within the context of a larger political and religious agitation for reform and renewal. By 1641 agitation led to open civil war between the king, Charles I, and Parliament. Indeed the language of astronomy, partly indebted to Copernicus's "revolutions of the heavenly orbs," provided vocabulary for the profound changes of the 1640s and 1650s. By 1660 the terms "revolutions and commotions" had become commonplace.[1]

Prior to 1640 religiously inspired reformers hurled their demands against what they saw as an obstinate monarchy, a corrupt court, and an insufficiently Protestant church. This confrontation first produced armed revolution, finally regicide, and the concomitant social upheaval affected the shape and direction taken by the new science. On the basis of modern scholarship the verdict has been rendered that the English Revolution shaped both science and its integration, not only in England but as English science spread also in much of the

Western world. The revolution contoured the natural philosophical thinking of Robert Boyle (1627–1691) and Isaac Newton (1642–1727) in ways that assisted in the development of their purely experimental and mathematical interests. Out of their achievement, a synthesis of philosophy and experimental method we describe as discernibly modern science came to exist. In addition the English Revolution raised the fundamental issue of the social uses of the new science; indeed scientific progress of a Baconian sort was central to the revolutionary vision of the Puritans. By the late 1650s they had failed to achieve their goals, but in the process they rendered science and natural philosophy vital elements in any alternative social ideology. By 1660 and the end of the first phase of the English Revolution, the prosperity of the English state came to be seen as linked—at first tentatively and then decisively—with the development of science and technology. The linkage between prosperity imagined and science improved, with technology as the assistant to both, remains part of the Western vision to this day.

Predictably many historians writing over several decades have demonstrated that the English Revolution, understood as a crisis stretching from the 1640s to the late seventeenth-century revolution of 1688–1689, bears close relation to the development of science. In the 1930s the American sociologist Robert Merton illuminated the links between the progenitors of the first revolution, namely the English Puritans, and the origins of modern science.[2] He provided convincing evidence to show that the Puritans, indebted as they were to the Calvinist doctrines of striving and predestination, were also particularly attracted to scientific inquiry. This evidence further strengthened the case for the linkage between European Protestantism and the rise of science, a case that can be documented not only in England, but also in the Dutch Republic after its liberation from Spain (1585).[3] More recent British and American historians have further developed the linkage between science and the Puritan reformers. They have stressed the importance accorded to science in the 1640s by a circle of leading reformers led by Samuel Hartlib. Francis Bacon never intended his vision to be used by either Calvinists or revolutionaries, but his wishes could not stop his books from being read and discussed by Puritan reformers.

Bacon and the Puritans

The Puritan reformers seized on the writings of Francis Bacon as their guide to the new scientific spirit and its empirical methodology. They interpreted Bacon not as the humanist and state builder he was, but rather they emphasized the millenarian and reformist aspects of his thought as revealed in *The New Atlantis* as well as *The Advancement of Learning*. Perhaps the most unique aspect of reforming Protestant thought, and one that sets it off from many Continental religious movements that embraced the new science, was its millenarianism. Put quite simply, English Puritans believed literally in the scriptural prophecies about the final days and the end of the world. God directs the course of human affairs, just as he directs the course of nature. At some point in time, one that could be determined by careful scholarship or even possibly by a gnostic illumination, history and nature would synchronize as God destroyed the world in a cataclysmic upheaval that would precede the institution of a thousand-year reign of the saints, the millennium. Almost every important seventeenth-century English scientist or promoter of science from Bacon to Robert Boyle and Isaac Newton believed in some version of the approaching millennium, however cautious they may have been in assigning a date to its advent.

The extraordinary link between new ideas about nature and millenarianism ought to alert us to beware of the simplistic categories—scientific versus magical, rational versus irrational—as having much relevance to the ideology or value systems of seventeenth-century science and scientists. Contrary to the retreat from worldly concerns or the conservatism that we might expect fundamentalist millenarianism to inspire, the Puritans' vision made urgent the work of worldly reform and renewal. Science or natural philosophy, as well as medicine, figured as critical elements in these Puritan projects. Hartlib and his friends promoted Baconian schemes for classification and improvement that led them to advocate universal education and medicine for everyone (a physician in every parish was the goal). They embraced the new philosophy as the cornerstone of all scientific inquiry; they championed mechanical experiments intended for improving the output of labor, and chemical experimentation for agricultural improvement.

With the outbreak of civil war against Charles I, the 1640s became an exhilarating time for social reformers as well as for natural philosophers and scientific experimenters sympathetic to the parliamentary cause. From that period we can trace the inspiration for the eventual founding of the Royal Society of London in 1662, as well as plans for the founding of new colleges, grammar schools, and academies, for a multitude of technological innovations in everything from mining to banking, and for a central "office of address" for the communication of useful knowledge. We can locate the 1640s, too, as the time of the earliest chemical experiments of the young Robert Boyle, an associate of Hartlib and his circle.

Within the context established by revolution and civil war, modern science in its English guise was perceived in terms of its social usefulness and linked to a larger vision of reform and enlightenment. The Puritan reformers assaulted the old monopolies of the physicians and the universities, and they took up the latest in scientific inquiry—everything from the theories of Paracelsus to the writings of Galileo, Bacon, and Descartes.[4] Had they succeeded in the 1640s in establishing even a modicum of these reforms, particularly in the areas of medicine and scientific education, the more humane aspects of scientific inquiry, as distinct from military and commercial adaptations, might have prevailed. More attention might have been given to medicine than to mechanics. But history cannot be written about what might have been, and by the 1650s, after the execution of the king in 1649, a very different, less militant mood had come to prevail even among many of the landed or propertied reformers.

The greatest fear of any early modern elite focused on the danger of popular unrest. The widening gap in seventeenth-century Europe between rich and poor, or simply between the relatively prosperous and the indigent, coupled with the absence of effective policing mechanisms, made even the possibility of lower-class revolt, whether by peasant or artisan, the most dreaded of all possible occurrences. In France one of the mainstays of monarchical power became the expected ability of the king and his bureaucracy to ensure an adequate supply of bread in the cities and the countryside. Failure to do so threatened the possibility, or the actuality, of civil unrest among the lower orders. But after 1649 England was governed without a king, by Parliament and increasingly by the army made up of lower-class volunteers.

As early as 1641 policing mechanisms in publishing and mores largely dissolved. The censorship of books, a function of established Anglican church censors, largely disappeared. Parliament dismantled the privileges of the established church and failed to put in place an equally effective alternative.[5] As a result more books and pamphlets were published in England from 1640 to 1660 than in the entire rest of the century put together. More dangerous than the ideas in books, however, were the various new sectarian movements that came to the surface abruptly in the 1640s. The theology of many of these religious groups can be traced back to the more radical phase of the sixteenth-century Protestant Reformation, when doctrines such as the priesthood of all believers, or the enthusiastic notion of the "inner light" within every man and woman that licensed private beliefs, came to justify the religiosity and religious freedom of lower-class, often illiterate, reformers.[6] That sort of radicalism was not what the Puritan reformers had in mind when they urged their reforms on Parliament.

In its war against the king, Parliament came to be dependent on the New Model Army, and within its ranks could now be found many radical sectarian movements and their most articulate leaders. Suddenly the dialectics of revolution threw to the surface a threat more serious even than the absolutist policies of Charles I against which Parliament had initially revolted. The radical reformers and their inchoate sects—the Levellers, Diggers, Ranters, Quakers, Muggletonians, and Socinians, to list only the most prominent—demanded a range of reforms, from one man, one vote, property redistribution, and complete religious toleration, to the right of women to preach, the end of church tithing, and the curtailment of the privileges of licensed physicians and lawyers. Some radicals mocked the sober lifestyle of the Puritans, their dedication to the work ethic, and their haughty strictures against swearing, smoking, drinking, and sexual freedom. Women stood up in chapels and preached against ministers; one man rode into Bristol on a donkey proclaiming the coming of the Messiah. Other heretics asserted the unity of God and said that Christ had been a good man, but not divine. Still others argued that the soul falls asleep at death and there can be no hell. Robert Boyle seems to have been particularly offended by such Socinian and mortalist views. Against every established institution, orthodox doctrine, or propertied interest unlicensed reformers threatened to turn the world upside down.

The radical threat posed to the moderate reformers included a direct challenge to the kind of science they would promote. Radical reformers also wanted a new science; they too repudiated the Aristotelian learning entrenched at

Oxford and Cambridge. But in place of Aristotle they would put Paracelsus or the naturalism and magic associated with the Hermetic tradition. According to John Webster, a surgeon and for a time chaplain in the New Model Army, the radicals wanted "the philosophy of Hermes, revived by the Paracelsian Schools"; he would have installed in the universities "true Natural Magicians, that walk not in the external circumstance, but in the center of nature's hidden secrets." They must have "laboratories as well as libraries" to pursue their alchemical and medical experiments.[7] The moderate promoters of the new mechanical philosophy took up against the natural magicians and simple "mechanicks" (or artisans), or more precisely against the radical and sectarian forces that would "turn the world upside down" by using the "science" of the people presumably in their interest.[8]

The moderates who had begun the revolution as Baconians and scientific reformers now found themselves outflanked by the radicals. The moderate scientific reformers and theorists—such as chemist Robert Boyle, the political economist, William Petty, and the linguist reformer, John Wilkins, as well as Henry More, the Cambridge Platonist and later tutor to the young Isaac Newton—used the new mechanical philosophy, particularly of Descartes and also of the French atomist and priest, Gassendi, against the naturalism of the magicians. The moderates were forced to articulate an intellectual and social stance that now sought to conserve as much as it sought to change. There may have been important differences between the philosophical commitments of More and Boyle, but there was no disagreement between them on the need to control change and temper extremism.

Natural Philosophy and the Reaction Against Sectarian Radicalism

One group of moderate natural philosophers led by Boyle, Wilkins, John Wallis, John Evelyn, Christopher Wren, and others went on to become the founders of the new Royal Society of London.[9] Their goal was to promote the organized pursuit of experimental science, but to detach it from any attempt at the radical reform of church, state, the economy, or society. They did not cease entirely to be reformers but rather these founders of the new private society couched their reforming sentiments in vague terms about improving man's health and estate through the disciplined work promoted by science. When they did become more specific it was, for example, to indicate the ways by which experimental science might be deployed to control the lower orders; curb the excesses of the great while increasing production, especially food production; and promote good health and commerce.[10] Scientific progress should occur without altering the existing social arrangements in the direction of greater leveling or redistribution of property. The founders of the Royal Society did possess modest reforming aims; they just sought urgently to avoid any further democratization of politics and society.

As if to symbolize their moderacy, many of the increasingly conservative natural philosophers left revolutionary London and retreated into Oxford colleges away from the social and political turmoil to pursue their thoughts in quiet contemplation. And when radicalism threatened to jeopardize their freedom within the university, they stepped in to oppose. The moderate Oxford reformers made a point of avoiding all religious and political questions while discussing science. This, however, does not mean that they were unaffected by the outside world and did not have their own opinions about it. Rather the reformers consciously distanced themselves from radicals who saw in science a powerful tool for promoting religious, political, and social revolution. In the radical vision science could justify democracy in church and state; it might also be used to extend popular education in schools and universities and to build a new society that would be more just and rational.[11]

The moderate reformers did more than simply retreat to Oxford and defend it against the proposals of the radicals. What was at stake now after the defeat and execution of the king in 1649 was nothing less than the survival of social order and property. The kingless country was now governed by Parliament and army, and the threat of lower-class and sectarian radicalism seemed everywhere apparent. In the moderate vision of natural philosophers like Robert Boyle science tempers extremism, both the zeal of the radicals and the arrogance of the old aristocracy, many of whom had doggedly fought on the side of their king.

What must be grasped in the crisis of the 1650s is the important role played by philosophies of nature in giving expression to human goals and aspirations. Natural philosophical and religious language constituted the stuff of scientific discussions; it also shaped discussions about the nature of political authority, the rights of the church, and the relationships between master and servant, husband and wife, lord and commoner. To picture the cosmic order, "the world natural" as Newton put it, was to speak by analogy of "the world politick." For a body to rise toward heaven or fall toward the earth might symbolize the "rising and falling in honour and power" of states or individuals. On a more abstract level the relationship between God and nature, hence between spirit and matter, could express the beliefs of an individual or sect concerning the role of priestly or kingly authority. If God stands above his creation, if spirit clearly dominates matter, does this not justify the continuation of a similarly authoritarian structure in society and government, both ecclesiastical and civil? Or to put it another way, if the spirit of God dwells in nature, in everyone, what need is there for the heavy hand of priestly or magisterial authority?

Such questions impinged directly on natural philosophy and the new science; indeed they were to haunt the new science throughout the seventeenth century. The most pressing concern of its defenders centered precisely on how to define the relationship between spiritual forces and matter. The Cartesian legacy demanded that matter be seen solely as extension, as the physical protrusion in space of an infinitude of clustered corpuscles. Consequently matter in motion could be explained only by reference to contact with other matter. Such a mechanical explanation could easily counter arguments that depended on inherent spiritual qualities or occult forces, which in turn justify a magical approach to nature. In one sense the new mechanical science was a perfect tool to hurl against the hermetic and the magical. But in another sense, mechanical arguments derived from Descartes came dangerously close to situating within matter the power to move itself. And once so endowed, how did the matter of the mechanical philosophers differ from the universe described by English radical philosophers and sectaries? Their universe was so filled with the spirit of God as to be moved by a pantheistic force open to the experience of every man and woman.

In England by the 1650s many pantheistic philosophers and prophets openly published and preached their beliefs. The Digger, and first English communist, Gerrard Winstanley, believed that God was in all things, that the creation stands as the clothing of God.[12] He also equated God with reason—an idea drawn directly from the Hermetic tradition with which the sixteenth-century reformer Giordano Bruno (see chapter 1) may also have been in contact. The Ranters believed the hermetic and pantheistic doctrine about spirit in the world, and it justified their self-conscious departure from Puritan morality. The spirit moved them to live as they pleased. One of the leading Quakers of the period, George Fox, admitted that he had nearly succumbed to the doctrine, also preached by Ranters, that there is no God and all creation comes from nature.[13] Another Leveller, Richard Overton argued for a pantheistic materialism and for the doctrine that the soul falls asleep at death; other radicals believed that it died with the body. All these ideas were part of a metaphysic intended to lay the foundation for a new religiosity and a new society, one less inequitable, freer, less rigid in its social and economic divisions, less priestly and magisterial in its system of authority. Such ideas constituted a direct challenge to the authority of the propertied and educated elites.

Hobbes

Heresy during the English Revolution came dressed in many forms. It could even emanate from the scientific community itself. Thomas Hobbes (b. 1588) may be best known today as a political theorist; but in the 1640s Hobbes was at the center of scientific and natural philosophical discussions in Paris as well as in London. He was a committed mechanist, a mathematician, and a violent opponent of Aristotle and his followers. He was also a royalist, loyal in his way to the absolutism of the Stuart monarchy, and tutor to the exiled son of the king. Yet Hobbes had to survive in a political universe made perilous by civil war and revolution. He was in no sense a sympathizer with radical causes, and he did not want to burn his bridges with Parliament.

Hobbes's political philosophy as found in his most famous work, *Leviathan* (1651), rested on a version of the mechanical philosophy that ignored entirely the operations of spiritual forces in nature. In denying a role for spirit in human affairs he likewise denied any independent role to clergymen—"those bugbears who prick the sides of their princes," as he injudiciously described them. Of course the clergy were the traditional guardians and interpreters of the operations of spirit in the world. But according to Hobbes, the clergy's claims to be intermediaries between God and man were bankrupt and they should be reduced to mere functionaries of the civil sovereign. Hobbes represented heresy coming from within elite culture itself, from the best scientific circles.

As seen by the Puritan moderates and Christian scientists, or *virtuosi* as they were called, Hobbes and the radicals made a curious alliance in their heretical philosophizing. The first sucked spirit out of the world so as to rationalize and explain in material or mechanical terms the existence of human greed and self-interest; the radicals pumped it back in with the intention of shaking the material and social order. Between Hobbes, the sucker, and the various radical pumpers, moderates concluded there was not much to choose from. Hobbes was just as dangerous, if not more so because he traveled in circles of powerful men.

During the 1650s the scientific reformers modified their understanding of nature to answer the threat posed by both Hobbes and the radicals. In place of either pantheism or materialism, Robert Boyle offered what he called his corpuscular or atomic philosophy and rendered it the foundation of chemistry. This amounted to a Christianized Epicurean atomism that Boyle elevated to the status of a hypothesis to be tested by experiment. Cautiously, he said that atomism was not a dogma, rather it was a theory worthy of consideration. Boyle and his associates held with Epicurus that the world was made up of lifeless atoms colliding in the vacuum of space. But the Puritan philosophers and their associates, like Boyle, departed from Epicurus by denying that the world as we know it was the product of a long succession of random atomic collisions. Rather, they held that only a providential God, not random chance, was responsible for all motion in the universe. He determined the paths the atoms took and hence maintained the order of the universe. Not only was this a workable scientific hypothesis capable of being refined and elaborated by a Baconian program of experiment, it was also an attractive candidate for adoption because it was applicable to social issues.[14]

Boyle's Christianized corpuscular and experimental philosophy allowed the Puritan scientists to escape the taint of heresy associated with the occultism and animism of the radical sectaries. More important, it allowed the reformers to attack the radicals. The idea that matter is moved mechanically by the will and according to the intelligence of a supernatural God upheld the orthodox Christian dualism of matter and spirit in the face of the radicals' animism, their belief that all matter was endowed with soul and that spirit was immanent in nature.[15] Nor was dualism merely a victory against false theological doctrine; it also had religious and political ramifications. The vitalistic or pantheistic idea of nature provided the metaphysical grounds for an attack on traditional authority in church and state. If spirit lay within people and nature, radicals had a strong argument against organized churches, supported by tithes and learned ministries. Traditionally clergymen claimed superior spiritual wisdom and separate spiritual authority—the power to teach, discipline, and punish.[16] Vitalism, with spirit diffused equally in the material world, could also be used to support the notion of human equality and to justify in cosmic terms antimonarchical and even democratic political ideas. The natural philosophy of the radicals tended to dissolve hierarchy, while hierarchical social order found support in the Christian dualism newly shored up by the corpuscular philosophy of the reforming Puritans like Boyle.[17]

The inductive or experimental aspect of the new corpuscular philosophy worked out in the 1650s also bore an ideological message meant to counter the radicals. Scientific progress would come through painstaking inquiry, the collection of evidence, and the testing of hypotheses. Knowledge then was not, as the sectaries with their emphasis on magic and the occult maintained, the result of mystical experience or God's direct revelation to the saints. God instead revealed himself indirectly by two means, nature and Scripture, his work and his word, and both required close study to bear fruit. This emphasis on patient, industrious scrutiny was directed against the antinomian theology of the radical sectaries, which insisted that God revealed himself immediately to the saints so that they might achieve perfection or at least perfect wisdom in this life. The fruits of salvation were accessible here and now as well as in the world to come. For the Puritan reformers, on the other hand, the effortless pleasures of salvation were deferred to the next life; in this life rewards would come only through reason and industry. Science, the new philosophy, was the model; knowledge would come to men of merit not through visions or divine illumination but rather through a searching and sustained inquiry into nature, the humility and dedication of the experimental philosopher. Nor was this update on the work ethic only directed against the illuminism of the sectaries; it was also seen as an instrument of social control to curb the excesses of the great. As Boyle insisted, hard work would keep men too busy to contrive heresy, to plot social revolution, or to dissipate their talents. Science would be particularly valuable in this regard because the practical application of its discoveries would create more employment.[18] The ethical posture of science as the pursuit of the truly meritorious originated in the mid-seventeenth century. In chapter 6 we will see its effects on one eighteenth-century family, the Watts of steam engine fame. When the ethic was formulated its originators saw it as a panacea for social peace. The corpuscularianism and the experimentalism of the reforming natural philosophers were designed to combat two threats, heresy and social insubordination, at the same time.

The reforming Puritan savants also deployed their corpuscular philosophy against Hobbes and Hobbesists. Indeed after 1660 Hobbesism was increasingly identified with subversion—and with good reason.[19] Hobbes's arguments denied the newly restored Anglican church an overarching authority in society and government. By contrast, corpuscularianism preserved a role for spirit in the universe; namely, immaterial forces imparted motion to matter and gave shape to the world through providential design. The corpuscularians, against Hobbes's materialist surgery, supported the imposition of order from above, hence the clergy's authority as interpreters of God's ways and will. The experimentalism of the Puritan savants also offered the way to knowledge through induction and the testing of hypotheses—not through Hobbes's deductive rationalism based in part on mathematical reasoning. Hobbes had advocated a purely geometrical science because he saw it as less contentious, as safer in securing consensus among the great. It was compatible with his absolutism. It required no separate space wherein voluntary societies would meet to conduct experiments, thus further securing the state's hold over its subjects. Boyle and his followers feared the development of monarchical absolutism as it existed in the great Continental and Catholic monarchies. The Christian *virtuosi* wanted a church separate from constant scrutiny by the court. Science in so far as it allied with Protestant Christianity enjoyed a comparable freedom. It also required relatively free exchanges among properly trained and cautious experimenters, hence the necessity for an independent, private experimental space.

Under the threat of radical sectarian and Hobbesist challenges the once Puritan philosophers grafted their reforming and experimental science onto an ideology that sought to reestablish order and stability in church and state. Science would not only alleviate man's material condition; it might also cure the excesses of revolution. The natural philosophers like Boyle and Wilkins who created the new ideology of practical science kept the original reforming aims of the Puritan scientific vision, particularly when they were easily accommodated to, or even promoted the larger political and religious goal. So they continued to argue for science as a means to greater private profit, national wealth and power, because science, to the extent that it increased agricultural production, trade, and shipping, would also foster domestic peace. The scientific protagonists equated a science made practical with growing prosperity, social order, and the public good.[20] The scientific culture that blossomed throughout the eighteenth century had its roots in the intellectual formulations of the 1650s, as refined by Christian and Anglican *virtuosi* in the period after 1660.

The Anglican Origins of Modern Science

The initial Puritan reforming vision of the 1640s thus survived in a continuing belief in the material benefits of science. By the late 1650s, however, this belief in science as an instrument of material progress was wedded to a new Anglican theology, one no longer essentially Puritan, but rather, liberal or latitudinarian. Its central tenets were the repudiation of the Calvinist doctrine of predestination, a concomitant emphasis on free will and striving as the keys to salvation, and an almost obsessive concern for design, order, and harmony as the primary manifestations of God's role in the universe. Evolved during the 1650s, this liberal Anglicanism relied on the reforming vision of the new science to verify both God's order in an unstable world and the superiority of cautious scientific inquiry over the illuminations of the spirit.[21] During the Cromwellian Protectorate the hope among men like Boyle and Richard Baxter was that this view could be translated into a church settlement based on proposals for moderate episcopacy. There would be bishops, but they would be tolerant and their power over other sincere Protestants limited.

These hopes, of course, were never realized. But they survived the Restoration of church and monarchy in 1660, and a scientifically grounded latitudinarianism, as it came to be known, received its classic formulations in the works of Robert Boyle published after 1660 and in Thomas Sprat's famous *History of the Royal Society* (1667), masterminded by John Wilkins. It was thus adopted as the public stance, if not the official ideology, of the Royal Society.[22] Briefly stated, the early version of latitudinarian or liberal Anglicanism designated science as a unifying force among all moderate Protestants, sought a way to bring repentant Puritans back into the Church, and gave complete but not slavish support to the restored church and monarchy. Moderates like Boyle wanted order and stability, monarchy and church, but not Continental absolutism.

Liberal Anglicanism distinctively assisted in the integration of the new science into the mainstream of English and eventually European thought. Churchmen of liberal persuasion began to base theological positions on scientific knowledge. They preached about the order and stability of nature and invented a scientifically grounded social ideology, as well as a new religious piety. It endorsed experimentalism and material progress based on science in a way in which no other contemporary social or religious vision had done. Anglican liberals made science a fit subject for pulpit discourse and in so doing rendered it more immediate to daily thought and experience. English churchmen made science far more relevant to worldly concerns than did the weighty tomes of the French Cartesians. The new mechanical philosophy as articulated by Boyle and his circle managed to escape the trap set by Cartesian dualism or Hobbesist mechanism; it was resolutely antimaterialist, not to mention antimagical, and hostile to sectarianism. The importance of this late seventeenth-century English synthesis permits us to speak of the Anglican origins of modern science as being not opposed to, but as superseding the Puritan origins of modern science. Liberal Anglicanism provides the ideological continuity between the science of Boyle—that is, the experimental method of modern science—and the science of Isaac Newton. With Newton, the new science achieved maturity. As Newtonian science spread it became less theological and more practical. Experimenters and lecturers made mechanics into an understandable and usable synthesis by which the physical order may be explained and exploited. Gradually Newtonian science shed some of its Anglican associations and became a cultural resource taught and applied by non-Anglican dissenters like the Watts who were of Puritan origin, as well as by marginally religious secularists.

English Science and Society Before the Principia (1687)

The cultural milieu of the English Revolution (1640–1660) sharpened the social implications seen in the varieties of discourses about nature available to early modern Europeans. To capture nature's power magically conferred power on self-proclaimed priests and prophets alike. Hence once again, as in France in the time of Descartes, naturalistic and hermetic doctrines had fallen into the hands of ordinary people. Only now in England radical Protestantism provided doctrines like the inner light and the priesthood of all believers, which further justified a heady arrogance toward established authority. The older scholasticism remained the true ally of Catholicism; and for English Protestants, with plenty of evidence to point to on the Continent, Catholicism meant absolutism. The scholastic alternative supported Catholic doctrines like transubstantiation, another reason why it was not viable. In the face of both Catholic and radical postures toward sovereign authority and the state, Hobbes offered a basic materialism as the only sane response. But the Hobbesist alternative sanctioned only mathematical and not experimental science; he also allied his materialism with an entirely secular form of absolutism. Of course, Hobbes was no Catholic, indeed he despised the power of the clergy, any clergy whether Protestant or Catholic. Moderate Protestants like Boyle and the Cambridge Platonists saw him as no friend to either the English church or state, nor did he support the kind of science that they wanted. Experiment required a separate space, an audience as well as specially acquired skills and technology, a civil society separate from the state, hard to police as a result. Hobbes's mathematical way, he said, was politically safe and if instituted, could easily have been policed by an absolute monarch.

Amid the absolutists, radicals, and Hobbesist materialists, by the late 1650s landed gentlemen as well as commercial venturers of Protestant persuasion wanted to retain a state religion and to ensure material prosperity. Increasingly they turned to what Boyle and later the Royal Society had to say. Boyle's air pump, the advanced technology of its day, was eagerly purchased and improved on. In London and the provinces a new audience, larger probably than what Descartes and the Cartesians courted in France or the Dutch Republic, found science attractive. Liberal Anglicanism countenanced a clergy docile to the landed and propertied, and one therefore not given to supporting the middle and lower orders. The revolution had unleashed a democratic impulse that had to be resisted. Likewise Boyle, Wilkins, and the leadership of the Royal Society supported a private, voluntary, and genteel engagement with nature, separate from both state and church but hardly hostile to them. Within this context and in the face of so many unacceptable alternatives, after 1660 Puritanism gave way to liberal Anglicanism, and the mantle of science passed to a new generation of intellectual leaders. Out of that generation came the metaphysical and religious assumptions that made possible the Newtonian synthesis.

The Social Elements in the Newtonian Synthesis

Before the extraordinary Newtonian synthesis could be achieved powerful reasons had to be found for a complete repudiation of Cartesianism. Among the Continental proponents of the new science in both France and the Netherlands such a total rejection of Descartes seemed unnecessary, if not bizarre. While there were problems clearly evident in aspects of Descartes's cosmology and physics, his unrelenting insistence on mechanisms and contact between bodies seemed the only viable alternative to the occultism of the magicians or the qualities and forms of the scholastics. Consequently, as we saw in the previous chapter, Cartesianism made slow but steady progress in the universities of Continental Europe, particularly in Protestant countries, and indeed in England and Scotland. Only in Cambridge in the 1660s among the liberal Anglican opponents of the radical sectarians and Hobbes, did Descartes's system come to be seen as untenable because of its materialist implications and nonexperimental style. The English Revolution and the reaction to it created the ideological climate that in certain circles undermined Cartesian certainty. Newton could not have laid the metaphysical foundations for universal gravitation and remained a Cartesian. In that sense we may say that while the culmination of the Scientific Revolution is unthinkable without Newton, Newton is unthinkable without the English Revolution.

If we focus attention very precisely on the Cambridge colleges of the 1660s, on the moment when the young Newton came to Trinity as an undergraduate (1661) and on the extant printed and manuscript evidence available, we can witness the intellectual revolution to which the young, but brilliant student of natural philosophy was exposed. In the 1650s the main tenets of Puritanism had been firmly repudiated among certain philosophers and college fellows who nevertheless wished to retain the new science. They too had repudiated scholasticism and turned, therefore, to the Christianized Platonic tradition of the Renaissance to search for explanations of nature that would counter Aristotle while preserving the basic tenets of Protestant Christianity and the immediacy of God's presence in his creation. Neo-Platonism, it was believed, when wedded to the new science, would preserve mechanical action while retaining spiritual forces in nature.[23]

The leaders of this Cambridge school were Henry More and Ralph Cudworth. In More's earliest published work, a collection of poems, *Platonica* (1642), he sought to articulate a Platonic sense of spiritual forces in nature that could be understood scientifically. At first he was also powerfully attracted to Descartes's writings and even corresponded with the French philosopher. In this same period More observed with horror the dislocations produced by the civil wars and the interregnum. He came in turn to despise enthusiasm and Puritanism—"such horrid errors, that they seem the badges of the kingdom of darkness"—just as he detested Catholicism.

By 1653, however, More was also voicing reservations about the Cartesian system; in 1665 he registered to Boyle his complete rejection of Descartes based on his fear that Descartes's system, just like the systems of Hobbes and Epicurus, led directly to atheism. More believed that true atomism required the assertion of spiritual forces in nature and a rejection of a purely mechanical and random material causation. Because he was a teacher of Newton, More's rejection of Descartes provided the backdrop for his pupil's theory of active principles and his life-long preoccupation with the role of the spiritual and immaterial in nature. Without the belief in active principles, Newton could never have postulated the existence of universal gravitation as an immaterial force operating throughout the universe independent of any direct, mechanical contact action between bodies. In the early 1660s the Cambridge Platonists sought, in their own words, to give the new generation of undergraduates an alternative to both Descartes and Aristotle:

> And seeing that they will never return to the old Philosophy, in fashion when we were young scholars, there will be no way to take them off from idolizing the French Philosophy and hurting themselves and others by some principles there, but by putting into their hands another body of Natural Philosophy, which is like to be the most effectual antidote.[24]

In Newton's earliest student notebook of 1663 we can discern that search for the antidote to materialism of a Cartesian variety. He is drawn to the atomism of Gassendi at this early stage, and he is repelled by Descartes's definition of matter as an infinitely extended plenum; the young undergraduate reasons that if all the universe were filled with matter, then there would be no room for motion. Atomism, on the other hand, permits vacuity between the particles, and it became one of the cornerstones of Newton's mature philosophy of nature. Newton's early student jottings show him abandoning Aristotle and encountering Descartes. But the young Newton found his natural philosophy (although not his mathematics or the new science per se) wanting.

Consequently Newton embarked on an intellectual odyssey totally dominated by contemporary scientific problems. His notebooks also reveal that he was in touch with natural philosophical matters frequently discussed within select circles at the university. Throughout his notes he responded to standard tutorial questions as well as to those philosophical issues. We know that in this period Hobbes and Descartes were being read at the university, even though in 1667 the vice chancellor of Cambridge University publicly condemned the reading of Descartes by candidates for the baccalaureate.[25]

With the restoration of the monarchy in 1660, the polemic against Hobbesism, enthusiasm, and naturalism began in earnest. While Boyle and Joseph Glanvill pounded at the naturalists in print, More and Cudworth in Cambridge worked out a variety of attacks against Hobbesism, Cartesianism, enthusiasm, and yet another version of materialism coming from the Dutch Republic in the form of Spinoza's pantheism about which we will hear more in the next chapter. Yet it should be noted that many of these same reformers still retained one vital element in the old Puritanism. While repudiating predestination and the "reign of the saints"—that is, the men and women who had sought in the 1650s independence from ecclesiastical authority—Boyle and the reformers based in Restoration Cambridge continued to believe in the possibility of a millenarian paradise. Indeed their millenarianism, conceived to include no alteration of the existing system of ecclesiastical and political authority, did nevertheless postulate an earthly paradise wherein the righteous would rule.

Newton's private writings from the 1660s echo much of the same polemical rhetoric. His manuscripts and notebooks from the period when he formulated metaphysical positions that came to rest at the foundation of his science—positions that stayed with him until his death—reveal his millenarianism. Most important, they make use of rhetorical formulations of natural philosophy directly relevant to the ideology of the new Anglicanism. Shortly after Newton's death, his associate John Craig, who many years earlier had been the intermediary between Newton and the young Newtonian polemicist, Richard Bentley, wrote that the reason for Newton's "showing the error of Cartes' Philosophy, was because he thought it was made on purpose to be the foundation of infidelity."[26]

A close reading of Newton's manuscripts spread throughout the Restoration period confirms Craig's view. The language he employed was remarkably similar to the Anglican polemics with which he was surrounded.

Newton repudiates Descartes's definition of body as extension because it does "manifestly offer a path to Atheism"; likewise he repudiates "the vulgar notion (or rather lack of it) of body . . . in which all the qualities of the bodies are inherent" because it too leads directly to atheism. Newton, like Boyle, wanted to construct an alternative to Aristotelian ("vulgar") matter theory because its implications were heretical and specifically because they chimed with the vitalistic and pantheistic notions of "the vulgar" (also another word for the people) spawned by the radical sectaries during the revolution. As Newton says in his manuscript, "Indeed however we cast about we find almost no other reason for atheism than this notion of bodies having, as it were, a complete, absolute and independent reality in themselves." In short Newton saw a profound danger in the specter of atheism, whether in the mechanistic version he read Descartes to be supporting or in the "vulgar" form that denied differences in substance between mind and body, in effect denying "that God exists, and has created bodies in empty space out of nothing." The basic definitions of the post-*Principia* Newtonian natural philosophy are clearly present in the pre-*Principia* manuscripts: the power of divine will to move "brute and stupid" matter; the independent, absolute existence of space and time; and, most essential to the formulation of the concept of universal gravitation, the notion that "force is the causal principle of motion and rest," which operates on bodies in a vacuum.

The wholesale repudiation of Descartes was essential before Newton could employ his brilliant mathematical skill to formulate precisely the law of universal gravitation. That repudiation began in the 1660s, and it was only late in the 1670s and again in the 1680s that Newton turned his attention again to the problem of gravity; the full-scale formulation of the law of universal gravitation emerged to be published in the *Principia* (1687).

The argument presented here does not presume to say that religious and ideological factors explain or account for Newton's scientific brilliance or his achievement. Rather the framework permitted his work to flourish in the direction that it did. At some point the historian must acknowledge the presence of a creative power, particularly in mathematics, of unprecedented force. We can only speculate as to how far religious beliefs and ideological concerns, so particular to the period after 1660, compelled the young Newton to search for evidence of divine efficacy in every aspect of the material order, in effect to develop as a natural philosopher and scientist. The religious Newton was never at odds with the scientific Newton; quite the reverse.

Newton was the most private of men. He chose to publish his science only when pressed. His religiosity lies to this day buried in voluminous private manuscripts found everywhere from California to Israel. In them the historian can glimpse his millenarianism, his hatred for Catholicism, his very liberal ideas on church government, his anti-Trinitarianism (one reason why he remained so private in these matters), and not least, his alchemy. Part of the reason for this "secret" Newton was simply a personal style that was slightly paranoid; part of the reason also relates to the age in which he lived. It was a time when the university had become "a machine to render maximum service to the state."[27]

The Restoration was a dangerous time for anyone who held to ideas associated with unorthodoxy. Newton practiced alchemy for most of his life; it had once been a cause among the reformers of the 1650s. For Newton, alchemy confirmed his sense of spiritual forces at work everywhere in the universe; indeed these spirits could decompose metals, and once refined, such matter "if it meets a suitable piece, quickly passes into gold." But it would not do to publish such beliefs. His alchemical beliefs and experiment also lie buried in a multitude of Newton's manuscripts.[28] Even his most singular contributions to the new science stayed buried for a time amid his private and youthful papers. Perhaps the greatest period of his creativity occurred in the mid-1660s, when he discovered the calculus; formulated the inverse square relationship between the sun and the planets, the earth and the moon; and through experimentation with light filtered through a prism, determined that colors are not complex modifications of light, but rather each color is unique and possesses its own degree of refrangibility. We can summarize these discoveries of the mid-1660s in Newton's own words:

> In the beginning of the year 1665 I found the Method of approximating series & the Rule for reducing any dignity of any Binomial into such a series. The same year in May I found the method of Tangets of Gregory & Slusius, & in November had the direct method of fluxions [in other words the rudiments of his calculus] & the next year in January had the Theory of Colours [Newton's work in optics] & in May following I had entrance into ye inverse method of fluxions. And the same year I began to think of gravity extending to ye orb of the Moon & having found out how to estimate the force with which a globe revolving within a sphere presses the surface of the sphere from Keplers rule of the periodical times of the Planets being in sesquialterate proportion of their distances from the centres of their Orbs. I deduced that the forces which keep the Planets in their Orbs must be reciprocally as the squares of their distances from the centres about which they revolve; & thereby compared the force requisite to keep the Moon in her Orb with the force of gravity at the surface of the earth, & found them answer pretty nearly [in other words, the law of universal gravitation]. All this was in the two plague years of 1665 & 1666. For in those days I was in the prime of my age of invention & minded Mathematics & Philosophy more than at any time since.[29]

From any point of view, that was quite a year the young Newton experienced. When he was older he pursued alchemy, theology, and church history with the same avidity that he had once brought to mathematical and natural philosophy. Indeed in the 1680s Newton, like so many other Anglicans, became once again obsessed with the meaning of the scriptural prophecies, with the final days of the world.

The Revolution of 1688–1689 and the Newtonian Synthesis

While the Anglican natural philosophers of the Restoration had successfully beaten back the threat to orthodoxy and the hegemony of the church once presented by the radical sectaries,[30] other dangers lurked around every corner. The sophisticated materialism of Hobbes could be used to justify a thoroughly godless but absolutist state. The republican legacy of the 1650s continued to attract elite as well as plebeian followers. In the early 1680s there were plots against the king and in 1685 even a brief open rebellion.

But the more serious challenge to Protestant ascendancy came from the monarchy itself. In the 1680s the specter of monarchical absolutism returned in the person of the new king and brother of Charles II, James, duke of York, soon to become James II. He was a devout Catholic; his brother also believed him not to be very smart. He was certainly intractable. In his private writings Newton saw him as a tyrant. When it became impossible to exclude him from the throne, his Catholicism deeply troubled pious philosophers like Boyle and Newton.

Suddenly James's person and his policies threatened the hegemony of the Anglican Church. All the other clerical institutions concerned with education or welfare, such as the Oxford and Cambridge colleges, also felt the cold scrutiny of a new king intent after 1685 to install Catholics in high places. Equally treacherous were James's policies, which tried to court the non-Anglican dissenters, Presbyterians (like the Watt family), and even Quakers like William Penn and his friends. The Dissenters spied, as one of them put it, "a snake in the grass," but they nevertheless built churches around the country and tried to come out from under the decades of persecution to which they had been subjected during the Restoration.[31] The Royal Society, like so many of the Oxford and Cambridge colleges, had lent its support to royalism and Anglicanism during the Restoration; evidence suggests that at various critical moments in the late 1670s and early 1680s Fellows of the Royal Society wrote in support of monarchical authority. Interested largely in the well-being of his Catholic subjects, James II seemed singularly unimpressed by their loyalty or their efforts.

In 1685 James inherited from his brother a court that was not only absolutist by inclination but also notorious for its private libertinism. Yet it was also open to the intellectual interests of its day. The French Epicurean Saint-Evremond had a following within it, while Charles himself, although largely ignorant of matters philosophical, had offered his protection to the Royal Society. Within this context established by royal patronage and the fear of political instability, the Royal Society attempted throughout the Restoration to chart its fortunes and those of the new science. It sought, as one modern commentator has put it, "to bring rationalization and order to all areas of national life."[32] Great emphasis was placed on technological improvements, on mechanical devices intended for industry and agriculture, and on learning from artisans, not to elevate them but to use their techniques in the service of theory. The inspiration for these projects was Baconian, or in some cases the motivation came from direct requests by government agencies for the Society to assist in one or another project.

Individual fellows, including Somerset clergymen like Joseph Glanvill, had direct ties with their rural parishes and the needs and interests of the local gentry. In early modern England the rural economy had come to include industrial development—mining for coal and minerals in particular, but also light manufacturing powered by horses or water wheels. In the records of the society from the 1680s we find evidence of interest in the earliest steam engines; and most important, the society was receptive at the time to what was to become a socially revolutionary argument. The fellows discussed the notion that mechanical devices could, and indeed should, save labor, in effect decrease rather than increase employment. At the time of those discussions it was extremely difficult to get a patent from the government for any device if its inventor argued that it would *save* labor. Indeed until the late 1720s patents may have been rejected if an applicant argued such a case. Yet in the minds of Restoration natural philosophers associated with the Royal Society we can find a mentality discernibly industrial in the modern meaning of that term and, most important, an eagerness to promote their vision of industrial progress whatever the immediate and, from the government's point of view, undesirable social consequences.[33] The alliance forged during the Restoration between the new science and the landed and commercial elite (whose interest prospered from the late seventeenth century onward) possesses historical implications that stretch right down to the late eighteenth century and the Industrial Revolution.[34] One piece in the cultural origins of the first Industrial Revolution was put in place as early as the 1680s.

But before genuine economic progress could be institutionalized political stability would be essential. Having allied itself with the search for order, stability, and the growth of commercial and industrial enterprise, the Royal Society was a quasiprivate institution dependent on the dues of its fellows and on monarchical support for its continuing respectability. It is little wonder then that in the late 1680s, when James II's absolutist policies threatened to destabilize the political order, undermine the Anglican church, and plunge the country into a new civil war, the Royal Society cast about to secure its interests as well as to remind the new king of his obligations.

At just that moment (in 1687) the *Principia* of Isaac Newton was published under the *imprimatur* of the Society. This is a singularly important date in the history of Western thought. From 1687 onward we are able to speak of the public formulation of a Newtonian synthesis; an ensemble of scientific laws, specifically, the law of universal gravitation, proven mathematically and in turn capable of being illustrated experimentally by the use of mechanical devices; a particular natural philosophy, neo-Platonic in origin and antimaterialist in intention; a polemically hostile anti-Cartesianism; and, just as important for the Anglican church, a series of social and political explications to be drawn by clergymen using the Newtonian model of cosmic order. As we have seen, the emergence of that synthesis bears relation to the ideological struggles we associate with the English Revolution. The timing of the publication of the *Principia* may also owe something to the return of political uncertainty. It was to be resolved by the flight and constitutional expulsion in 1688–1689 of James II.

In light of what is now known about the political activity of the Royal Society during the Restoration, we should at least consider the question of why the *Principia* appeared when it did. The standard story is that Edmond Halley, a fellow of the Royal Society and a friend to Newton, prodded the reticent and otherwise preoccupied genius into writing and publishing his magnum opus. Throughout the instabilities of the 1680s, it should be remembered, Newton seems to have been particularly preoccupied with the rise and fall of ancient monarchies and with the apocalyptic texts of the Old and New Testaments.[35] But Halley persuaded him to leave aside his historical and alchemical studies when he brought news about debates in London on the phenomenon of universal gravitation. The result of that digression was, of course, the famous *Principia*. It bore on its title pages the *imprimatur* of the Royal Society and as its representative, the name of Samuel Pepys in bold type. Pepys in this period was avidly seeking favor at the court of James II, and indeed he paid dearly for his sycophancy in the postrevolution wilderness to which followers of James II were consigned after 1689.

There are, however, difficulties with the Halley–Newton story, attractive though it is. For one thing it too closely resembles George Ent's description of his role in prodding William Harvey to allow his *De generatione animalium* to be published in 1651.[36] That, of course, does not make the story untrue in the case of Newton. But if there is something to the hypothesis that the publication of Newton's *Principia* during the reign of James II was inspired by political motives, to which Newton may very well not have been privy, we would expect some oblique indication of that covert design, some hint dropped—either in Halley's admiring ode on Newton and his achievement prefixed to the *Principia* or possibly in Halley's fawning and explanatory letter on that achievement addressed to James II and later published in the *Philosophical Transactions of the Royal Society.*[37]

Halley's ode does indeed supply some very interesting hints. It employs Epicurean language to woo admirers toward Newton's achievement. Recall that Epicurean ideas were fashionable in court circles, and Halley's ode made use of Lucretius's atomistic poem on the nature of things, *De rerum natura*. The poem was (and is) a major source for Epicurus's ideas. Halley begins by reminding the *Principia*'s readers that "the pattern of the Heavens" is based on "Laws which the all-producing Creator, when he was fashioning the first-beginnings of things, wished not to violate and established as the foundations of his eternal work." After this brief mention of the eternality of "Law" and the role of the "Divine Monarch" as its creator and preserver, the poem goes on to glory in the power unleashed by Newton's intellect, which "has allowed us to penetrate the dwellings of the Gods and to scale the heights of Heaven." Couched entirely in Epicurean language, Halley's ode commends the new science sponsored by the Royal Society as the means by which "we are truly admitted as tableguests of the gods."[38] In short, Halley may be seen as trying to win over the Epicureans associated with the Stuart court, to tell them that what the Royal Society has to say in science is novel and worth hearing. It was a message sent at an absolutely critical time for Anglican natural philosophers and churchmen who had been systematically excluded from James's court.

Perhaps we can now better understand why, after the church's hegemony had been reestablished in the early 1690s, Newton wrote a seemingly bizarre letter to Pepys in which he asserts, almost hysterically, "I never designed to get anything by your interest, nor by King James's favour." If indeed the *Principia* had been published in an attempt to be ingratiating, by reestablishing the supportive role that scientific knowledge had played for the monarchy during the Restoration, then either Newton was innocent of these motives or, as a firm supporter of William III, he became almost paranoid after the Revolution in his concern that his name not be associated with Pepys, who was by that time suspected of Jacobitism (that is, of still supporting James II).[39]

If Newton was in all probability naive in 1686, he was not so by 1692. He had led the anti-Catholic opposition to James II in Cambridge, writing in his private papers that "men of conscience" should care not "for their preferments but their religion and Church."[40] He then gave his wholehearted assent to the Revolution of 1688–1689, which deposed James II. As a member of Parliament representing Cambridge, Newton urged his parliamentary constituents to do likewise. Newton, the commoner, like Boyle, the gentleman, had always feared the excessive powers of absolutist kings such as James II had sought to become.

But the Revolution of 1688–1689 undid more than the Stuarts. It secured the church's constitutional place, yet vastly weakened its legal and moral authority. Dissenters received limited but real toleration, and church courts were abolished. Soon censorship would also largely disappear. As staunch supporters of the revolution, the latitudinarian faction now ascended to positions of leadership within the church's hierarchy, and its problems became theirs.[41] After 1689 Newton's natural philosophy served the latitudinarians as the underpinning for the social ideology preached by the intellectual leadership of the church in response to the Revolution Settlement. The Newtonians once again resumed the polemical assault against philosophical and political radicalism, and they did so in language characteristic of Restoration Anglican science. They spoke, fittingly, from the podium established by Boyle's last will and testament (1691). With Newton's assistance and approval the Boyle lecturers—Richard Bentley, Samuel Clarke, William Whiston, and William Derham—brought Newton's "system of the world" to bear against the radical Whigs of the 1690s and beyond. Their republican tendencies were as odious as their heterodox religiosity, which owed much to their reading of Hobbes and Spinoza as well as to Bruno and Servetus, to the extreme pagan naturalism of the late Renaissance. Indeed the Boyle lecturers did precisely what Newton had indicated to a friend in late 1691 could be done: "A good design of a publick speech (and which may serve well as an Act) may be to shew that the most simple laws of nature are observed in the structure of a great part of the Universe, that the philosophy ought there to begin."[42] Newton had formulated his theology in the context of the Restoration church's attempt to recreate its own sense of legitimacy; now his disciples would do the same within the post-1689 revolutionary context.[43]

From Boyle's endowed pulpit and in their writings the clerical Newtonians preached to London-based and exceedingly prosperous congregations. They extolled the virtues of self-restraint and public-mindedness while at the same time assuring their congregations that prosperity comes to the virtuous and that providence permits, even fosters, material rewards. The nation must acknowledge God's providence by the cultivation of virtue, by the pursuit of what Newton's tutor, Isaac Barrow, had called "sober self-interest," and by support for Anglican hegemony. The same God whose laws of motion Newton had discerned in the natural world would also inevitably ensure order, prosperity, and the conquest and maintenance of empire in the political world. Adopting simple, nontechnical language, Newton's first-generation advocates used his science, as Restoration Anglicans had used Boyle's, to support the social ideology and political goals of liberal Anglicanism and constitutional monarchy bound by law—both had been rendered supreme within the recently secured church. Gradually and only after 1714, the liberal Anglican Newtonians became supporters of the Whig party, although in Scotland many Anglican Newtonians remained Tories.[44]

With the lapse of the Licensing Act in 1695 and the intensification of party rivalry between Whigs and Tories in the late 1690s, the liberal Anglican establishment, along with court and monarchy, found itself under assault by the radical and republican Whigs. With the freethinker John Toland in the vanguard, they put forth materialistic and pantheistic—to use the word invented by Toland in 1705—arguments to justify the rule of Parliament over court placemen and standing armies, of civic religion over the established church, and of religious pluralism over a narrowly circumscribed toleration. From their pulpits the Boyle lecturers, with the Newtonian Samuel Clarke as their most philosophically gifted spokesman, put forth counterarguments to justify order and stability, to maintain the hierarchical and providential interpretation of the constitutional settlement.

But if Anglican hegemony after 1689 now owed so much to Newtonian science, what did Newton's science owe to its religious and ideological roots? On the crucial level of matter theory, on Newton's insistence that universal gravitation must operate through immaterial forces in the universe and not as a property inherent in matter, it seems plausible to argue that Newton had accepted the central arguments of the Anglican virtuosi as formulated during the 1650s and beyond. Certainly his private manuscripts from as late as the 1690s repudiated the materialistic arguments by which "the vulgar" described the world and lashed out at those who postulated an impotent deity—a "dwarf-god," as Newton put it.[45] Newton's insistence on a mechanical philosophy that relied heavily on spiritual forces led him to adopt a baroque and neo-Platonic ontology that to this day has puzzled those purely philosophical commentators intent on unraveling its complexity. The approach taken here does not seek to minimize that complexity, but it does offer one explanation for its existence.

If we date the origins of the European Enlightenment to the 1690s in England, then it seems clear that English science from Boyle to Newton sponsored but one version of Enlightenment, a moderate and theistic, occasionally deistic, movement. Given what we now know about the institutional and ideological relations of the new science, in short about its Anglican origins, it must be acknowledged that the Newtonian Enlightenment was intended by its participants as a vast holding action against materialism and its concomitant republicanism, against what is best described as the Radical Enlightenment.

As we shall see in the next chapter, the Newtonian Enlightenment cast its light in a variety of directions. Overwhelmingly its practical applications were mechanical, but Newtonian physicians also argued that "the mechanism of the body [is] conducted by the same law that support the motions of the greater orbs of the universe."[46] Later in the century social theorists such as Adam Smith took inspiration from Newton's physical laws and sought their analogue in the behavior of the market.[47] The invisible hand that keeps order in the market owed much in its formulation to the Newtonian synthesis. From England Newtonian science spread quickly onto the Continent, largely assisted by the French-language press at work in the Dutch Republic. There, as in England, its advocates first attacked the Cartesians. As it became increasingly legitimate, partly because of the work of French Newtonians, the science of the *Principia* slowly penetrated into the schools and universities. Although the penetration occurred much later on the Continent than it did in Britain, Newtonian science with its mechanical applications, appealed to industrial promoters as well as to philosophers and social reformers. The model of order based on knowable laws embodied in the Newtonian synthesis offered a powerful alternative to a variety of other belief systems, not least to the doctrines of the scientifically naive clergy. With the dissemination of the new science in the early eighteenth century through lectures, sermons, journals, and textbooks, all educated people were expected to know something about science. In this one area, the break between high culture and low culture was now complete. For those European elites who also embraced science the goal became enlightenment, and England and its science became the model of order, stability, and progress.

4

The Newtonian Enlightenment

The cultural ascendancy of science in late seventeenth-century Europe and its colonies—from a body of knowledge once promoted by select devotees in Florence, Paris, Leiden, or London, to the cornerstone of progressive thought among the educated laity—occurred with extraordinary rapidity. We can date the transformation in the role of science in Western culture from the 1680s to the 1720s. Within one generation, largely in northern and western Europe, the transformation was complete. Mechanically based science left the hands of the mathematically adept and went into the everyday conversations of journalists, learned societies, coffee house lectures, and church sermons. As a result, science altered the way urban merchants, progressive aristocrats, literate gentlemen, some gentlewomen as well as artisans and tradesmen, understood the physical world around them.

The assimilation of science was so rapid, and its impact so great, that historians since the 1930s have identified the period in European culture from the 1680s to the 1720s as one of profound crisis. Out of the crisis emerged a mentality discernibly modern, a new cultural moment called in retrospect the "age of enlightenment." At that moment, high culture armed with scientific acumen distinguished itself completely and irrevocably from the culture of the untutored or semiliterate people. Science became essential to educated discourse; nature mechanized provided analogies and metaphors for every aspect of human experience. Nature now imagined as knowable also fueled a new heterodoxy. A new rationalism, "active, zealous and intrepid" as one leading historian of the period has described it, became a weapon against Christian orthodoxy and piety, as well as against established authority.[1]

Among the educated, new religious persuasions appeared as did new forms of social interaction. All eventually acquired a debt to scientific renderings of nature, although the sources of the creeds and practices often derived from contemporary practices or ancient philosophy: Socinianism or Unitarianism, which denied the doctrine of the Trinity and in England eventually consolidated into a new religious sect; deism, which would have God be remote, the great clockmaker; pantheism, which defined nature as God; freemasonry, which made gentlemen into brothers meeting in secret "upon the level"; and of course, freethinking, which could mean anything from atheism to skepticism or anticlericalism.

The most virulent heresies, pantheism and freethinking—words first used in English in the early 1700s—amounted to a disdain for all forms of organized religion. Some bold freethinkers then proclaimed nature as the object of their worship. All these beliefs and practices amounted to a massive shift away from the religious toward the secular, toward living in a "timeless" world without a biblically described beginning or knowable, appointed end. Science played a role in inaugurating the secular; in the "wrong" hands it was also used to confirm heterodoxy.

In making possible the shift toward this world and away from the next, the new science from Descartes to Newton offered a radically altered picture of nature. Science made nature lawful, and as the definition of creation changed so too did the human conception of the Creator. A new religious outlook was being invented. "Natural religion" and "natural theology" became passwords to a distinctive religiosity.[2] Miracles and divine interventions became rarer; being religious began to mean thought rather than prayer. A vision of order and harmony, God's work, replaced biblical texts and stories, God's word. But in the hands of freethinkers science also permitted the first articulation of a coherent universe without any creator. The roots of our uniquely modern ability to examine nature and society as self-contained entities and to offer explanations totally natural, that is entirely human, lie in the crisis of the late seventeenth century. By the end of the eighteenth century philosophers began to articulate branches of knowledge focused on society, government, and the human psyche. The beginnings of the modern social sciences lie in their efforts.[3]

In consequence of the Enlightenment so many commonplace beliefs about natural science have also been inherited: a faith in its progressive nature leading to constant improvement of the human condition; its supposed superiority to mere beliefs, opinions, and subjective judgment; the heroic role of the scientist; the presumed need for all other disciplines, however social their focus, to be "scientific"; and not least, the absolute right of free scientific inquiry as an extension of freedom from censorship—a freedom demanded regardless of the social or moral consequences of the inquiry. With this sort of cultural legacy, it is extremely difficult to use our modern historical imagination, itself an outgrowth of science, and realize that these assumptions emerged as dominant in Western culture only early in the eighteenth century. Their rapid acceptance in northern and western Europe was provoked by a European-wide cultural crisis primarily political, yet also social in its origins and dimensions.

By the 1680s the new science, whether in its Cartesian or Newtonian form, if coupled with a tolerant version of Christianity, seemed the only alternative to the political rigidity and religious intolerance increasingly associated with the absolutist state and its clergy. This new synthesis of science and Christianity articulated by liberal clergymen in England and the Dutch Republic had many uses. It served to combat varieties of intellectual radicalism, the old naturalism of the people, the new naturalism of the literate proponents of materialism and pantheism, and the sectarian enthusiasm of popular religiosity. A religious sensibility rooted in thought more than in prayers and ceremonies could efface the differences between Protestants and Catholics, the source of hostility and persecution.

The promotion of science as the new guarantor of a rather cerebral religion owed much to the ideological struggles waged across Europe by the seventeenth-century prophets and promoters of science. From Galileo through to Gassendi, Descartes, Boyle, and the Cambridge Platonists, they had all enlisted the mechanical philosophy against the culture of "the vulgar." To a man they envisioned some sort of alliance between the established church, whether Catholic or Protestant, and the state on the one hand with science on the other. Galileo hoped for just such an alliance before the Inquisition got the better of him. As we saw at the end of chapter 2, the French Cartesians similarly offered their services to the absolutist state. But by the 1680s in both France and England an alliance of progressive science and the state—with the freedom of inquiry and practical application that it promised—seemed doomed by the political ambitions of absolutist monarchy, by Louis XIV in France and James II in England. The Enlightenment was born of a crisis that was as political as it was intellectual.

The Threat of Absolutism

In 1685 the French king Louis XIV revoked the Edict of Nantes and sent over 100,000 French Protestants into exile in search of religious toleration. Those who remained either converted to Catholicism or faced persecution and imprisonment. The prison records of Paris show Protestants incarcerated with common criminals, sellers of illegal books, even alchemists. Simultaneously with this assault on the domestic religious practices of a minority, Louis XIV embarked on an aggressive foreign policy that threatened the territorial integrity of the Dutch Republic and the Spanish Netherlands (later the Austrian Netherlands and eventually called Belgium), as well as the western German cities and principalities. In England James II, as we saw in the last chapter, sought to install Catholics within the army and the universities, that is, to undermine the Anglican church established by law. He believed that Catholics would be his allies as he attempted to abolish Parliament and rule with absolute authority vested in himself and his court alone. Because his rule lasted less than four years before he was toppled by a revolution, we will never know if the policy would have worked. All the major English scientists, beginning with Boyle and Newton, and much of the church, opposed him.

By the late seventeenth century rigid censorship was already an established fact of life in much of Catholic Europe, as was clerical control over the universities.[4] In Naples proponents of atomism, an essential element in the new mechanics, found themselves on trial in 1688. At the same time the absolutism of the Hapsburg monarchy in Spain was so taken for granted that astute and hostile observers largely missed its relative decline. Suddenly the 1680s in western Europe resembled the 1580s. Continental Protestants feared for their survival, monarchs once again sanctioned religious persecution, religious refugees crowded into the urban centers of the Low Countries, and French Protestant intellectuals were driven in pilgrimage to London, Amsterdam, Berlin, and Geneva. Within the context of repression came the crisis out of which a new secularism emerged.

Predictably, given the immediate political causes of the crisis, its impact came first in the area of political beliefs and values. Beginning in the 1680s we see a rapid disintegration of confidence in the doctrine of the divine right of kings, an increasing emphasis among political theorists on the rights of subjects rather than on their duties. To justify the theoretical attacks on absolutism, natural law theorists of a previous era (such as Hugo Grotius) were invoked, and arguments for the rule of law rather than the will of the sovereign became fashionable. The prevalency of such arguments among the political opponents of absolutism may indeed have encouraged a predisposition to the new science, a desire for the experimental rather than the simply memorized or doctrinal, and a sympathy with general theories operating according to predictable laws and not whimsical forces. The assault on absolutism and Catholicism gave credibility to the rhetoric of probability over that of absolute certainty. It is hardly accidental that one of the most subtle historians of Protestantism and a supporter of the Revolution of 1688–1689, the Anglican polemicist Bishop Gilbert Burnet, argued for the probable certainty of knowledge, scientific and otherwise, against the claims to eternal, and hence absolute, authority made by Catholic historians.[5] Similarly Protestant apologists for limited monarchical authority were drawn to scientific arguments to reinforce the existence of a natural harmony and order, which would make absolute authority less necessary. By contrast, the supporters of Louis XIV in France said that arguments drawn from nature would diminish the grandeur of the king.[6]

The crisis that began in the 1680s gave European circulation to events and political ideologies once germane only to an English setting, to English Protestants such as Burnet, Boyle, Newton, and their associates. In short, the political crisis of the late seventeenth century brought the legacy of the first of the great modern revolutions into the mainstream of European thought. Once internationalized, the English revolutionary legacy associated on the Continent with the writings of John Locke merged with indigenous traditions of anticlericalism, philosophical heresy, and antiabsolutism. The English Revolution of mid-century had produced a body of political, religious, and scientific thought so rich and complex that, once discovered by the European opponents of absolutism, became a major source for the new synthesis we describe as enlightened.

The English version of science, whether in the form of Hobbes's materialism or Boyle's Christian atomism, as we saw in the previous chapter, had been inextricably bound up with the search during the 1650s for an alternative to rigid Puritanism, to radical sectarianism, as well as to the pretensions of absolutist monarchy supported by an independent and culturally dominant clergy. The science and natural philosophy of Boyle and Newton in that sense may be seen as the complex, exceptionally rich by-product of a revolution against the established clergy and the absolutist state.[7] It is little wonder that English theorists—from the liberal Anglican promoters of science to Hobbes and his enemies, the republicans—were favorably received by a Continental audience suspicious of Louis XIV and his legacy. The audience included exiled French Huguenots, Dutch lawyers and doctors, French poets of minor aristocratic background (like Voltaire), and an entire generation of Protestant refugee journalists. Exiled in the Dutch Republic and armed with their native command of French, these journalists used the freedom of their presses in a vast campaign against absolutism.[8] They were aided by caricaturists who used print culture to lampoon the French king and his clerical sycophants. Cheap engravings flooded the market and they portrayed kings as arrogant and clergymen as sycophantish fools and knaves. It was a short step from satire to turning against all forms of organized religion. Through translations and journalistic explications the same refugee writers and publishers introduced educated Europeans, literate in French, to English science and culture.

The Dutch printing presses published in various languages engravings that showed the persecution of Protestants as wrought by priests eager to please the French monarch. (With permission from the Teyler's Museum, Haarlem.)

The Failure of the Old Learning

Coupled with the political origins of the crisis were other cultural factors, by the late seventeenth century more cumulative than traumatic in their effect. The increase in European traffic to non-Western nations had produced a travel literature richly descriptive of customs and beliefs that were totally non-Christian, yet "curiously" moral. Although much racism and Christian chauvinism were mixed into the Western response to the non-West, by the late seventeenth century the cumulative effect of the travel literature had been to call into question the absolute validity of religious customs long regarded, especially by the clergy, as paramount. The French Catholic Church had argued that all people have in their hearts a belief in God; the travel literature turned that claim into nonsense. And not least, a century of Protestant versus Catholic polemics about the biblical authorization for either version of Christianity had, willy-nilly, rendered the Bible into an historical document. Once reduced to human scale, its contents were open to skeptical scrutiny. Such scrutiny when offered to the literate could only render the task of instruction more difficult for the clergy.

Simultaneously, literacy increased in England and Scotland (and possibly in the Dutch Republic). On both sides of the Channel by 1700 probably over 50 percent of males were in some sense literate. In France the figures were nowhere as high, but after 1700 they were increasing, not stagnating or declining. In Protestant Germany literacy, in the sense of reading, appears to have been common, although by no means a mass or majority phenomenon by the late sixteenth century. The presence of active literacy in early modern Europe is notoriously difficult to calculate, but it would seem to have been increasing among urban men and possibly women after 1680.[9] The increase may have been coupled with a decline or stagnation in the rural or poorer areas of Europe, thereby further widening the gap between elite and popular knowledge. The crisis of Western culture that gave birth to the widespread assimilation of the new mechanical science was profound precisely because of the presence of a large literate citizenry, greater than any ever assembled in the West since ancient times.

The crisis had also resulted from the failure of the older, scholastic culture to deal effectively with the challenges presented by new experience and empirical data. Throughout the seventeenth century, from the Jesuits' attack on Galileo to the discomfort about Cartesianism among French and Dutch theologians, elite culture had been badly served by the philosophical guardians of religious orthodoxy. In the sixteenth and seventeenth centuries both Catholic and Protestant theologians had based the metaphysics of such doctrines as transubstantiation (the belief that the priest transforms bread and wine into the body and blood of Christ), consubstantiation (both bread and the body of Christ are present), and the Trinity, on scholasticism. Aristotle's doctrine of form was a key ingredient in explaining how the bread of the Host continued to be sensed as bread, but the essence or form had changed into Christ Incarnate. Scholasticism rested on a Christianized Aristotle. The clergy taught it in their elite schools—hence its name. Yet as early as the 1630s, after Galileo's confrontation with the church and the publication of Descartes's *Discourse on Method* (1637), it was clear that Aristotle and Ptolemy no longer adequately described the operations of the natural world, either celestial or terrestrial. Yet the clerical promoters of scholasticism failed to find any alternative to a philosophy inherited from a previous age and badly in need of revitalization.

The threat to the metaphysics of Orthodox Christianity was real and immediate, yet the clergy of the schools resolutely clung to the old scholastic explanations. "Forms" still animated matter, not atoms or bodies in collision. By the 1680s the Aristotelians occupied powerful positions in any school in almost every country, but in western and northern Europe they were on the defensive. They now desperately—sometimes mindlessly—sought to maintain doctrinal orthodoxy in the face of the destruction of Aristotelian natural philosophy. Not surprisingly by 1700 scholasticism was in retreat. The issue became what body of philosophy would replace it and still maintain essential Christian positions.

Liberal Christianity

Eventually the more ingenious clergy, largely of Protestant Europe, realized that a new Christian religiosity was required, and they found its foundation in the new science. The synthesis of science and religion emerged first among moderate Anglicans, who had been forced under the impact of the English Revolution to rethink the relationship between the natural order, society, and religion. Simultaneously all progressive European Christians, from the German philosopher Leibniz to the Cartesian priest Malebranche, were being forced to restructure the philosophical foundations of Christianity to conform to one or another version of the new science. It is hardly surprising that liberal Anglicanism, wedded as it was by the 1690s to Newtonian science, took the lead in this enterprise. The Boyle lectures by the major Newtonians were quickly translated into a variety of Continental languages, and Samuel Clarke remained throughout the century the leading theologian of the godly version of the Enlightenment. Later in the eighteenth century Rousseau invoked the teachings of Clarke, whereas French materialists such as Baron d'Holbach saw him as one of their prime enemies.

In tandem with the liberal version of Christianity came the science of Boyle and Newton. In stark contrast to the doctrinal rigidity of the French Catholic church, or of fundamentalist Calvinism, the English theologians in the tradition of the Cambridge Platonists preached a natural religion founded on reasonable expectations of salvation in the afterlife and reward in this life. Doctrines like transubstantiation, or even the Trinity, as well as belief in the existence of hell, quietly slipped away. The laws of science vindicated God's existence. The instilling of belief in order, both social and natural, took precedence over complex and increasingly controversial doctrines like the existence of hell. Suddenly a version of Christianity emerged that focused on achievements in this world, on a Christianized self-interest; and this version also embraced the physical universe delineated by the new science. We can find the doctrines of design and harmony being preached in the pulpits of prosperous London churches as well as in books written by progressive Dutch Protestants.[10]

After 1689 liberal Christianity of English origin became associated in the minds of Europeans with two extraordinary developments. The first was a successful and bloodless revolution in 1688–1689 that removed an absolutist king, James II, established parliamentary sovereignty, and forced the Dutch stadholder, William of Orange, to accept a Bill of Rights as one of the conditions of his ascent to the English throne. The Revolution of 1688–1689 also established a limited religious toleration for all English Protestants, although in theory not for Catholics or anti-Trinitarians. The second innovation was Newtonian science. In the 1690s liberal Anglican clergymen championed both the political settlement of 1689 and the Newtonian synthesis, and related one to the other.

Suddenly the new consensus forged in England stood in stark contrast to the capriciousness of Continental absolutism. A viable national and Anglican church remained amid limited religious toleration, leading clergymen offered justification for revolution and constitutional government, and experimental science had uncovered previously hidden and universal laws. The Newtonian system of the world could be championed as the model for the stable, harmonious, moderately Christian polity ruled by law, not by an arbitrary and capricious will. This polity was the creation of the parliamentary class: large landowners, prosperous merchants, voting freeholders. Political revolution against absolutism had been achieved without social upheaval, without an uprising of the lower orders. Not least, the alliance of England and the Netherlands against the French colossus proved effective. By 1710 Louis XIV had been humiliated on the battlefield; he had lost his conquered territory in the southern Netherlands and his treasury also stood empty. In this period we should never underestimate the cultural implications of military victory or defeat.

The International Context

Yet before we explore this triumphant Newtonian Enlightenment, as well as the radical alternatives offered to its theism by materialists, pantheists, and atheists—who were themselves enamored of the new science—we should examine the variety of uses to which scientific knowledge was put during the crisis of the late seventeenth century. The psychological epicenter of the crisis lay predictably within Protestant culture. In the face of persecution and migration the traditional responses of piety, prayer, and biblical prophecy came to be seen as increasingly inadequate.

We may take as typical of the older Protestant tradition under challenge the mentality of a Protestant Dissenter from southern England, one Samuel Jeake (b. 1652). He was a highly literate merchant who read widely, whose family favored the Puritan side during the civil wars, and who had supported the revolution of 1688–1689. Indeed had he been Scottish, Jeake would look a great deal like his contemporary, John Watt, whom we shall meet in the next chapter. Unlike Watt, Jeake was more a merchant than a scientific teacher and craftsman. From an early age Jeake began to interpret the events in his life, as well as revolutions in the polity, in astrological terms. This is not to say that he knew nothing of the new science; he certainly read William Harvey's treatise on anatomy and the circulation of the blood. But the culture of the Royal Society or the publication of Newton's *Principia* (1687) passed him by completely. In the early 1690s he found himself on the defensive as he tried to justify astrology "experimentally" and to show that events in 1688–1689 conformed to the radical alterations of the planets in those years.[11] Despite the growing confidence in science, his faith in astrology never wavered; nor did his belief that the events in his lifetime had somehow been foretold by the scriptural prophecies. Like John Watt, who was uncle to the more famous James Watt, Jeake probably knew the astrological predictions of the radical prophet, John Pordage, but Jeake never evinced the interest in science found among the Watts.

In the 1690s prophetic responses to dramatic events were still commonplace throughout Europe. The Huguenot minister and refugee Pierre Jurieu and his followers identified Louis XIV as the Antichrist of biblical prophecy and, not surprisingly, predicted his demise. Jurieu cast a cold eye on natural religion or indeed on any version of Protestant rationalism that denied the clergy an independent and dominant role in the state. He verbally persecuted another Huguenot refugee, the journalist Pierre Bayle, who in turn used his encyclopedic *Dictionnaire historique et critique*, (1697) to mock the doctrinal rigidity of those who predict the future, as well as to scorn absolutism.[12]

Laymen like Bayle, so typical of the end of the century crisis, embraced the new science—in his case in its Cartesian form—as an antidote to the scholastic pretensions of the orthodox clergy, both Catholic and Calvinist. Of course the new encyclopedic mind with its passion to order and classify was deeply indebted to the Baconian method of classification and collection. Bayle was the foremost encyclopedist of his age, and where we find his *Dictionnaire* being re-edited or imitated in the eighteenth century we will also find journalists with a keen interest in the new science.

The vision of Francis Bacon, his call to classify all knowledge, did not require the science of Boyle or Newton for its survival or application. It appealed to the organizers of knowledge, those directly connected with the printed word who by the late seventeenth century faced the monumental task of simply keeping track of all that was now being published. The line of influence from Bacon to the great encyclopedia of the Enlightenment, Diderot's *Encyclopèdie* (1751), lies through the world of journalists like Bayle and refugee publishers in the Dutch Republic, who were forced to devise cataloging and classificatory systems just to keep abreast of their inventory.[13] Their receptivity to the new science lay partly in its salability, but it also grew out of their own sense of the necessity to order the world around them. The inordinate number of Huguenot refugees—many of them once Parisian booksellers—drawn after 1685 to the freer presses of England and the Netherlands meant that presses receptive to science were also the centers for antiabsolutist and anti-Catholic propaganda. Almost singlehandedly the refugee publishers invented the French-language literary journal intended for an international circulation.

Among the most important Continental promoters of the new science, particularly in its Newtonian form, was the liberal Calvinist minister and journalist, Jean Le Clerc. His journal, *Bibliothèque universelle et historique*, published in Holland, disseminated news of the *Principia* to thousands of French readers, and it also championed the liberal Christianity of the Anglican moderates. In addition Le Clerc embraced the epistemology of his friend, the English philosopher, John Locke.[14] While as an exile himself from repression enforced by the Stuart kings, Locke worked on his *Essay Concerning Human Understanding* (1690). It was a radical statement of how human beings know, one that gave little credit to innate ideas or the force of tradition. It proclaimed the senses as the starting point of all knowledge. Locke had come to intellectual maturity deeply under the influence of the new science as explicated by Robert Boyle, and not surprisingly his philosophy laid emphasis on the external, physical world as the starting point of all that is worth knowing.

Locke and his place within international Protestant circles are perfectly symbolic of the European crisis and its resolution. In the 1680s as an English opponent of royal absolutism he fled for safety to the Dutch Republic. There living under an assumed name he became closely associated with liberal Calvinists such as Le Clerc as well as with English refugees of radical background, such as the Quaker merchant Benjamin Furly. Together they discussed every aspect of the contemporary scene: the threat from France, the danger of invasion, and Locke's ideas on parliamentary sovereignty, which he had worked out primarily in the early 1680s when James II's ascent to the throne appeared inevitable. Not least, he and Furly were familiar with the latest medical theories and with medical reformers who attempted to apply both mechanical and hermetic theories to their practice.[15] In Locke and his circle we see the confluence of interest in the new science and hostility to doctrinal rigidity and absolutism—in short, the Enlightenment in embryo. Yet even Locke could not work his way through the mathematics in Newton's *Principia*. Without the new explicators of Newton's system it would have remained esoteric knowledge for the gifted. A member of Locke's political party in England, a prominent Whig and nobleman, wrote to Le Clerc in 1706 about the new era they were witnessing in thought and culture: "There is a mighty light which spreads itself over the world, especially in those two free Nations of England and Holland . . . it is impossible but Letters and knowledge must advance in greater proportion than ever." All that might spoil the new enlightened age, he told Le Clerc, would come from religious fanatics or atheists.[16] As time would show, there would be plenty of atheism in the age of Enlightenment.

Likewise in Locke and Le Clerc's international circle during the 1690s we can see the confusion of possibilities open to the educated person in search of alternatives to rigid orthodoxy and authoritarianism. The influence of hermeticists, such as the alchemist F. M. van Helmont (d. 1698?), was still in evidence; indeed Furly himself believed in the mystical doctrine of metempsychosis—that is, the migration of souls after death.[17] By contrast Le Clerc and liberal Dutch theologians argued for science, theism, and toleration, while young refugee journalists, who would later edit Bayle and become radical materialists, found Furly's circle and his library an interesting place to congregate. All were drawn together by the wars against France and the very real fear of a French invasion. In their midst English radicals of freethinking inclination such as the young John Toland visited and sought converts; the radical version of the Enlightenment mingled in its early years with moderacy.

At this stage in its cultural history scientific knowledge was still very much a matter of philosophical principles, cosmologies, and rules of reasoning. It was part of the search for an alternative synthesis among educated laymen, doctors, merchants, journalists, politicians, and liberal clerics, for a way out of the crisis provoked by clerical and monarchical authority. Newtonian science had not yet become, as it would by the 1720s, a body of learning for laymen to master, practice, and apply. Yet there were extraordinary uses to which natural philosophy could be put within the context "of an entire philosophical liberty," as Le Clerc's correspondent described the intellectual atmosphere.

The new light that spread in northern Europe focused on magic and popular superstition, as elite culture defined it. A Dutch clergyman of rationalist persuasion, Balthasar Bekker, launched what became a famous argument against witchcraft and magic. From his vantage point as a citizen of a vast seafaring empire, Bekker compiled a massive catalogue of the superstitions and magical practices found both at home and abroad. Introduced to rationalism and the new science by reading Descartes, Bekker embraced both while retaining his own version of Christian orthodoxy. Where the Bible speaks in the language of the people, as for example in asserting geocentricity, Bekker simply dismissed its cosmology as rhetoric necessary to keep the attention of the common folk. Likewise he denounced the Catholic doctrine of transubstantiation as simply unreasonable.[18] Where the Bible clearly speaks in the voice of God, as in the prophecies that describe the time and circumstances of the end of the world, it must be taken literally. Fideism and Cartesianism mixed in Bekker's mind—so typical of the transition we are describing—in such a way as to permit him to render a major theoretical assault against magic and still remain a millenarian of sorts, a believer in the Biblical prophecies.[19] His Dutch book, *De Betoverde Wereld* (*The World Bewitched*, 1691), was dedicated to a mathematician and the *burgermeester* (mayor) of his native city, Franeker, and it pitted the mechanical philosophy of Descartes against all *tovery en spokery* (witchery and spookery). Bekker sought "to banish the devil from the world and bind him in hell so that the King Jesus might rule more freely." It also labeled the Catholic church as the kingdom of the devil.[20] When he translated his textbook against magic and the power of devils into French, Bekker muted the outright assault on Catholicism and confined himself to attacking the superstitions of "popish" priests. Bekker's French text took its place among a number of such assaults on popular religiosity emanating from French rationalist circles. It provoked a flood of criticism, much of it sponsored by the clergy, who saw Bekker as an extreme rationalist, ignorant of the true power of demons and witches.

Bekker's book enlisted the new science against the pagan naturalism of the people, and it became a standard and widely read work of the early Enlightenment. We may see Bekker as a transitional figure, not unlike Newton,

In this pro-Calvinist engraving Bekker is driven from the church by devils. He rides away on his book against witches. (With permission from the Teylers Museum, Haarlem.)

who combined scientific rationalism with intense religious piety and a rather cerebral penchant for the prophecies. Yet both thinkers addressed themselves to the laity or to liberal clerics; indeed as might well be imagined, Bekker quarreled with other Calvinist clergy. Remember that Newton cautiously limited his anti-Trinitarianism to discussions with those clergy who were his followers and, possibly, with John Locke.[21]

In France Cartesianism had been used by clerical supporters of absolutism to render glory to the Sun King. Indeed, a French Huguenot was one of the first to attack those scientific mandarins and to argue that science should serve other, more humane ends.[22] And in the hands of Dutch Protestants such as Bekker we can see where Cartesian science might have led had it not been for its negative associations with French absolutism and for the lingering fears about the materialistic implications of Cartesian matter theory. Descartes's prescription that thinking defines existence could empower men and women to think for themselves and in the process challenge centuries of fear and superstition.

Not surprisingly throughout the seventeenth century doubts lingered among the clergy about the meaning of Descartes's radical separation of mind from body. Dutch anti-Cartesians had been vociferous in the 1640s about the danger of materialism. The Cambridge Platonists of the 1660s had sought for similar reasons to lead a new generation of students away from the French philosophy. As we saw in the last chapter their teachings shaped the thinking of the young Isaac Newton. As late as 1671 Scottish Cartesians in Edinburgh actively taught Descartes while still warning against atheistic attempts to use the mechanical philosophy to undermine religion.[23] All those warnings, however dire, only preceded the heretical impact made by the Amsterdam philosopher Benedict de Spinoza (1632–1677). He was a nightmare come true for theologians.

Spinoza and Spinozism

Born into a family of recently immigrated Portuguese Jews and the son of a merchant, Spinoza read Descartes as part of his school education. As one contemporary biographer put it, from Descartes Spinoza learned "that nothing ought to be admitted as true, but what has been proved by good and solid reason."[24] Spinoza combined his reading of Descartes with deep knowledge of classical and Hebraic texts. Out of that mélange he forged a solution to the Cartesian separation of mind and body that possessed devastating implications for all forms of organized religion. The fears of a century were actualized in Spinoza. He constructed a naturalistic, philosophically anchored version of the human and physical worlds that John Toland, the English radical and follower of Bruno, labeled pantheism.[25] Briefly stated, Spinoza asserted the existence of one infinite substance in the universe, namely Nature or God. He argued that it was illogical or contradictory to posit two kinds of substance, as did all traditional Christian metaphysics, in other words to posit the infinity of God and the separate finiteness of matter. In true Cartesian fashion Spinoza pursued his reason to its clear and distinct conclusion. In the *Tractatus Theologico-Politicus* (1670) he presented his pantheism in an eminently readable fashion and linked it with a philosophy of total freedom from intellectual constraint and with his republicanism.

In the midst of the crisis of the late seventeenth century, spinozism proved the most virulent heresy, and its debt to the new science was inescapable. Spinoza accepted all the Cartesian and mechanical definitions of matter and motion. He then perversely collapsed matter into spirit, God into nature, and a nightmare for Christian natural philosophers became a reality. First Hobbes, then Spinoza—very different, to be sure, in their philosophies of government—and both were so comfortable with purely naturalistic, materialistic, pantheistic explanations of man, society, and nature. Whatever adjective we use should not obscure the adjective most commonly used by contemporaries: atheistic.

To this day the spinozism of those decades possesses a murky history. To its opponents it was everywhere; yet just try to find an avowed spinozist. When the authorities did and he had been unwise enough to publish, they locked him away. The Dutch Republic spawned Spinozism, and there it can be found as an early and radical version of the Enlightenment, located among very private circles of professional men, merchants but also publishers and journalists. They invented and circulated clandestine treatises that proclaimed Jesus, Moses, and Mohammed as imposters; they championed all science and taught themselves mathematics. They were republicans and critics of monarchical authority; and they had little use for the clergy or doctrines of the Dutch Reformed Church, which had the right to interrogate heretics and force the authorities to do something to silence them. In their private letters Spinozists described the providential God of Christianity as "the god of the lazy."[26] In short they took care of themselves in a competitive world and never resorted to traditional piety for solace. We know of a postal official and servant of the Austrian administration in Brussels in the early eighteenth century who was a Spinozist. It all sounds harmless enough from this distance until we realize that this official almost certainly assisted his publisher friends to ship clandestine and heretical literature into France, to undermine the authority of both its church and state. Eventually, in the 1740s, he lost his job because he could not resist publishing and circulating himself yet another heretical piece, a French Jansenist work that attacked absolutism.[27]

Jobs and careers could be lost in this period if an individual was accused of irreligion, especially of Spinozism. In 1668 an Amsterdam lawyer and doctor who belonged to Spinoza's circle and publicly blasphemed the Judeo-Christian tradition got a sentence of ten years in prison; he died there the following year. A liberal Calvinist theologian of international reputation, Philip van Limborch, stepped gingerly as his clerical colleagues harassed and interrogated him regularly about his views.[28] In the early eighteenth century Tyssot de Patot, a professor of natural philosophy and mathematics at Deventer in the Netherlands, lost his position for holding heretical views and was ostracized from polite society. Not incidentally he had known Toland in The Hague, where they exchanged clandestine manuscripts—a form of communication for heretical thought that became commonplace during the Enlightenment.[29] An English deist, Thomas Woolston (d. 1733), who challenged the authority of the Bible in matters miraculous and prophetic, died in prison. Prisons were very unhealthy places. In Paris during the 1720s the authorities closed down an aristocratic club, L'Entresol, because its members toyed with Spinozism and freethinking. The leading Continental Newtonian of the first half of the century, Willem Jacob s'Gravesande, was accused of Spinozism—a heresy to which he did not subscribe but of which he might be accused by devout Dutch Calvinists simply because of his intense involvement with the new science. His successor to the chair of natural philosophy at the University of Leiden, J. N. S. Allamand, was similarly accused although he too successfully protested his innocence.[30] We will meet him again in chapter seven as a rather mediocre professor of physics. In Leipzig, one of the cultural centers of Protestant Germany, official censors persecuted publishers and booksellers with special zeal when they were suspected of distributing Spinozist literature, but also for selling that antimagical text of Bekker. The orthodox clergy regarded any attack on the power of spirits to be tantamount to undermining all spirituality.[31]

The implications of Spinozism threatened secular as well as clerical authority, and not simply in the absolutist monarchies. The specter of leveling—as remembered from the English Revolution—lurked about in the early eighteenth-century circles of London deists and freethinkers, many of whom adopted the naturalism of either Hobbes or Spinoza. An anonymous freethinking poem of the 1730s wittily summed up one aspect of the Spinozist legacy: Add "mind" or spirit to "Nature" and "this Mighty Mind shall be/A Democratic Deity/ . . . all of which we behold is God,/From Sun and Moon, to Flea and Louse/And henceforth equal—Man and Mouse."[32] Freethinkers of this period could hold to ideas with democratic implications while still having little use for the people or their clergy. Naturalism or materialism of scientific origin also justified the new erotic literature of the age and by the 1740s a lively clandestine trade had developed in pornography. Works like *Thérèse philosophe* and *Fanny Hill* preached materialism and anticlericalism while displaying the lascivious in graphic, engraved detail.[33] By the mid-eighteenth century not only could someone think as a freethinker, it was also now possible to live as one.

The New Scientific Culture of the Educated Elite

The crisis of the late seventeenth century brought to a head the long-standing tension between the new learning, particularly the new science, of the educated laity and the doctrinal rigor of the traditional clergy. By and large the latter lost the struggle. They could no longer control the printing presses, particularly in England and the Dutch Republic; nor could they eradicate the demand for books and learning, the ever-expanding market for knowledge. The crisis also starkly exposed the heterodoxy to be extracted from the new science. Indeed, as a result of the crisis a new *persona* emerged, first in England and then in western Europe: the literate gentleman who read the periodical press, attended literary and philosophical lectures or clubs for the purpose of being cultured, and remained vaguely Christian, generally Protestant, but explained his beliefs in terms of the order and harmony of creation. He might be a merchant of the city or a landed gentleman of the country; he might even be a shopkeeper, a doctor, or a lawyer. He believed in educating his children; his wife, although generally more pious than he, was almost certainly literate and a reader of books, especially novels.[34] By the 1720s, particularly in England, such a gentleman or merchant could have increasingly easy access to applied science as taught by the Newtonian lecturers. Both he and his wife might attend the increasingly fashionable scientific lectures complete with mechanical demonstrations. By the 1760s his son might be investing in industrial ventures or, as we shall see in chapter six on the Watts, possibly even be an industrial entrepreneur himself. Liberal Protestantism and science made it possible for such men to explain nature to themselves and to feel comfortable in it; eventually, applied mechanics also made it possible for them to exploit it.[35] The weight of atmospheric air powered the new steam engines of Newcomen and Savery and if properly used, they could do the work of many horses or men.

Occasionally a scientifically literate gentlemen might slip over the edge into outright atheism, generally into pantheism or materialism. We can be sure that when such a conversion occurred it was made easier by familiarity with science. It may even have been prompted by taking Descartes's method of reasoning for oneself too literally or by assuming, as did Toland, that Newtonian gravity was a sufficient explanation for the workings of the universe and that no deity other than nature was necessary. In England such a radical departure from the prevailing wisdom was often accompanied by opposition to the ruling oligarchy or to any version of the old order as it manifested itself at home or abroad. Where we find such radical groups in late eighteenth-century Britain they will be frequently at the forefront of industrialization. For them science was tied to a larger vision of social reform through the application of machinery to production. Such radical gentlemen like the materialist and physician, Erasmus Darwin, or the Unitarian, Joseph Priestley, embraced capitalism perhaps even more willingly than did their more moderate counterparts. For those who could control it, capitalism of an industrial and mechanical sort represented, on both sides of the Channel, an effective means of destroying the monopolies exercised by the old landed aristocracy. An unreformed, entrenched clergy would fare equally badly.

Whatever one's private religiosity or politics became, in the eighteenth century the religion of the educated increasingly bore less resemblance to the expressive piety of popular Catholicism or to the rigorous demeanor of orthodox Calvinism. The century saw on both sides of the Channel a marked revivalism among the lower and middling classes. In England Methodism provided an outlet for intense religious devotion; in Germany Pietism sprang up within many Protestant churches.

Any kind of sectarian "enthusiasm," the public preaching of millenarians who announced the end of the world or the ecstasies of parishioners who thought they had discovered a saint in their midst—such an event occurred in Paris in the 1720s—was an object of scorn and derision on the part of those who identified themselves as enlightened. The scorn for "the inferiour herd of people" was endemic to enlightened culture; "our people of the lowest rank, for want of due care to instruct them, are worse than Hottentots," as one smug freethinking English journal put it. The only remedy was to instill "the most familiar and evident truths in natural philosophy . . . some of the fundamental maxims of a free-government . . . and practical precepts of religion and morality." These alone might "dispose the people to virtue, without which we can never long continue a flourishing nation."[36] The new scientific learning allowed some educated Europeans to judge the illiterate or the foreign with a superior gaze. The tendency to superiority induced by science was among the least attractive aspects of the new cultural dispensation.[37] But other apostles of scientific progress like Joseph Priestley turned instead to the study of history in an effort to understand developmentally and comparatively the sources of Western prosperity. In the process they laid the foundations for written history as we know it today.

Throughout western Europe the market for scientific learning coming from England, the science of Newton and the Royal Society, steadily widened. In 1700 Pierre Bayle urged a bright young Huguenot refugee with an interest in science to go to England: "It is the one country in the world where profound metaphysical and physical reasoning is held in the highest regard."[38] By that year Newtonian science had begun to attract followers in the Dutch Republic, especially at the University of Leiden as well as at the French-language presses run by Huguenot refugees or Dutch Arminians. The Leiden professors offered a new, yet moderate synthesis that eschewed materialism but implied a tolerant and progressive way out of the crisis in confidence that since the 1680s had afflicted elite culture. Among non-Calvinist Protestants who emphasized the right of the individual to find his or her own salvation, the Mennonites were particularly receptive to the new science throughout the late seventeenth and eighteenth centuries. The leading liberal Mennonite theologian of the Dutch Enlightenment, Johannes Stinstra, adorned his wall with a portrait of the Newtonian philosopher Samuel Clarke, whom he had translated.[39]

But the French-language press of the republic, edited by the Dutch and Huguenot journalists, led the way in scientific learning. The pages of the *Journal littéraire* (1713–1732), the *Nouvelles de la république des lettres* (1700–1710), *L'Histoire critique de la république des lettres* (1712–1718), *Nouvelles littéraires* (1715–1720), and the *Bibliothèque raisonnée* (1728–1752) are bursting with English culture, but particularly with explications of liberal Anglicanism and the latest scientific publications.[40] In addition the Dutch theologian Bernard Nieuwentyt wrote one of the most important textbooks on liberal and Newtonian theology, *The Religious Philosopher* (1715), which after its translation became a standard text in English schools; it was also popular in French and German translations. It was unrelenting in its attack on spinozism, and it presented a mélange of science and religion—known at the time as physico-theology—that emphasized the harmonious and hierarchical order of nature and society. Significantly the English translation done under Newtonian auspices removed Nieuwentyt's extensive references to the Bible.[41] A century of doctrinal quibbling had convinced liberal Protestants that science was a better anchor for religion than either of the Testaments.

In England the first generation of Newtonians—Richard Bentley, Samuel Clarke, John Derham, and William Whiston—took Newton's science into the pulpit. Yet as early as the 1690s Newtonian science, or more precisely the new mechanical science as synthesized by the *Principia*, was also unveiled in far more secular settings. In coffee houses and printers' shops, Newtonian explicators such as John Harris, Francis Hauksbee, and William Whiston assembled audiences and gave "a course of Philosophical Lectures on Mechanics, Hydrostatics, Pneumatics [and] Opticks."[42] Such lectures frequently received aristocratic patronage and became very much a part of the culture of the ruling Whig oligarchy.

Indeed the linkage between the promotion of Newtonian science and the interests of the Whig oligarchy was by no means accidental. After 1714 the latitudinarian hierarchy of the established church, much to the horror of the lower clergy, gave its blessing to the triumphant Whigs. The scientific ideology of order and harmony preached from the pulpits complemented the political stability over which they sought after 1689 to preside so comfortably. At the Royal Society the followers of Newton, partly as a result of his direct influence, were firmly in control and kept antigovernment or Tory dissidents out of positions of authority. By the 1720s and the accomplishment of the Hanoverian

Succession (1714)—which ensured the survival of Protestant monarchy, the Whig party, and the established church—a new generation of Newtonians had come to prominence and very much set the terms of the moderate Enlightenment in England.

At the Royal Society under the leadership of such Whigs as Martin Folkes and Sir Hans Sloane scientific application to industry and commerce—always a part of its mission—took on increasing prominence. Likewise we see an easing of the doctrinal preoccupations of the first generation of Newtonian clergymen; indeed Folkes and his friends appear to have had little interest in organized religion.[43] Newtonianism supplied all the answers they needed to live lives of relative comfort amid the prosperity and political stability enjoyed by the upper classes—and some middling folk—in the Hanoverian state. The elegant gardens of Queen Caroline at Richmond contained busts of Newton, Locke, Clarke, Boyle, and the liberal theologian William Wollaston, which expressed her faith in Newtonian science and natural religion.[44]

Typical of this Newtonian culture, with its emphasis on practical science, is a text like Henry Pemberton's *A View of Sir Isaac Newton's Philosophy* (1728). It is a much more straightforward and succinct account of Newton's philosophy of nature, his definitions of matter, space, time, the vacuum, and the law of universal gravitation, than that found in the Boyle lectures. Christian apologetics have been deemphasized in favor of a general, but constant, emphasis on the power of the deity, on a straightforward explanation of Newtonian physics. Whenever Pemberton enters into polemics, it is only against the materialists: those who, like Toland, assert that gravity is essential to matter, those who would have the immortality of the world, and those who deny the supremacy of God in every aspect of creation. This fashionable Newtonian and providential "deism" had now replaced the doctrinal exactness of the early Newtonians.

By far the most famous transmitter of Newtonian culture to the Continent was the French poet and philosopher, Voltaire. When he arrived in London in 1726 he learned Newtonianism directly from Samuel Clarke, and for Voltaire it took on the force of a new religion.[45] His *Lettres philosophiques* (1733), an immensely popular paean of praise to English government, social mores, and science, linked the achievements of Newton to a milieu of intellectual liberty such as existed, he claimed, only in England. He offered English science and society as a universal model for enlightenment, and in the process he further secularized Newtonianism. He insisted on the existence of Newton's God; but in Voltaire's hands the concept becomes largely impersonal; its function could be described as simply social. The deity maintains order and so too should monarchs and governments. The English aristocracy is praised precisely because of its willingness to be educated, to mix with men of learning and science. Even the London Stock Exchange became for Voltaire a symbol of how the ever-expanding market promoted toleration. He said that on its floor men of different religions mingled and traded. We now know that Voltaire could make that observation because of what he actually saw on the floor of the exchange. There men congregated by their religious identities (p. 168), and occasionally by occupation. The market may have forced them to be more tolerant, but it did not render them void of social identity. Voltaire was so taken with the relative toleration he witnessed in England that he wanted to imagine it as the end of cultural differences. It is better seen as a new and different form of culture, more modern, hence expansive, than what could be seen anywhere else in Europe at the time.

Like the English Newtonians, Voltaire repudiated the science commonplace in educated French circles, the science of Descartes, and he did so for similar reasons. Not only did it not explain celestial motion as well as its Newtonian counterpart, but Voltaire's private notes tell us that Cartesianism led directly to materialism and atheism.[46] He got this from Samuel Clarke. Voltaire, the deist, made Newton and his science fashionable throughout Europe, and he linked both to his rabid anticlericalism and his denunciation of superstition and intolerance. The new science, he proclaimed, was the alternative to priestcraft and bigotry. The argument became famous by the 1740s.

Beginning in the 1690s English and then Continental Newtonians undertook a vast propaganda campaign against Cartesian science. For someone like the Dutch doctor and Leiden professor Boerhaave, the science of Descartes was insufficiently experimental, for others the primary fear was that Cartesianism led directly to materialism. Voltaire put his objections succinctly:

> With regard to the pretended infinity of matter [for Descartes matter is extension], that idea hath as little foundation as the vortices. . . . But what are we to understand by an infinite matter? For the term [indefinite], used by Descartes, either must be explained by this, or it signifies nothing at all. Do they mean, that matter is essentially infinite in its own nature? If so then Matter is God.[47]

Voltaire's deism rested on the assumption that "God the General in the Universe gives orders to different bodies."[48] Without those orders there can be no order. Voltaire believed that without God nothing would restrain kings or impose order on the masses. Any explanation for the triumph of Newtonian science in the early eighteenth century that ignores or underestimates the force of these social and ideological considerations misses the context within which science, like any other body of knowledge, had to be mediated.

Voltaire popularized Newtonian science, but of even greater importance in its spread among the scientifically literate was the Dutch scientist, Willem Jacob s'Gravesande (1688–1742). His *Mathematical Elements of Natural Philosophy* (Latin edition, 1720–1721; English, 1720–1721 and five subsequent editions; French, 1746–1747) gave a sophisticated and highly mathematical explication of Newtonian science in textbook form that was never surpassed in the first half of the century. In 1717 at his inaugural lecture for the professorship in astronomy at the University of Leiden, a post secured for him through Newton's intervention, s'Gravesande defended mathematicians from the accusation of atheism and irreligion. He also lashed out at "those men who have never thought that their very existence and that of the things around them would not be possible without the effects of a powerful and a very wise Cause . . . and against those who are only occupied with religion as it is an object of their indecent railleries." s'Gravesande always maintained the Newtonian objection to materialism. But with that caveat in place s'Gravesande contributed to a generally more secular version of Newtonianism than what was commonplace among Newton's immediate and Anglican followers. He eschewed their clerically inspired polemics, and concentrated his attention on explicating the *Principia*.

S'Gravesande used mechanical devices in his lectures and pioneered on the Continent the more applied version of Newtonian mechanics. His impact, particularly on Dutch higher education, as we shall see in chapter 7, was profound. S'Gravesande educated an entire generation of students at Leiden, and his collection of mechanical instruments and illustrative devices was one of the finest in Europe. A woman visiting his university in the 1720s saw proudly displayed in its library "a fine brass sphere which shows the motion of all the planets according to the Copernican system and is moved by a pendulum."[49] Indeed with devices and textbooks such as these, especially when combined with the easier texts in Newtonian science that became increasingly commonplace, the *Principia* could be safely ignored by those in search of a basic scientific education. What surprises the historian about s'Gravesande and Dutch Newtonianism lies in the failure of both to extend their influence outside of academic circles. Yet for the mathematically serious, s'Gravesande's *Mathematical Elements* was the book to master. As late as the 1780s James Watt was forcing his recalcitrant son to do exercises out of s'Gravesande's text.

The Birth of European Freemasonry

The new cultural synthesis based on science, religion, and social ideology was preached from the fashionable London pulpits and published in fancy editions partly financed by lawyers, merchants, and Whig members of parliament.[50] Newtonianism also underwrote a new form of social gathering complete with ritual and costume. British freemasonry began in 1717 as a speculative, gentlemanly club quite different from the older masonic guilds from which it originated. The practicing masons, along with their culture of itinerant work, were displaced—indeed the concept of the guild protecting the wages of its workers is self-consciously repudiated by the new masonic *Constitutions* (1723). In place of practicing stonemasons came the scientific devotees with one out of four freemasons in the 1720s being fellows of the Royal Society.[51] The most active freemason in the early years of the London lodges was the Newtonian scientist and experimenter, Jean T. Desaguliers. We will meet him again in the next chapter where he becomes the leading Newtonian teacher of his generation. In his masonic garb, Desaguliers helped spread the lodges from London to the English provinces and to the Low Countries.[52] His background as a French refugee probably paved the way for his ease of access throughout the cities of Western Europe.

At the masonic gathering, the quintessential popularization of enlightened culture, literate gentlemen of substantial means (one had to afford the dues) worshiped the "great Architect," the god of the new science, and gave allegiance to any religion they cared to name: "to the religion of that Country or Nation, whatever it was, yet 'tis now thought more expedient only to oblige them to that Religion in which all men agree, leaving their particular opinions to themselves."[53] Armed with the principles of geometry as well as "the Mechanical Arts," "several noblemen and gentlemen of the best rank, with clergymen and learned scholars" constituted lodges where "all preferment" is based on "personal merit only." In some of the earliest British lodges the prosperous and meritorious could watch scientific experiments performed by visiting lecturers.

The lodges as they spread to both sides of the Channel were never primarily centers of scientific learning. They were social clubs that gave ritualistic expression to a fraternity of the meritorious and encouraged them to improve their literacy, education, and decorum. The lodges sometimes kept libraries or sponsored reading societies; not accidentally, freemasons in eighteenth-century Europe were active in promoting scientific education in excess of their numbers. When Desaguliers lectured on mechanics in Rotterdam, Amsterdam, The Hague, and Paris—speaking in English, Latin, or French—he undoubtedly attracted men who in turn sought out membership in his fraternity. Women were generally excluded from the lodges, although in late eighteenth-century France, lodges for women were popular and they advocated that women learn science so as to better equip them to search for equality.[54] For dissidents of church and chapel, for opponents of established authority and social reformers, the masonic lodges offered an alternative society where any heresy might be freely discussed. Not accidentally, the leading Amsterdam freemason of the 1730s and 1740s was a self-proclaimed pantheist who adored the new science and believed that "Nature places us willy-nilly on this earth, not forever but for a limited time, whose extent and final end are alike hidden from us; this is the universal order to which everyone, but especially men of reason, do well to submit themselves."[55]

Such a new and extraordinary faith in the order and reasonableness of nature, as proclaimed and mediated by Newtonian science, might also make political radicals out of those who took it too seriously. Eighteenth-century society and government, especially on the Continent, was at best profoundly oligarchical and at worst rigidly hierarchical and totally unrepresentative of mercantile or industrial interests and values. Inevitably there are links between the synthesis of science and religion that resolved the crisis at the beginning of the eighteenth century and the revolutions coming later, first in the American colonies (1776), then in Amsterdam and Brussels (1787), and finally in Paris (1789). A progressive faith borne out of the new science and sustained by its achievements rendered enlightened people potentially impatient, even rebellious, in the face of practices or old elites disinterested in improvement or in economic growth based on the freedom to trade, or worship, or experiment. Late in the eighteenth century reformers in Germany imitated the masonic lodges and turned them into new societies of *illuminati* where their radicalism and impatience could gain expression. The new groups probably had little resemblance to the lodges founded earlier in the century by Desaguliers. Yet in ideological terms they remind us that the progress promised by the new science might lead to expectations never intended by its original exponents.

The Application of Newtonian Science

Newtonian science in the hands of the laity was, however, more than ideology or the inspiration for new quasi-religious rituals. Increasingly it became practice. Prior to the assimilation of the *Principia*, mechanics certainly existed as a body of science and craft capable of application; but what was lacking was any overriding theory or set of principles, a natural philosophy and set of laws to give it coherence. We can compare a pre-Newtonian textbook of good practical mechanics with what came immediately after it. Such manuals were frequently anti-Aristotelian but could offer no coherent alternative explanation of gravity, although they were perfectly adequate in explaining how levers, wedges, and pulleys could be employed.[56] As one historian of science has put it, "the parallelogram of forces, the law of the lever, the principle of virtual work, the action of contact forces, and the principle of energy had extensive earlier histories," but all the parts of classical mechanics "were to be absorbed in or united to the Newtonian stream."[57] From the more general perspective of Western culture, mechanics also received an unprecedented public exposition after the publication of the *Principia*.

In the first Newtonian lectures ever given, Francis Hauksbee explicated "the general laws of Attraction and Repulsion, common to all matter." As proclaimed in the *Principia*, the laws "establish . . . the true system of Nature, and explain . . . the great motions of the world." Then follows a detailed description of Boyle's air pump as a machine "for giving a swift motion to bodies in vacuo." Hauksbee possessed a particular interest in the phenomenon of "action at a distance," of which electricity was the most fascinating and spectacular example. The attractive electrical forces are defined as essentially an aspect of the overall "power in nature by which the parts of matter do tend to each other"—in short, another illustration of Newton's principles. Throughout these lectures mechanical devices are used to illustrate the laws of Newtonian science, and the emphasis is on perfecting mechanical devices.

In Hauksbee's lectures no direct industrial application of the demonstration machines was made, although tables were given for the specific gravity of stone and coal found commonly in the mines of the Midlands.[58] The coal mining of Britain was by 1700 the most advanced in Europe. The output of coal in France at the end of the seventeenth century probably did not amount to more than 75,000 tons a year, which was less than what had been mined on a single north country manor prior to the English Revolution.[59] On the Continent only Belgian coal production came close to the English figures, and predictably both s'Gravesande and Desaguliers were active in Belgium (the Austrian Netherlands) in the 1720s attempting to install steam engines, probably of the Newcomen or Savery type, to drain deeper coal mines.

The application of the new science could hardly be resisted; indeed it had been encouraged by the scientists of the Royal Society as early as the 1680s.

The earliest applications were, however, more desired and dreamed than possible. Nevertheless, the commitment to render science useful to trade and industry became a part of English science from the 1660s, if not earlier. After 1700 ideology came to bear relation to reality, and simultaneously the lecturers of the London coffee houses moved into the provinces: to the north, Newcastle-upon-Tyne in 1711–1712, Derby in 1728, to the midlands, Peterborough, and Stamford, in the 1730s. The provincial academies, the schools of the non-Anglican Dissenters, also eagerly took up that science and replicated the lectures in their classrooms. By 1730, not incidentally, over 100 steam engines were at work in Britain. As we shall see in the chapters ahead, the scientific culture that gave pride of place to mechanics permeated Britain more broadly and deeply than any other place in Europe.

The steam engine cannot be severed from the diffusion of the English Enlightenment, from the science that lay at the heart of that cultural transformation. The Enlightenment could foster industry just as easily as it could instill a rather cerebral piety. It could edify and instruct the genteel; it could also appeal to provincial entrepreneurs more interested in capital profits than cultural sophistication. For them the engine worked as a symbol and it also just worked—in mines and factories. Such men possessed a sense of what was happening in the world around them and why it was necessary to educate oneself in science. They bought scientific books and attended scientific lectures in increasingly large numbers. The ease with which Newtonian science could now be taught made irrelevant the ideological struggles and metaphysical disputes that had once dominated natural philosophical discourse in seventeenth-century Europe. By 1720 a family like the Watts could be scientifically interested and imagine itself as being in the vanguard of a new, if controversial, cultural movement.

Late in the eighteenth century, at the height of the British Industrial Revolution, mechanical science and the ideology of progress it promoted seemed to the leaders of mechanized industry the answer to all human misery. It would secure their wealth and power eternally while eliminating the excesses of poverty still commonplace to the majority of men and women. Industrialists presumed that "the application of steam to the various purposes contemplated [will not be] very difficult," that there will be new machines "with greater the velocity and less the expense." They proclaimed their faith in words best taken from a leading industrialist writing to Maria Edgeworth, an early nineteenth-century moderate feminist:

> in the stupendous effect which the application of mechanical science is on the eve of bursting upon the world—when, in the transport of ourselves—as well as the enormous masses of other hands [i.e., the workers]—time, distance, and expense shall be almost annihilated. This will be laughed at now, as was Sir Richard Arkwright a half century ago, when he predicted Cotton Yarn and Cloth would be sent from here to the East Indies. [60]

The new industrialists like the Strutts or Josiah Wedgwood, gloried in mechanical science; they made heroes out of entrepreneurs like Arkwright and

Watt (when they were not competing against them). As did the Watts, they sent their children only to Edinburgh and Glasgow, to universities and Dissenting (non-Anglican) academies where they were sure the most up-to-date state of the art would be taught.[61] Armed with science and new technology they believed it possible "to ameliorate the condition of the great mass of the people not in Europe only but in the World—the rising generation are soon to form that Mass, some will rule and some will obey, but all will in one way or other have influence in the management of affairs." Armed also with a science significantly divorced by the early eighteenth century from the people and their immediate necessities, the first industrialists (not unlike their modern successors) believed that somehow they could retain a social order that primarily rewarded and enriched themselves while still improving the human condition. The dream goes back to Francis Bacon. Its widespread acceptance among the educated elite began only in the early eighteenth century, and so too did modern scientific culture organized under the banner of the Newtonian achievement.

II

CULTURAL AND SOCIAL FOUNDATIONS

5

The Cultural Origins of the First Industrial Revolution

Overture

Sometimes a single life or lives within a single family manage to embody the major themes of a book. Such is the case with the Watts—uncles, fathers, wives, sons, spanning three generations in Scotland and then England from roughly 1700 to 1800. All were interested in science; all turned to independent, entrepreneurial business and then to mechanized industry. James Watt (1736–1819) became world famous because he modified and improved the simpler steam engines of the eighteenth century and made them into the most advanced technology of the age. With his modifications patented in 1775 the engines provided unprecedented power from water and coal, replacing men and horses. They could drain deep mines and fill tidal harbors. Fitted with a patented rotary device, they ran the new cotton factories, potteries, and breweries. The steam engine became both symbol and reality of industrial changes that by the 1780s in textiles such as cotton were beginning to be seen as revolutionary.

Before James Watt became world famous he was the son of a little-known Scottish merchant, James Watt of Greenock (1698–1782) and nephew to two uncles, John and Thomas. All were in one way or another mathematical practitioners and knowledgeable about instruments and machines. One uncle, John Watt of Crawfordsdyke (1687–1737), short-lived and struggling, left the outlines of a life that, along with what is known about his more famous relatives,

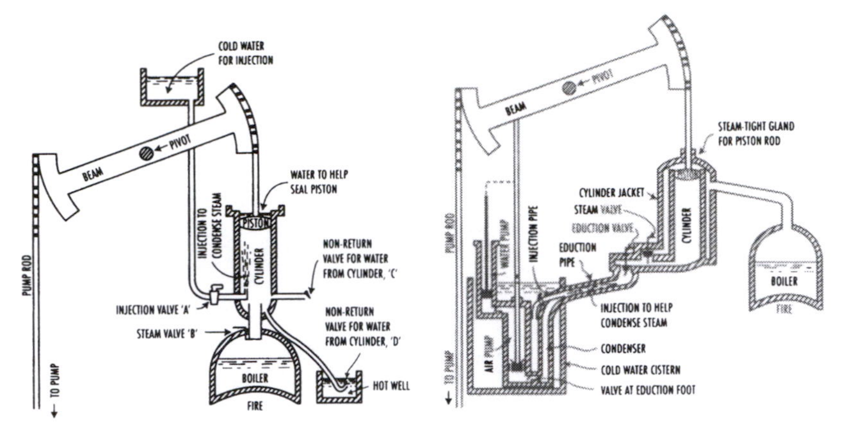

A sketch of the basic mechanism of the Newcomen atmospheric engine (left); Watt's separate condenser (right).

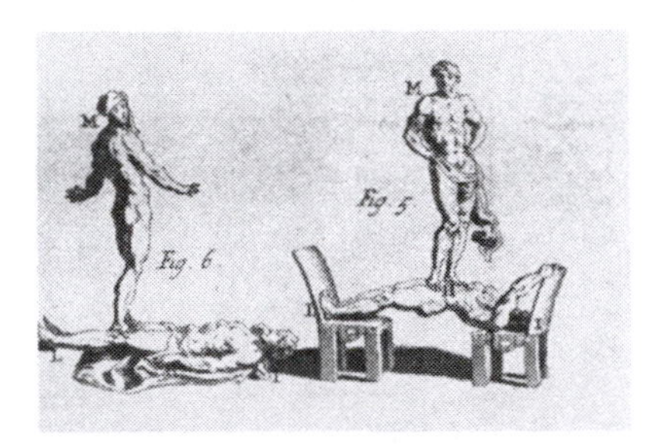

Illustrative drawings of men using leverage and balance to enhance strength; taken from Desaguliers's textbook on mechanics. (Courtesy of Van Pelt Library, University of Pennsylvania.)

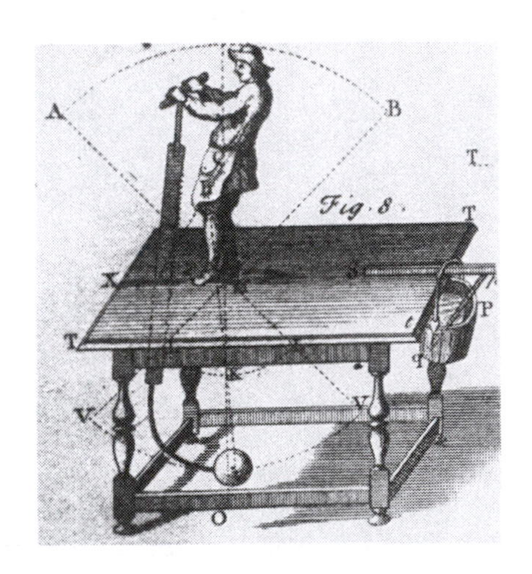

enliven any history book. In his hand-written notebooks, inherited from someone of the previous generation and shared with his brother, Thomas, John Watt recorded the intellectual and conceptual tools he learned from the new scientists, from Copernicus right through to his brilliant contemporary, Isaac Newton (d. 1727). Watt also inscribed his debt to the intellectual ferment associated with the English Revolution and with reforming Puritanism as it made its way after 1660 into Dissent. The Watts were all Calvinists of sorts; in Scotland and England that generally meant being a Presbyterian.

The intellectual roots of the Industrial Revolution are rudimentarily there in the jottings of John Watt, obscure artisan, self-made teacher, small-time entrepreneur. We would probably never have known about him had not his nephew, James Watt, become famous and been a compulsive saver of letters, indeed of every scrap of paper. In Britain by 1720, as we shall see in subsequent chapters, there were many artisans turned educators like John and Thomas Watt. All were obscure and made their living from applied science and mathematics. They did not have an easy time of it.

John Watt's surviving business cards are dated both 1730 and 1732 and contain a short self-portrait: "a yong man come to the Cost-side that professeth to teach . . . Mathematics . . . square and cube roots, trigonometry, navigation, sailing by the arch of a great circle, doctrine of spherical triangles with the use of both globes, astronomy, dyaling, gauging of beer and wine, surveying of land, making of globes, and these things he teacheth either arithmetrically, geometrially or instrumentally." For the date it was written the English used is old-fashioned, betraying the Scottish roots of John Watt. But his artisanal learning is prodigious and he is used to explaining things by instruments for those who possess little mathematics. Like his brother, the shipping merchant in Greenock, John Watt made a business from both land and sea, and like his brother, his handwriting suggests a man who is literate, but just.[1] Making a living as a scientific lecturer was harder in 1730 than it would be in 1780, when so many more men and women saw the value of such learning. By then, however, a kit of scientific instruments would cost about 300 pounds, a sum that John Watt probably did not see in an entire year of work.[2] Before his death a few years after he printed his business cards, John Watt got into financial trouble. We do not know why. His nephew, James Watt of engine fame probably inherited his books and used the mathematical exercises and mechanical lessons when he too learned surveying and the making of globes and quadrants.

The just-literate uncle was learned in higher scientific culture, but in his way. Aside from being literate—only slightly more than half of all Scottish men and even fewer women were at the time—he had an acquaintance with the teachings of Kepler, Copernicus, Tycho Brahe, Newton, and the mechanical philosophers. "Kepler observes that ye pulse of a strong healthful man beats about 4000 strokes in an hour . . . 67 times in a minute," Watt taught, and knowing how to count a pulse beat, a navigator at sea without a clock could roughly estimate time. One manuscript exercise book that John Watt owned started up in the 1680s; it too was probably inherited from a relative of the previous generation. It gave the phases of the moon supposedly from William the Conqueror right into the reign of Charles II (d. 1685) "whom God grant long to reign over us." Then came another page with the dates of full moons from 1687 to 1690. This book had been started sometime after the English Revolution, during the Restoration of the established church and king (1660–1685).

To show the position of earth, moon, and sun, the maker of the Watt book gave both the Copernican and the Tychonic systems. Living after 1660 he was savvy enough to know that the geocentric model of Ptolemy was, as Descartes put it in the 1640s, "now commonly rejected by all philosophers."[3] While natural philosophers in the Royal Society at this time were sure enough about the Copernican system of the sun in the center of the universe, there was still some doubt among everyday scientific practitioners. So this fellow hedged his bets and learned Tycho's system, which still put the earth in the center with elliptically orbiting planets around the sun. He also understood the completely heliocentric system of Copernicus with earth and planets revolving around the sun. For the purposes of navigation, either would do. Indeed what interests us is how this teacher of seamen and navigators was up on the latest theories about the structure of the heavens. By the 1680s the Ptolemaic system, with the earth in the center and perfect circular orbits made by planets and sun around it, was simply no longer believed. The Watt brothers were better at science than they were at history. Their knowledge of Copernicus was sketchy, perhaps recorded from memory: "Copernicus a famous astronomer of Germany, who lived in the year 1500. . . ." Actually he was a Pole who published his famous work in 1543. But no mind, the details of Copernicus's "system" were accurate enough in John Watt's handbook of applied science.

The new mechanics that evolved in the seventeenth century along with the new astronomy was synthesized into English-language textbooks written generally after 1700 by the followers of Robert Boyle and Isaac Newton. This new science, as we have seen, presumed on seeing the world, everything from air to water and earth, as composed of particles possessed of weight and measurement. In addition rational mechanics as it was developing did not abandon the traditional function of the discipline; it too organized local motions and made them more usable with the assistance of levers, weights, pulleys, and rotary motion.

Somehow John Watt and his brother Thomas had learned enough of the new mechanics to make drawings of inventions intended to be used at sea to measure the distance traveled by ship. Presumably they were the inventors. Wheels of graded circumference turned one into another, powered by the weight of water against a wheel that protruded into the sea. Carefully calibrated, each wheel reduced the feet into inches traversed, as would a series of connected pendula, and the final wheel mounted on a cabin wall would show (in ten movements of a hand on a circle) that the ship had traveled 10 miles. One drawing bears the signature of Thomas Watt, and it was still more sophisticated: "The great wheel which is to turne about once every 100 part sailing, turn yet second wheel 6 times about, and this turn ye ballance wheel 6 tymes about . . . ye index wheel turning about once in 10 tymes of this which will make a 10th part of a day. . . ."[4] It was an extremely cumbersome device, easily dislodged by the rocking of a ship. It probably never made it to the patent stage.

The drawings prove that mechanical invention occurred in the family and that early in the eighteenth century the Watts could think about the weight of water and the calibration of movement proportionally. They could also think about the smallest particles of air possessing weight as a result of their motion and they did exercises to determine "the weight of smoke that is exhaled of any combustable body." In a separate notebook probably dated 1722–1723, John Watt left a treatise on mechanic principles full of axioms and definitions: "The Center of Gravity of a Body is the point thereof about which the parts remain in equilibrio . . . velocity . . . by which a body runs a given space in a given time is the ratio of the space to the time. . . ." Watt was learning his Newtonian mechanics, possibly using a French treatise by the Dutch Newtonian s'Gravesande. In the same book he went on to apply the principles to weight balancing on a lever, to wheels, gears, etc. He is also reading physico-theology.[5]

Although learned in the latest mechanical science neither Watt uncle had even a modicum of success at inventing. Although they were teachers of mechanics, navigation, and fortification, the astrological predictions they also inherited may have meant more to them. Their notebooks contain what is described as the 1681 writings of the radical astrologer, John Pordage. What the astrologer had to say may have appealed to the precariousness of their lives, both personal and as Dissenters, political. Why else would someone in the family have copied the predictions?

Pordage was no run-of-the-mill astrologer. From the 1650s onward he was a radical in both philosophy and politics who sided with the enemies of absolute monarchy and regularly predicted dire fates for kings and potentates,

An engraved portrait of James Watt in his successful years. His somewhat grim affect is consonant with the depression he often described in his letters. (Courtesy of the Mansell Collection.)

even for bankers and clergy: "The conjunction of ye sun and mars will have a strange effect in some countreys in Europe & some prince perhaps last from England finds its true lot . . . some moneyed men shall suffer loss, & that from ye breaking of some great banker or bankers in or about ye city of London; some clergyman may be frowned upon by his prince."[6] The authorities of church and state never liked the Pordages of their world, and after 1660 outlawed the Dissenters (non-Anglican Protestants) who were especially drawn to preachings associated with radicals like Pordage.

The year 1681 was bad for Dissenters and, as far as we know and as far back as anyone of the next generation could remember, the Watts were Dissenters. Although more numerous in Presbyterian Scotland than in most places in the kingdom, they faced persecution and now the prospect of a Catholic king. In 1681 the movement led by Whigs to exclude James, duke of York and brother of Charles II, from the throne had failed utterly. Since 1660 Dissenting clergy—Presbyterian, Congregational, especially Anabaptist and Quaker—had been jailed or fined and many had migrated to the new world or to the Dutch Republic. Although granted liberty after the Revolution of 1689, people like the Watts would remain second-class citizens throughout the eighteenth century. Not surprisingly, the same Watt notebook with the predictions contains considerable information on the colony of Pennsylvania where William Penn and the Quakers had granted everyone full religious liberty. Being attracted to the subversive preachings of Pordage and having an interest in Pennsylvania bespoke a degree of religious, if not political, radicalism in the roots of this entrepreneuring family. A hundred years later it would surface again in the revolutionary decade of the 1790s when the grandnephew of John Watt, James Watt, Jr., sided with the French revolutionaries.

A full century earlier, reading the astrologer Pordage along with the Scriptures also denoted a devout Protestantism. As Pordage said in predicting by the stars: "we do not thereby pervert ye true meaning of ye Scriptures & tho we are forbid in ye holy Scriptures to be afraid or dismayed at ye signs of heaven, viz. to be possessed with such a fear as is inconsistent with our confidence in God, or as disturbs us in performing ye duties we owe as creatures to our great creator." Another searcher of the Scripture, Isaac Newton, who preferred to take his millenarian predictions directly from his own reading and nominally an Anglican, could not have agreed more.

The Watts of Newton's lifetime illustrate the way in which we must understand science in his time, like dark thread entwined in a tapestry of many colors, a whole cloth made up of religious and secular values crisscrossed with scientific learning. Once people were literate they had resources that went from the Bible to astronomical tables; once they had some capital and some commerce they could try to take shortcuts in industrial ventures by using levers, weights, and engines. We separate science from religion, science from technology, theories from practices. They did not.

John Watt left a legacy of scientific learning and disciplined striving that never deserted the Watts for a hundred years. In the course of the eighteenth century other Europeans would arrive at the same knowledge with different values and assumptions: devotion to kings or Catholic clergy, or an aristocratic dislike of business and commerce, or a good eye for commerce and no particular interest in applied mechanisms. Of all the ways science could be woven into a wearable cloth, the way the Watts spun will remain the focus of this book. Their success was not, however, in the stars despite their interest in astrology. The economics of their situation could not predict their eventual triumph, although having access to capital was clearly essential. By the mid-eighteenth-century consumption and international commerce had given the British a precious commodity in the eighteenth century, surplus capital. They also had coal, iron, and cheap labor. As we are about to see, they also possessed a distinctive scientific culture that now needs to be factored into the economic setting.

The Turn to Mechanized Industry: The Setting of Engineers and Entrepreneurs

Purely economic models traditionally assume that if people have coal, capital, and cheap labor they will see it as being in their best interests to industrialize. If they need any specialized scientific or technical knowledge to do that, they will just go out and get it. Such arguments about the way human beings change, make choices, or even recognize what choices are available, presume a particular definition of the way people are. Their free will prodded by their economic interests creates the advantageous cultural setting required, or free decultured agents simply transcend any restraints that culture may impose. Rationality means always choosing what is perceived as being in one's best interests. Put somewhat crudely, offer someone the chance to make a profit—in this case to industrialize—and they will perceive progress, do anything, invent or innovate as needed, try and try again until they succeed.[7]

What is missing in the story of early industrialization to date is any convincing cultural paradigm—a set of recognizable values, experiences, and knowledge patterns possessed among key social actors—that offers insight into the formation of the industrial mentality of the late eighteenth century.[8] According to David Landes, for the West "work has barely begun on the nonrational obstacles to innovation, on the negative influence of institutional, social, and psychological attitudes."[9] The economic model of human actions finds little of interest in the differences among the various scientific cultures that emerged in eighteenth-century northwestern Europe. The model points us elsewhere, solely to supplies of capital or cheap labor, to explain Britain's extraordinary leap forward in mining, transportation, and manufacturing. The role of culture—imagined as the tinted spectacles that enhance or impede individual perception and choice, or that sharpen short-range or long-range vision—has no place in traditional economic explanations. This book seeks to remedy a deficiency in our own cultural knowledge.

Showing the marked differences between the scientific cultures found in Britain in comparison to France or The Netherlands tries to recreate the different universes wherein entrepreneurs actually lived. From there the cultural model presented here suggests that mental universes played an historical role that was important. In this chapter we will concentrate almost entirely on Great Britain in the eighteenth century, on institutions and attitudes that worked in favor of innovation. Later chapters will explore the culture of science that can be seen in other western European settings. Laying emphasis on culture should never be seen as an attempt to supplant economic factors. In a sophisticated historical account cultural and economic life should be seen as they are experienced by human beings, as intrinsically woven together.

The eighteenth-century British civil engineer or mechanician, barely a professional figure, often self-educated and self-fashioned by pioneers like Jean Desaguliers, John Smeaton, and James Watt, is the key figure in the cultural side of the story discussed in this chapter. Indebted to the scientific culture established in England by 1700, such men acquired the necessary learning to do the more advanced calculations needed to move heavy objects over hilly terrain or out of deep coal mines never before tapped. British engineers and entrepreneurs who sought to build or improve canals and harbors and invent, as well as use, steam engines, had to be able to understand one another. Too much was at stake for their partnerships to fail (as was so often the case despite their best efforts). Scientific culture anchored around the Newtonian synthesis provided the practical and increasingly accessible vocabulary.

As it turned out, both engineers and entrepreneurs were well served by knowledge of applied Newtonian mechanics. After 1687 and the publication of the *Principia* mechanics, pneumatics, hydrostatics, and hydrodynamics had all been regularized and systematized by the Newtonian synthesis. Its eighteenth-century explicators, beginning with Francis Hauksbee and Jean Desaguliers, then wrote textbooks, which by 1750 made applied mechanical knowledge available to anyone who was highly literate in English, soon in French and Dutch.

Access to the mechanical knowledge found in the textbooks was critically important, yet the depth and breadth of its European diffusion differed widely. By the 1720s mechanical knowledge was more visible in Britain (in both England and Scotland) than anywhere else in the West; by then the British had invented what Larry Stewart calls "public science."[10] On the Continent the spread of specifically Newtonian and applied scientific knowledge to the larger public was inhibited—but not stopped—by various factors. High among them was the power of the Catholic clergy at work in the various educational establishments found, for example, in France and the Austrian Netherlands (Belgium).

In mid-eighteenth-century Britain industrial entrepreneurs in partnership with engineers merged in a preexisting setting conducive to innovation. It fostered trial and error through a common mechanical language and through relatively egalitarian interaction among and between them.[11] Both the language and the setting guaranteed trial and error, and it was (and is) absolutely essential to technological development. Engineers needed to have a hands-on familiarity with the site intended for development while speculators or local improvers also had to possess a meaningful understanding of applied mechanics to communicate with them. Such an understanding was best learned through touching or watching mechanical devices from table-top models to the real thing. Installing the wrong engine could bring bankruptcy. Applied mechanics taught by lectures, textbooks, and schoolmasters, served as the lingua franca when coal mines needed to be drained or harbors dredged or canals installed, or mechanical knowledge transferred from one industry to the next. As we saw in John Watt's notebooks, eighteenth-century textbooks of applied science slid effortlessly into technology, if for nothing else than to illustrate with weights and pulleys the principles of local motion and how they related to planetary motion. Decades before we can date the onset of industrial development fueled by power technology, its rudiments lay in the Newtonian textbooks available to literate people.

Historians once assumed that "much of [British] technical, scientific and organizational elements were international property before 1750."[12] But the evidence drawn from formal and informal educational sites from Rotterdam to Lyon suggests that the Continental diffusion of the culture of applied mechanics was much more sporadic and uneven than has been previously imagined. In some European cases the scientific element defined as a set of laws memorized or mathematically formulated was available, but the technical elements and organizational circumstances—the informal learning, the mechanical illustrations, the hands-on use of devices, the relatively egalitarian philosophical society, the cultural "packaging" of science—differed enormously.

British scientific culture further rested on relative freedom of the press, the property rights and expectations of landed and commercial people, and the vibrancy of civil society in the form of voluntary associations for self-education and improvement. In early eighteenth-century Britain these structural transformations worked for the interests of practical-minded scientists and merchants with industrial interests. Using Newtonian science taken from those parts of the *Principia* pertaining to the mechanics of local motion, the scientists created and the merchants consumed curricula and books applicable to technological innovation. In some cases engineer scientists also developed pumps and steam engines specifically intended as early as 1710 to enable "one man to do the work of a thousand" and aimed at the marketplace of entrepreneurs.[13]

In the Royal Society of London, but especially in numerous provincial scientific and philosophical societies from Spalding to Birmingham and Derbyshire, mechanical learning formed the centerpiece of discussions, demonstrations, and lectures. Into a setting of formal, but just as important informal institutions for applied, yet experimental scientific learning, came eighteenth-century entrepreneurs, would-be engineers, governmental agents, local magistrates, even skilled artisans—all faced with economic and technological choices and receptive to new knowledge systems promising new solutions. The route out of the *Principia* (1687) to the coal mines of Derbyshire or the canals of the Midlands was mapped by Newtonian explicators who made the application of mechanics as natural as the very harmony and order of Newton's grand mathematical system.[14] As we shall see in chapter 9 when we examine British settings as diverse as coal mines or select Parliamentary committees investigating the plans submitted by engineers or canal companies, after 1750 technically literate laymen and civil engineers communicated through a common scientific heritage.[15] Their cultural universe had fashioned the "mental capital" of the first Industrial Revolution.[16]

The cultural approach emphasizes not simply the intellectual component in the British setting, the books and lectures, but also its public and social nature, how and by whom it was absorbed and deployed. The British scientific societies were populated by men of land, business, and finance. They made science innovative in application, but not necessarily in original achievements. The social and cultural setting of British science after Newton helps explain the relative absence of originality by comparison with French science.[17] Taking note of the aristocratic character of French scientific institutions and examining how it reenforced their theoretical and mathematical bent (as we will in greater detail in chapter 8) throws the British model into sharper relief.[18]

Within an applied framework the Newtonian mechanical tradition laid particular emphasis on mechanical experimentation and actual demonstration with levers, weights, pulleys, table-top replications of engines, and so on. When turned toward application the practical and investigative style was critically important for encouraging industrial development. It tied science to machines as well as to an accessible method capable of being used by technicians and engineers who eagerly embraced the discipline and style of replication and verification. They in turn brought these practices to technological problems. Such men could simply not have understood the sharp distinction made in modern times between the scientific and the technological.

A letter of 1778 from the civil engineer, John Smeaton to James Watt concerning his steam engine, illustrates the interaction of scientific method with trial and error industrial innovation and, not least, with profit. As part of his normal way of proceeding, Smeaton explains that "to make myself master of the subject, I immediately resolved to build a small engine at home, that I could easily convert it to different shapes for Experiments. . . . I determined to prosecute my original intentions of finding out the true *Rationale*. . . . The fact is . . . I have no account upon which I can depend, of the actual performance upon a fair and well attested experiment, of anyone of your engines. . . . If you can shew me a clear experiment . . . I should think it no trouble to go to Soho [Watt's workshop] on purpose to see it."[19] If Smeaton became convinced of the value of Watt's innovation, then the engine could be built into plans or consultations for which Smeaton was being commissioned by canal or mine developers.

With these disciplined methods of verification and replication British engineers imagined themselves to be scientists or their imitators. They could move from hands-on knowledge of machines to the application of theories drawn from mechanics, hydrostatics, or pneumatics. In addition, science and mathematics occupied their leisure and informed the education of their children, and they bought books and instruments in all fields from optics to astronomy and telescopes.[20]

In some middle-class households technical knowledge was shared by both husband and wife, as the letters between James and Annie Watt illustrate.[21] He invented the separate condenser for the steam engine; she was a chemist in her own right who sought to perfect bleaching techniques and to replicate the experiments of the French chemist Berthollet who had produced chlorine.[22] Women's participation in scientific culture, given the inequality of their status throughout the West, can be turned into one important index of its spread. From the 1730s onward there was a European-wide effort led by Newtonians like the Italian, Francesco Algarotti, to find a female audience for science. British periodicals appeared specifically aimed at making science accessible to women. This may also have had something to do with their use of surplus capital. A 1775 guide to the London stock exchange said that stockbrokers developed to assist women making investments and to represent them on the exchange floor.[23] In Birmingham where the Watts lived, mechanics appeared in the curricula of girls' schools by the 1780s.[24]

By the 1780s many of the girls in Birmingham must have come from families where manufacturing and machines were commonly discussed. The mental posture of such mechanists or engineers with entrepreneurial interests might best be described as a merger of theoretical science and highly skilled artisanal craft. They knew machines from having built them, or from having closely examined them, and what is important from our perspective, they knew that machines worked best when they took into account mechanical principles learned from basic theories in mechanics, hydrostatics, and dynamics. Once learned, the theories could then be laid to one side for as long as the basic skill in metal working or mathematics remained. As the great engineer William Jessop told his inquisitive employers in the Bristol Society of Merchant Venturers, "in the earlier part of my time [I] endeavored to make myself acquainted with these Principles [respecting the discharge of water over cascades], and having been once satisfied with the result, I have, as most practical men do, discharged my memory in some measure from the Theory, and contented myself with referring to certain practical rules, which have been deduced therefrom, and corrected by experience and observation."[25] One needed the principles as well as the practices. A good workman, as Matthew Boulton put it, should "have brains as well as hands." As a frustrated French teacher of physics said in the 1790s when his school was too poor to buy the machines and devices: "Here it will be impossible to supply by [mathematical] figures in the absence of machines . . . verbal descriptions are truly insufficient in the sciences where one can only instruct by continual manipulations [of the devices]." Or as another teacher in the same national system of secondary schools put it, without the machines "I am reduced to teaching only theory."[26] In another one of these same schools, where the commitment to instilling industrial application had become part of revolutionary ideology, a French translation of the 1740s British textbook of Desaguliers was being used in the late 1790s.

In exactly the 1790s, when the French were bringing their educational system in science closer to the British model, the Society of Civil Engineers was established in London. It embodied the marriage between theory and practice, which reformers and industrialists on both sides of the Channel advocated.[27] Its membership consisted of a "first class" of engineers, a "second class" of

"Gentlemen . . . conversant in the Theory or Practice, of the several Branches of Science necessary to the profession of Civil Engineer," and a third class of "various Artists, whose professions or employments are necessary and useful to . . . Civil Engineering." Into each class fell men whom we shall meet again: in the first, James Watt and William Jessop, civil engineers (and seven others); in the second, Matthew Boulton, Watt's genteel business partner, and Sir Joseph Banks, president of the Royal Society; in the third, men about whom another book should be written, a geographer, two instrument makers, land surveyors, a millwright, an engine maker, and a printer. Although different in their "classes" (both within the society and in the larger social universe), all shared a common technical language that mechanical manuals and textbooks had helped to codify and disseminate. Only in the society did the engineers come first, ahead of their genteel betters. By the 1790s they had become leaders in the newly emergent industries.

Applied mechanics also required some mathematical training, especially in basic geometry. As British evidence suggests both engineers and entrepreneurs needed it, but they also required skilled workmen at their industrial sites who in the words of Matthew Boulton, "can forge, file, turn and fit work mathematically true."[28] Men with a basic mathematical knowledge were everywhere scarce, but rarer still on the Continent where mathematical education had not permeated as deeply into the general population as it had in Britain.[29] British schools were teaching basic mathematics, for example, algebra, geometry, surveying, mechanics, and astronomy, in some cases as early as the 1720s. Arithmetical and mathematical texts doubled during the first half of the century, their numbers peaking in the 1740s.[30] When the engineer James Watt gave instructions to his son for his education he said that "geometry and algebra with the science of calculation in general are the foundation of all useful science, without a complete knowledge of them natural philosophy is but an amusement, and without them the commonest business is tiresome."[31] He also wanted him to master physics and mechanics along with bookkeeping.[32]

The mechanical and mathematical knowledge possessed by British engineers, entrepreneurs, and even by artisans such as those who belonged to the Society of Civil Engineers, came from courses given by traveling lecturers, from patient study of textbooks based on the *Principia*, from handbooks for practical mechanics or textbooks used in private academies intended for artisans, or from regular attendance at the proceedings of voluntary societies like the Lunar in Birmingham, the Literary and Philosophical in Manchester, even the Royal Society in London.[33] While on the lecture circuit Desaguliers alone addressed hundreds of men and women each year who attended ten-week courses generally at a cost of two guineas. This most famous Newtonian of the 1720s and 1730s, former official experimenter for the Royal Society, then finally gathered his texts together and published *A Course of Experimental Philosophy* (1744, expanded from 1734). It put the bulk of the new mechanical knowledge into two hefty and beautifully illustrated volumes. They began with calculating the distance needed to offset disparate weights balanced on a beam, went through levers, weights, pulleys, pumps, and steam engines, and ended with a verbal and pictorial description of the Newtonian universe as explicated by the law of universal gravitation. The accessibility of the British to the new mechanical knowledge can be put concretely. A young artisan like John Watt possessed as early as the 1720s a good working knowledge of rudimentary mathematics and mechanics.[34] Similarly a school master in Bristol in the same period offered his young pupils "A Train of Definitions according to the Newtonian Philosophy."[35] Even Oxford and Cambridge taught Newtonian mechanics and basic mathematics to young gentlemen while the Dissenting academies were hotbeds of scientific learning throughout much of the century.[36]

Women and the Culture of Practical Science

The industrial process seen as a culturally configured series of applications dependent on knowledge and technique might be considered an entirely male or masculine venture. The cultural history of the first Industrial Revolution should not, however, be gendered so exclusively. We should not miss the attitudes and values that by 1800 women were beginning to bring to scientific learning. These are hard to get at because published texts germane to the education of mechanists and entrepreneurs are overwhelmingly by men. Aside from periodicals like the *Female Spectator*, the known attendance of women at lecture courses in mechanics and electricity and their subscription to underwrite textbooks, their independent role in economic life was often not visible. Even Annie Watt has largely stayed hidden from view, her private letters only now revealing how active she was in James Watt's business life.

But early in the nineteenth century women's relative silence is more easily penetrated. Margaret Bryan broke it by publishing a textbook in mechanics, *Lectures on Natural Philosophy: The Result of Many Years' Practical Experience of the Facts Elucidated* (1806). It grew out of her years as headmistress of a girls' school outside of London. Its subscribers' list of people who put up money to finance the publication was generally filled with the names of elite women of the aristocracy and also with many unmarried women whose London addresses suggest wealth. There were, however, other female subscribers about whom little is known. The book was dedicated to Princess Charlotte of Wales and the naturalist, Charles Hutton, who encouraged the project. There is more physico-theology in the text than was usual in other comparable texts by men, and its purpose was explicitly to arm women and all readers "with a perpetual talisman," which will "guard your religious and moral principles against all innovations."[37] The truths of religion and natural philosophy possess a deep affinity, or so it is argued, and the purpose of the text was to teach girls about physics as well as "to impress them with a just sense of the attributes of the Deity." But typical of the new industrial vision, Bryan's intention was to be "not merely mechanical, but really scientific" and therefore to associate "the theoretical with practical illustrations." She presents herself as "merely a reflector of the intrinsic light of superior genius and erudition" who is translating and moderating knowledge for anyone without "profound mathematical energies." Male writers and lecturers often said similar things. She confesses to being a follower of William Paley's version of natural theology. He stood in a long tradition of Newtonian clergymen that began with Samuel Clarke,[38] and to a man they used the Newtonian universe to illustrate God's providence and beneficence.

Like Haukesbee and Desaguliers nearly a century before her, Margaret Bryan begins with Newtonian definitions of matter and gravity in the process of introducing students to the history of the new science, beginning with Galileo and on to Boyle and Newton. She then moves to fire, evaporation, and steam. The engine described is by no means state of the art; it is a Savery engine. Yet the steam engine is presented as an instrument of progress: "But for this machine we could never have enjoyed the advantages of coal fuel in our time; as our forefathers had dug the pits as far as they could go." In predictable fashion immediately follow levers, weights, and pulleys with mechanics brought to its conclusion with "Of Man as a Machine," which despite its materialist sounding title, attributes the wonderful mechanism of the human body to divine artifice. From there she went on to air pumps, atmospheric pressure, pneumatics in general, hydrostatics, hydraulics, magnetism, electricity, optics, and astronomy (on which she had written another whole book); all were illustrated by experimental demonstrations. The scientific instruction is capped off by a preachy lecture on stoicism, obedience, cheerfulness, affection, and duty. Each stands in the service of politeness.

By 1800 the British mechanical vision neatly synthesized nature with a moral economy intended for young readers both male and female, of lowly or genteel origin.[39] Routinely teachers tied it to national greatness as exemplified by decades of technological advances. As historians now seek to understand the rise of British nationalism, the success of mechanical science needs to be added to the discussion. When British soldiers were captured by the French during the Napoleonic wars they were interrogated for their manufacturing and mechanical knowledge.[40] Such knowledge arose out of a century-long culture of science originally fostered by Protestant clergymen and scientists. By the 1790s footsoldiers could possess it. An education in scientific culture such as Margaret Bryan offered gave nationalistic pride to the daughters and sons of men who belonged, in whichever "class," to the Society of Civil Engineers.

The Anglo-Irish novelist and moderate feminist, Maria Edgeworth, was such a daughter. Richard Edgeworth, her father, mentor, and friend, belonged to the civil engineering society established in London as well as to the Lunar Society in Birmingham; both father and daughter revered scientific learning and utility.[41] They identified industry and applied science as the vehicles for improvement, particularly if they could be learned by their "backward" Irish tenants led by a patrician and educated but Protestant elite. Richard Edgeworth predicted in 1813 that "steam would become the universal Lord, and that we should in time scorn post horses."[42] Although William Strutt of the Derbyshire family of industrialists told Maria Edgeworth that mechanical learning was too dirty a business for women, he said to her that this was not for want of their ability: "Ladies are excluded . . . from Mechanics and Chemistry because accurate ideas on the subject can scarcely be acquired without dirtying their persons but in other things they are competitors."[43] As a lady she would have been the first to agree. Her novels such as *Belinda* (1801) painted gallantry as a vice and utility as a virtue; her private correspondence to the Watt family shows her keen interest in construction devices and steam. Her eagerness extended to wanting to be among the first to try the new steam boat from Holyhead to Dublin, although she did take the precaution of writing to James Watt, Jr., to ask him if he thought it was safe. Perhaps a similar knowledge of mechanical processes may have led the far more radical feminist, Mary Wollstonecraft, to argue in 1792 in her famous *Vindication of the Rights of Woman* that now women lived in an age where brute force need not predominate.

The Cultural Argument Summarized

The cultural roots of industrial technology in Britain were long, deep, and early to multiply. By 1800 so pervasive was the new scientific learning that it fueled the imagination of British entrepreneurs and feminists alike. The Royal Society of London as early as the 1680s discussed the labor-saving value of machines. Yet for an inventor or entrepreneur to get a patent in Britain up to the 1740s the bias of the authorities was overwhelming toward the argument that a device would put the poor to work, not enhance profits by reducing labor costs.[44] Indeed Desaguliers's 1744 textbook in mechanics while discussing the steam engine contains the first instance when anyone, writing in any language, spelled out in print (vol. II, p. 468) the critical insight that mechanization undertaken by engineers could enhance the profit of entrepreneurs precisely by reducing labor costs. Desaguliers's understanding of entrepreneurial industrial practice was consonant with what earlier seventeenth-century English theorists of political economy such as William Petty had explained. They looked to the marketplace as the model of human freedom. But they equated free choice with the ability to sell commodities, not with the selling of one's labor for wages, and certainly never with leisure or idleness.[45] By the 1730s an ideology of commercial development had come to be linked in the minds of some entrepreneurs with mechanical applications and Desaguliers' writings appealed directly to them. English science in the form of Newtonian mechanics directly fostered industrialization. It was not simply or merely its handmaiden as an older historical literature once claimed.

In eighteenth-century Britain the behavior and power of the landed, propertied, mercantile, and manufacturing was understood as the natural condition of all humankind. As Paul Langford puts it, "In a society dominated by property nothing could be more inimical to prevailing values than distinctions unconnected with property."[46] By 1700 within scientific circles an ideology with Baconian roots, aimed at the propertied and mercantile, had been developed. It was distinctively favorable to industrial and entrepreneurial activity. A partnership between wealth of any kind and applied science had been forged. The economic and technological wherewithal needed to make the ideology work most effectively would, however, take many decades to emerge.

When teaching about mechanics and experimentalism, the scientific lecturers of the eighteenth century reenforced the entrepreneurial interests of the middling (often higher) men and women in their audience. They officiated at the earliest marriages of convenience formed between engineers and entrepreneurs. Desaguliers interspersed mechanical practices with a discussion of the profit to be made from doing them correctly. But the mechanician had to be mathematically and naturally philosophically learned: "The contriver was a curious practical Mechanick, but no mathematician nor philosopher; otherwise he would have been able to have calculated the Power of the river." Had the trained engineer correctly calculated the volume and hence weight of the water, Desaguliers concluded, the management of power would directly have reduced costs and increased profits.[47]

Although deeply identified with propertied interests, engineers and industrial entrepreneurs had to have different skills from those of the traditionally landed or mercantile. And predictably, as Stanley Chapman has shown, eighteenth-century merchants and industrial manufacturers were not by and large the same people. British industrial entrepreneurs either had to possess technical skill or they had to be able to hire and converse with people who did.[48] They needed to assimilate applied scientific knowledge along with business skills and the Protestant values of disciplined labor and probity. As we shall see with the Watts, enlightened notions of progress and improvement also played a distinctive role in the value systems of late eighteenth-century entrepreneurs. Improvement became the watchword of the age. Its achievement relied on the coercive power of Parliament to guarantee patents or promote turnpikes and canals. In practice that meant having members of the two Houses who could understand what the engineers and entrepreneurs were trying to do.

Mechanical knowledge came to businessmen, as well as MPs, from a variety of channels. It was taught by scientific lecturers, by schoolmasters, and by self-help textbooks. It could be found even at Cambridge and Oxford. Mechanical knowledge became a centerpiece in the curriculum of the Dissenting academies that also laid great emphasis on ideologies of personal freedom, progress, property, and representative government, and on the writings of John Locke and Adam Smith.[49] A similar optimism and emphasis on "the improvement of our *Manufactures*, by the improvement of those *Arts*, on which they depend . . . chemistry and mechanism" was routinely discussed throughout the informal network of voluntary associations commonplace in towns and cities by the second half of the eighteenth century.[50]

Under the ideological umbrella pervasive in the philosophical societies emerged a new social space. The public culture of British science created, perhaps also required, a distinctive social ambiance among engineers and their employers. Collecting, experimenting at philosophical gatherings, as well as reading and discussions of literature, even the habits of sermon and lecture attending, gave engineers and entrepreneurs a common discipline and vocabulary. In this relatively egalitarian setting the civil—as distinct from the military—engineer achieved a newfound identity. He acquired skills of direct interest to men with capital to invest or commodities to move or manufacture more quickly and more expeditiously. At the same time the entrepreneur became remarkably literate in matters technical, applied, and occasionally theoretical. In explaining to a Russian count how to turn his visiting son into a manufacturer, Matthew Boulton wrote: "I also hope he will attend a course of Experimental and Philosophical Lectures . . . when he has attained some knowledge in these sciences, I beg he will allow me the pleasure of shewing him the application of some branches of them to the Manufactures and useful Arts and not return from Soho without seeing its manufactory."[51]

Brought together by a shared technical vocabulary of Newtonian origin, engineers, and entrepreneurs—like Boulton and Watt—negotiated, in some instances battled their way through the mechanization of workshops or the improvement of canals, mines, and harbors. Their mutual scientific literacy was also the source of much grief. British engineers frequently complained of the interference they encountered at an industrial site as entrepreneurs or investors proceeded to tell them how to go about their mechanical business. John Smeaton was particularly eloquent about his frustration: "The parties interfering suppose themselves competent to become Chief Engineers."[52] But Smeaton's frustration provides a critically important piece of information. By the mid-eighteenth century British entrepreneurs and speculators knew enough mechanics to think that they could stand on the river bank or at the mine shaft and tell the engineers how to do their job. For our purposes it is sufficient to know that by 1750 British engineers and entrepreneurs could talk the same mechanical talk. They could objectify the physical world, see its operations mechanically and factor their common interests and values into their partnerships.[53] What they said and did changed the Western world forever.

6

The Watts, Entrepreneurs

In the previous chapter, using the rosy spectacles of hindsight we credited the expansive public sphere of eighteenth-century Britain with helping to foster industrial development. Seen in retrospect networks of voluntary associations, lobbying in Parliament, voluminous publishing, and public lecturing made an ideal setting for cultural and intellectual advancement. Perched across the Channel and seen from a Continental European perspective, it was ideal. But not all British entrepreneurs and engineers immersed in the setting would have seen it that way. This was not a universe of limitless opportunity, neutrally open to all comers. Success in business—and civil engineering was also a business—meant cultivating one's interests at the expense of any and all competitors. Living in the highly competitive environment fueled by consumption and capital painstakenly accumulated made men long for monopolies. Indeed as partners Matthew Boulton and James Watt prospered, they were envied, despised, and imitated as monopolists. They shielded their pride by imagining themselves having a true reputation only among men of science. They knew that other men of business, as Watt said, "hate me more as a monopolist than they admire me as a mechanic."[1]

We want to look briefly at themes in the lives of these early, and now famous, partners, in particular at Watt and his family, engineers as well as entrepreneurs, for a more intimate, less abstract portrait of early British industrialists than the preceding chapter suggested. The Boulton–Watt partnership rested on mutual interest, unceasing desire for profit, and a shared network of intellectual and political interests. The partners also possessed a common body of technical knowledge that permitted each to communicate with the other. The partnership flourished within the framework created by voluntary association in the Lunar Society of Birmingham, and it depended also on a shared moral economy of values and attitudes best described as simultaneously self-aggrandizing and enlightened.

These were men of very different backgrounds. Boulton inherited his father's business and was something of a dandy, an Anglican who understood higher finance and the manufacturing of everything metal from buttons and buckles to watch chains. He loved fame, married well, and used his wife's capital as he needed it. While his firm struggled for many years and was overextended in debts before the success of the steam engine business, Boulton became a man of elegant tastes and social grace, at home in courtly society more than on the workshop floor.[2] By contrast we think of Watt as an engineer and inventor. He was also an entrepreneur, almost entirely self-made with family assistance, a dour and provincial Scot, inward-looking, repressive of himself and his family who had little regard for "the aristocratical." In private the Watts even laughed affectionately at Boulton's excesses, his eating and drinking, his conspicuous consumption of everything from coaches to garden chairs. Watt was also a compulsive saver and, as a result, he and his family left one of the largest archives ever assembled by a single family and business. All of it is now finally in the public domain at the Birmingham City Library. All the more reason why the Watt family is irresistible for any history focused on early industrial culture.

Not being readily able to form joint stock companies, British manufacturers resorted to partnerships. Together the dandy and the Scot created a business in steam that made their engine and their company paradigmatic of the profound change at the heart of the first Industrial Revolution. They died rich men. By 1800 they and their heirs were key players within an emerging industrial elite that had been unimagined a mere generation earlier. Watt was even hailed by some as the Newton of his age.

In the late 1760s Watt perfected the separate condenser that kept the engine's steam at a constant temperature and pressure and permitted the engine's cooling and condensation to occur in the condenser without affecting the steam in the cylinder. The condensation, then the refilling, of the steam in the condenser—not the external atmosphere—lowered or raised the piston of the engine without the other parts ever having to be cooled. Older engines like Newcomen's with one condenser had to have the steam (and hence its metal container) cooled by a spray of cold water, and the parts had in turn to be reheated by the next infusion of steam. Watt's innovation was elegant and brilliant. It drew on his exceptional talents as an instrument maker and his knowledge of drawing and mathematical exactness. His earlier work with clocks, watches, steel springs, levers, iron braces, brass fittings, and mathematical instruments gave him habits of precision and exact fitting that proved critically important for making his complex engine work.[3] His innovation with the separate condenser also aimed at economy; it allowed his engine to do, it was claimed, as much as five times the work for the same quantity of coal. When he was sure of his engine, Watt then applied to Parliament for an act to secure his patent and a parliamentary committee took testimony on the genuineness of his discovery. That testimony, as we shall see in the final chapter, required a reasonable knowledge of mechanics by the members of Parliament who oversaw patents.[4]

But mechanical knowledge was the least of what little men like James Watt were required to have. Patenting and private bills also required lobbying—licking "some great man's arse" was the way one of Watt's philosopher-friends put it. Or as Lord Cochrane coyly said when he wrote to Watt with the good news from Parliament: "I wish you all the success you could wish, notwithstanding that we coalmasters, have no reason to rejoice at any improvement that diminishes the consumptions of fuel." Nonetheless with Watt's patent secured, Cochrane ordered an engine for one of his new Scottish mines. So too did the president of the Royal Society who had mines on his estates.[5] Soon textile manufacturers in Manchester would also be ordering the engine equipped with a rotating device Watt invented to power the machines in their factories.[6] Within less than three years of securing its patent, the firm of Boulton and Watt had installed twenty-seven engines.[7]

Business and lobbying success in Parliament not withstanding, Watt and his associates took a dim view of all the imagined protectors of the public interest.[8] We may see the virtues of the vibrant public sphere that came into existence in England as early as the mid-seventeenth century. Watt could not wait to get away from "an ungrateful public."[9] In his letters to his second wife, Annie, written when he was trying to renew the patent to his engine, he was blunt: "We go to the House of Commons with no hope of victory; . . . I am held out as an extortioner, as a . . . man who claims rights to the inventions made by others before my days begun. It may so be deemed but if it is, I hope to live to see the end of a corrupt aristocracy that has not the gratitude to protect its supporters, nor the sense to uphold its own decrees." He never stopped worrying that someone, somewhere out there among the knowledgeable public would infringe his patent or that it would be revoked. Even when he had made 3,000 pounds profit from his engine, he lamented that "we have got so many pretenders now that I fear they will make us little people if we let them."[10] When he wrote to Boulton to express his fears he said that his enemies would cynically argue that the breech would be "for the good of the public." We may see public science as the great innovation of the age; Watt had little confidence in any aspect of the so-called public in part because the access it provided to innovation favored his competitors.

Watt's anxiety bred depression. Indeed he struggled with depression and severe headaches all of his adult life. Even as a young instrument maker and surveyor he was literally terrified by the risks that he was taking and success did nothing to alleviate his dark brooding.[11] The fierce competition ate at him. The debts he incurred launching the steam engine business made him "prey to the most cutting anxiety."[12] Watt said in his case that only science saved him from depression and the languor it induced.[13] Even when he had become successful, his son and heir of the same name had to beg him to "treat with the contempt they merit . . . the malignant cavils of your competitors in trade and the envious suggestions of the rivals of your abilities and reputation." Such was the nature of men that "in every age and country the wisest and best men have always suffered most by the petty calumny of those who had no other means to make themselves conspicuous."[14] James Watt, Jr., inherited his father's distrust of other people's interests, although as we shall see, he also had his idealistic and utopian side.

All the Watts suffered from various mental and physical aliments; indeed tuberculosis claimed two of the children. But there was nothing wrong with their sense of self, with their ability to project themselves effectively into the world, to undertake arduous tasks. The habits of striving, disciplined labor, and self-examination within a universe framed by piety and science can be seen in the family as early as 1690. The surviving papers of John Watt, James's uncle whom we met in the previous chapter, attest to the family's Protestantism of a Calvinist variety, its interest in mechanical science, its diligence at artisanal craft of a mechanical and mathematical sort. The hints we have of the family's politics at that time suggest a debt to Puritanism with radical digressions into Quakerism, Whiggery, and possibly revolutionary sentiments. The Watts may have begun as men of little property but they too, like the generation of Presbyterians who revolted against their king, were not to be trifled with.

James Watt inherited all these cultural debts. As a young man he kept his Bible with him wherever he went; he also always carefully maintained his accounts. He advised his son (even when abroad in Calvinist Geneva) to spend his Sundays in Bible reading. In his youth Watt saw an Anglican service in the great cathedral at York and he found it "ridiculous" for its ostentation. He was shocked by the talking of prebends and canons during the service.[15] A few decades later Watt had become something of a religious seeker, departing from the Presbyterianism commonplace to a Scottish childhood. Off in Cornwall surveying coalmines where his engine could be installed he even attended Anabaptist services—much to the chagrin of his second wife.[16] Only in courtship and early marriage (to his first wife, who died from the effects of childbirth) did his Scottish sobriety and earnestness give way to light-heartedness. He went to see a conjurer perform in London and mentioned to his new wife back in Scotland how he was charmed by such a "surprising fellow."[17] But once in London as a journeyman apprentice to a carpenter, and despite "rheumatism," he worked ten, twelve, or more hours a day—so hard that his hands shook with exhaustion.[18] Yet he kept meticulous records of his expenditures and was held accountable by his father, who acted as his creditor, banks being seen as a court of last resort turned to when a career or business was foundering.[19]

James Watt had no intention of failing or remaining a carpenter. Although he had had some formal education, applied science and mechanical craft were his ticket to a better future. His prosperous father, James Watt of Greenock (1698–1782), was a merchant and outfitter of ships who knew about mathematical instruments and navigational devices (as did the elder Watt's brother John and their father who taught mathematics). Watt's father was also an elder in the local Presbyterian kirk. The family had links to the academic and scientific community in Glasgow; young James supplied the professor of chemistry, Dr. Black, with cinnamon from one of his father's shipments. Clearly James aspired to higher things as well as practical knowledge and he was in London with the purpose of learning as much as possible about machines. He cut out numbers and letters in the shop of a clockmaker; he bought telescopes, compasses, and needles for his father and his father's friends; he learned how to make quadrants, mathematical and musical instruments, organs and flutes; he made globes; he was tutored by a schoolmaster, probably in mathematics; he taught drawing and map making and within two years he was hiring his own workers. By 1773 his first wife was addressing her letters to "James Watt, Engineer." With his instrument business struggling, Watt was in the field surveying terrain intended for commercially useful canals. By then he could assess the economics of a building venture as well as the hydraulics, discussing the projected savings to shippers in time and insurance money, and the value to investors in the building project as the result of profits to be made from the reduced cost of coal shipping.[20] When he first returned to Glasgow after his London apprenticeship his artisanal skill was such that he could work as a mechanic-instrument maker for the college where he was allowed to receive his mail.

The youthful Watt never relaxed. By the mid-1760s Watt had turned his attention to the most sophisticated mechanical devices of the day, to the cutting edge technology of steam and electricity.[21] He may not then have understood the principle of latent heat on which Joseph Black, professor of chemistry at Glasgow demonstrated that his engine depended, but he understood the effect of gravity or the force of inertia on its strokes.[22] He also came to know sometime in his youth about the properties of steam as an "elastic fluid" and the necessary geometry as well as practical mechanics needed for his engine.[23] He also had learned geometry and trigonometry, which he used in surveying, and had read the mechanical textbooks of Desaguliers and s'Gravesande. In 1763 he made his own steam engines of both the Savery and Newcomen kind.[24] So when asked to repair a Newcomen engine, he knew the mechanical principles at work as well as its strengths and deficiencies, and he knew how to work in wood and metal. The repair job turned into his life's work; he never stopped trying to improve his already vastly improved version of the steam engine. He did all of this without ever spending a day in a university classroom or a Dissenting academy.

Yet the university in Glasgow, with its fierce non-Anglican Protestantism and dedication to science, would remain a place of aspiration for the Watt family all their lives. There is even some evidence that in the early 1760s Watt knew about what Black was teaching on the latent heat produced by steam in his chemistry classes.[25] When James and Annie Watt's son, Gregory (b. 1777), showed literary and scientific talent, Watt proudly sent him to study there and Annie invited Gregory's professors and their wives to visit the household, which by then had become a centerpiece of Birmingham society.[26]

From youth to old age Watt, and his sons after him, retained their interest in science in part because it was an occupation undertaken by men of merit. Intelligence and hard work were all that was required. Science also cemented the camaraderie and values of Boulton and Watt and those of their entire circle. Matthew Boulton, a chemist in his own right, put the relationship between virtue and science succinctly: "A man will never make a good Chymist unless he acquires a dexterity, & neatness in making experiments, even down to pulverising in a Morter, or blowing the Bellows, distinctness, order, regularity, neatness, exactness, & cleanliness are necessary in the laboratory, in the manufactory, & in the counting house."[27] Scientists also belonged to the right political circles. As Watt explained to his wife, he "shall go to the Royal Society in the evening in hopes of meeting some friends who can be of use to us in Parliament."[28]

With Boulton and Watt's success in engines and business came their conviction that they too were scientists. Watt thought himself to be as smart as the French chemist, Lavoisier, indeed he and Priestley thought "Mr. L. having heard some imperfect account of the paper I wrote in the spring has run away with the idea and made up a Memoir without any satisfactory proofs. . . . If you will read the 47 and 48 pages of Mr de la Place and his memoir on heat you will be convinced that they had no such ideas thus, as they speak clearly of the nitrous acid being converted into air." Watt worried that no one would believe him as he was not, like Lavoisier, an academician and a financier.[29] But as far as Watt was concerned they were in the same league. Gradually Watt evolved from his Dissenting Protestant roots with their fundamentalist associations and legal disabilities into a secularist, a man of the Enlightenment.

At Glasgow in the 1790s, being given the opportunity for a university education, predictably Watt's son Gregory learned science and specialized in geology. But the ethos of practical science intended for progress and industry could also have a political analogue. Gregory did Greek and rhetoric while learning as his college notebooks tell us, that wealth and power produce "a crowd of servile sycophants" and that there are societies where "the haughty tyrant seated on his gorgeous throne . . . dreaded and obeyed by an abject people is for the time considered . . . at the zenith of human glory. The hand of death cuts him short in his career; he perishes in the midst of his splendor"[30]—dangerous teachings undoubtedly inspired by events in Paris. They harkened back to the seventeenth-century revolution and were hardly appropriate for young men destined for prosperity and business to be hearing.[31] But presumably the Watts found what Gregory was learning acceptable enough, at least for an intellectually gifted son.

For their part James and Annie Watt were as cautious and conformist as they needed to be, with Watt even advising his eldest son, James Jr., who was given to radicalism, to be submissive and respectful to powerful men.[32] In 1791 Watt told his friend, the radical chemist Joseph Priestley, that "while Great Britain enjoys an unprecedented degree of prosperity" and other countries are in the throes of revolution, it would be folly to risk "the overturn of all good government."[33] During the Birmingham riots against Priestley and other Dissenters the Watts were protected by their workmen (whom they generally regarded as worthless), and after the riots they were very careful and somewhat withdrawn politically. By 1793 Watt was laying great emphasis on his loyalism, but there is no evidence that he began to treat his workers any differently or better.

The Watts were far from content with the existing social hierarchy, but it was not the plight of their workers that concerned them. By the 1780s he and Annie had the time to follow contemporary national and international politics avidly, and politics seems to have been a regular subject of household conversation. In the 1780s Watt railed against excessive taxation and he saw little common interest between the landed who controlled Parliament and industrialists like himself. When frustrated in Parliament he would rail to Annie against "aristocratic scoundrels" and say that "a little more of this will make me an enemy to corrupt p.s [parliaments] and a democrat if democracy were a less evil."[34] He also helped to procure witnesses "to the cruelties practised by the slave traders." Indeed the Watts' vision of the political process was intensely social. Men of birth were the problem: "I hear a Society is formed at Freemasons Hall for shortening the duration of parliaments, but as the leaders are noblemen & gentlemen of great property, & I believe aristocratic there is no danger of their acting upon republican principles."[35] Such men, he said, "struck great terror into the supporters of the present corrupt system." But by the same token Watt said that he had nothing to do with them. He is never clear what alterations he would put in place of the existing system. But Watt is clear that only people of his class and interests can be trusted. Science, industry, reliance on family and kin made the Watts of the world harder, not easier, to govern. The divergence provided by work and profit did however keep them relatively passive.

Even as they were horrified by the violent turn taken by the French Revolution the Watts had no illusions about the power and pretensions of kings and aristocrats.[36] When the French threatened Italy Watt said that "if they spare the monuments of the arts the rest is only the retribution of divine justice on an excerable government." His hatred of tyranny and superstition is best illustrated by the willingness of James and Annie to send Gregory to Glasgow, where at the time the political learning was decidedly radical.[37] Not surprisingly, the attitudes learned in Glasgow took hold. When traveling in Austria Gregory wrote home that it was a land of "aristocracy, gluttony and imbecility. . . . Every wholesome regulation of poor Joseph [the II] has been annulled and Austria sunk a half century into Barbarism."[38]

But before the more spacious times that came with prosperity when the Watt family could indulge in politics or send their son to university or their children to the Continent and wider learning, there was much work to be done. From the time of Gregory's grandfather, if not before, the Watts were fiercely intent on upward mobility and they were harsh to indict any family member who could not or would not work. The rigor can be seen in Watt's father and it passed from generation to generation as well as laterally within each generation, only softening between siblings of the later eighteenth century. By then prosperity and the cult of sensibility encouraged kindness between brothers but especially between sisters and brothers. But in the 1750s Watt's father harshly judged his sons; young Watt in turn despised his brother, Jockey, when he had no work and urged his father to give him not a farthing.

Watt raised his own son by his first marriage, James Jr., with a similar harshness.[39] When Jamie wrote from the Continent to show his father his ability in

French, Watt could not fault the grammar so he attacked his style and penmanship. He sent his son mechanical problems from the Newtonian textbook by s'Gravesande while chiding him for using too much paper. His unaccountable harshness toward the daughter of his first marriage—after her mother died and he was settled in his second—extended to Watt's refusal to attend her wedding and he disdained James Jr. for wasting his time with sentimental journeys to see his sister.[40] As distinct from the affectionate, even pleading tone of some of his letters to his two wives and theirs to him, Watt wrote to his daughter on the eve of her marriage: "It is his province to order and yours to obey nor are you ever to dispute his will even in indifferent matters." He regarded her as "very dull and far from being accomplished."

It is little wonder that when the French Revolution broke out James Jr. lined up against the "crimes of tyrants," and much to his father's horror, supported the revolution right into the Terror. He told his father firmly that the monarchs of Europe "are in general so despicable that they are not worth attending to . . . by an enlightened age."[41] In his youth James Jr. displayed another version of the radicalism that cropped out from time to time in the family. He quarreled with his father, lectured him on politics, told him not to trust stories born of "aristocratical malice that only make me smile." He mourned the revolution's turn toward violence, but as someone who continued to believe in its principles. By 1794 James Jr. was in the thick of English Jacobin circles, had been denounced on the floor of Parliament by Edmund Burke, and feared returning to England. He did so only after the radical Thomas Walker was acquitted by the courts.

But even James Jr. believed in the common sense inculcated by experience and observation. By late 1794 he thought that "the revolution will remain more a terror to the friends of the people perhaps than to their enemies." Yet despite the disappointments brought by the French Revolution, he continued to expect and welcome profound change. James Jr. deeply believed that the changes in industry as they accelerated in the 1790s will "result in strange things . . . I have repeatedly said to father now that the machine is set in motion we may wait quietly the result."[42] Perhaps that prospect, plus the hope of inheriting his father's business, kept him from fulfilling his plan to expatriate to America with Priestley and his radical friends of the early 1790s.[43]

While we know a great deal about the politics of James Jr. and his brother Gregory, Watt senior played his political cards close to his chest. In his youth while trying to set his machine in motion, if Watt ever thought about politics he kept his views to himself, never even confiding them in extant letters to his first wife or to his father. Only infrequently did he vent his spleen against the great and well-born. In the 1790s his letters abroad were moderate and loyalist, but then he knew that spying had become commonplace and that the authorities were opening the mail of men in his circle, especially the known friends of Priestley.[44] Historians often say that industry and commerce in late eighteenth-century Britain diverted people who might otherwise have turned more radical. In the case of Watt and his circle, infused as it was with radical politics, there is truth in the argument.

It is not as if in the 1790s a timid James Watt made an abrupt about-face. Business and health were always perennial topics in all the Watt family letters. First and foremost this was a family in the business of getting ahead. Like fathers and sons, wives and husbands made partnerships based on worldly striving. Watt's first wife, Margaret Miller (Peggy), who was barely literate, still worked in his instrument shop replacing two of his "lads," and ran his business when he was in the field working as a civil engineer and surveyor. In the 1760s, in the critical period of Watt's inventiveness, Peggy's uncle loaned Watt money on the guarantee of his merchant father.[45] When Peggy died, Watt the widower and father was once again aided by his father and family, who looked after his children. When he remarried to Annie MacGrigor she was significantly more literate and learned than Peggy had been, as befit the wife of an engineer and inventor. By the 1770s she could enjoy consumption and a relative prosperity to which every member of the family aspired for several generations.

Annie Watt came from a bleaching family and had her own serious scientific and intellectual interests. She and her father were experimenters in bleaching techniques and in the 1780s worked with the newly discovered substance, chlorine, about which the French chemist, Berthollet, wrote at length to Watt. The partnership in marriage of James and Annie meant that they discussed her experiments. She also never hesitated to scold her husband or to lecture him affectionately on everything from his health to the curtain fabric he should buy on his travels. She understood his business both financially and technically. He could write to her for parts of engines and clearly she knew precisely what he was describing.

Annie Watt worshipped self-improvement—as she told her beloved son, Gregory, "you know we live but to improve"—but in that parental relationship her character softened, as did James Watt's, toward a sickly but talented son who would die at age 27.[46] In 1800 the elderly Watt even wrote to this favored son about the beauty of the seacoast and "its most romantic forms." In wealth and then retirement James and Annie Watt changed and mellowed, but only slightly. He became a true gentleman of science, branched out into chemistry and machines for medical treatment of lung and breathing disorders, took up an international scientific correspondence and espoused a version of the moderate Enlightenment. She carried on an intense friendship with her son Gregory, which included her opinions on what he should study and the virtues he must cultivate.

Yet despite his own debt to artisanal practice, Watt insisted that his sons have an even more rigorous and formal scientific and mathematical education, although it did include bookkeeping. Along with good morals James Jr.'s schooling both in England and on the continent was meant to provide him with a career, either as an independent mechanical engineer or a merchant depending on the direction his talents took.[47] Of these Watt took a harsh and dim view. Indeed James Jr. admitted that he had little mechanical skill and hence could do little to assist his father with a malfunctioning engine he was supposed to examine.[48] Nevertheless, the education Watt gave James imparted enough of the necessary business skill, science, entrepreneurial spirit, and international contacts that he eventually inherited the engine business and did well at it.

We can contrast James Watt's education of his son with that given by an exact French contemporary to his mercantile and industrialist son. The Oberkampfs were among the earliest Continental mechanizers of cotton and in 1780 father Oberkampf gave his son, Émile, detailed instructions on what he needed to know to succeed.[49] Many of the virtues required were remarkably similar to what Watt preached to James Jr.: solidity, scrupulous thrift with your capital, fairness, attention to detail, never taking anything on faith, and never trusting strangers. The Oberkampfs were also Protestants, a distinct minority in France. The father even left his son a list of every company in the world with which they did business, identifying each by religion only if it were Protestant, and outlining its virtues and its vices. But never once in these admonitions did father Oberkampf recommend the necessity for mechanical training. Had Émile Oberkampf been forced by the revolutionary circumstances of the 1790s to flee to England and try to keep his fortune, he would have discovered by 1800 dozens of steam engines at work in the cotton factories in Lancastershire, Cheshire, and Manchester. Despite the advantages that the Watts would have assumed a Continental education afforded, the exiled Oberkampf would have to rely on others to tell him in detail how the machines worked. There was not a cotton factory in France at that time where steam, and not men or water, powered the spinning.

Ironically, if you had asked the Watts, they probably would have seen Frenchmen like Émile Oberkampf as more sophisticated than themselves. Part of what both the Watts and the Boultons did for their sons was to ensure that they received a Continental education and that they were conversant with a wider cultural and intellectual universe than Birmingham or its environs. Although they were quintessentially provincial entrepreneurs, both Boulton and Watt in their way wanted to be cosmopolitans. Through the universe of scientific contacts they developed an international correspondence from the 1770s onward. Watt was proud of his French and like Boulton valued politeness and the skills only formal education could impart. Yet by this time and in retrospect we can see both Watt and Boulton as more cosmopolitan than the Oberkampfs. They traveled extensively but did not inhabit the same world of books and advanced scientific learning.

Yet there were important differences in emphasis between what Boulton and Watt valued for their male children. In young Boulton's case the education was self-consciously genteel; Boulton had large pretensions both for himself and his son. But Watt attended little to the elegance of the setting and was much more concerned that James Jr. learn science and mathematics and not waste his time on theater or novels. Annie Watt saw much greater value in cultural pursuits, in travel, poetry, and rhetoric. She begged Gregory to show her his writings and to be her friend. Yet both the Watts and the Boultons expected their male children to make their own way in the world, to be stiffened by the discipline of education, practical, applied, rigorous, and bookish. Only by such discipline would their business inheritance be properly managed.

Despite these varied social aspirations, first and foremost both the Boulton and Watt families instilled science in their sons. It was the key to personal and business success. For in the end above all else, both families coveted a place in a world that would make room for their interests and success. As Watt said when Parliament saved his patent and thus served his interests: "that two poor Mechanicks & the justice of their cause have more interest in his house [of Commons] than an Aristocrat, so be it always."[50] When his interests were placated and he could get on with being a scientific entrepreneur, then Watt was a loyalist. In his mind the world was divided between men of science, practice, industry, and merit, and everybody else, great and lowly. Any one of them could thwart the progress of one's industry.

The sources of the entrepreneurial spirit have been the subject of much historical discussion. Nearly a century ago the German sociologist, Max Weber, said that the worldly ascetism induced by Protestantism held the key to unlocking the new personae of the seventeenth and eighteenth century.[51] He found the new entrepreneurs most notably among Watt's associates and even used Benjamin Franklin, a distant colonial correspondent of his intellectual circle in Birmingham, as the prime exemplar of the early spirit of capitalism. Certainly Watt and Franklin were evenly matched when it came to frugality, obsessive saving, caution with regard to frivolity and luxury, all characteristics, Weber argued, of the self-made entrepreneur.

Much nonsense has been written by extreme champions as well as detractors of Weber's thesis. Among the errors has been the assumption that in naming Protestants as good capitalists, Weber meant to exclude Catholics or Jews. But Weber need not be read as having created "ideal types" trapped by their theology rather than molded by time and circumstances. Weber's point should be seen as an historical one: Protestantism with its emphasis on predestination induced an unceasing uncertainty about salvation. If it did not lead to despair, the Protestant ethic was slightly more conducive to worldly ascetism, to an almost mindless striving among larger numbers of literate non-nobles who were generally attracted to it in the first place. They were most commonly found in cities and towns where heresy was harder to eradicate. They had access to printing presses, and in urban settings were better able to practice artisanal and commercial crafts. The other major and older forms of Western religiosity carried with them historical baggage and associations with corporate, hierarchical, or ghettoized life that restricted individual freedom or frowned on aggressive expressions of self-interest coupled with risk-taking and enforced, even selfish savings. In the case of Catholicism an independently minded clergy responsive by custom and law to bishops and kings meant that it was harder to get the new values of self-made men preached from the pulpit. Where Protestants were in the ascendancy, as was the case in England, the Dutch Republic, Geneva, and parts of Scotland, not surprisingly mercantile life flourished. None of this means that Catholics could not or would not excel at business, but it does suggest that Protestantism succeeded in creating a capitalist ethos more easily and more effectively. Protestants need not have turned easily from mercantile to industrial capitalism. Many complex factors—among them market size, patterns of consumption, attitudes of elites, access to science and civic associations—had to be in place before, for example, the British and not the Dutch industrialized first.

Among the factors omitted by the Protestantism/capitalism paradigm has been the turn toward the secular within European cultures of the eighteenth century. We assign the term "Enlightenment" to this shift. As we saw in chapter 4, it had certain key components: dedication to experimental science generally of a Newtonian variety; an emphasis on the reform of existing institutions with a particularly cold eye cast on religious practices described as superstitious; and a glorification of print culture, sociability, utility, and merit. The Enlightenment also had a radical side. Atheism, materialism, and republicanism lurked among the values of its avant garde intellectuals. For people caught in the precarious world of the market radical values may have seemed more appropriate.

The enlightened voice was a universalizing one in that it resembled the Christian or clerical voice. But after that the resemblance ended. In the most extreme case the man or woman of the Enlightenment could live entirely for this world, dispense with Bible reading, fear of damnation, hope for salvation, church or chapel attending, and almsgiving. Charity, benevolence, sensibility, passions and interests, consumption and comfort, even luxury, as well as politeness in society, could preoccupy the life of the new secularist. Most devotées of the secular in Protestant countries where the clergy had been subjected to lay authority never went to the extreme of atheism or pantheism. They quietly shifted from Bible reading to the newspaper and drifted away from church-going except perhaps on special family occasions.

Something like that odyssey seems to have occurred in the lifetime of James Watt and his family, especially in the lives of his two male children. Gradually his own letters say less and less about Bible reading, or invoke God, or even send Christmas greetings. Annie Watt sent such greetings occasionally, but they were highly secular in tone even to her adored son, Gregory. Her greatest desire was for his longevity and their mutual friendship. His father counseled him to express disapprobation if "any of your companions expresses sentiments that are immoral or irreligious." But Watt did not spell out what those sentiments might be, and if the letters Gregory received from some of his friends are any indication, the advice went unheeded.

Watt may have held to one standard for himself in his family, another when relaxing with his cronies. When Watt and his first wife were courting, Peggy worried that he might think her too bold if she told of her feelings for him, suggesting a restraint on both sides. Certainly their extant letters are chaste and proper as are those between James and Annie Watt although they evince a love of possessions and comfort. Even in profound sorrow at the death of Gregory, neither has much to say about the will of God or eternal salvation. If sermons were heard by either, never were their contents remarkable enough to bear discussion in their letters. The same was not true for science.

In happier times when addressing Watt as "my dear philosopher," Watt's comrades saw him as far less proper than his family letters ever suggest. Writing to Watt quite colloquially, Dr. James Hutton could be bawdy and graphic: "A modern gentleman is not satisfied with simple action and reaction but when he goes to bed must have elasticity forsooth to work for him . . . a smith here has been consulting me about taking a patent for some improvement of a bed; I'm thinking of adding to it a machine which shall be called the muscular motion whereby all the several parts shall be performed of erection, intrusion, reciprocation and injections; this will become absolutely necessary in christian countries that do not allow the eating of children and where people will have pleasure at the easiest rate. . . . Please communicate [these temperature readings in my garden] to my friend Dr. Darwin."[52] Hutton seems to have made light of autoeroticism. The new, modern pornography of the age did likewise, and it also relied on mechanical metaphors, imitating the artifice of science to distinguish it from an older, bawdy naturalism.[53] It may be presumed that at the very least Watt was neither shocked nor offended by the highly secular conversation of the gentlemen "lunatics" like Hutton who met monthly in the Lunar Society supposedly to discuss only serious science and high culture.

Something of their "men on their own" mores may also have been communicated by father to sons. Certainly Gregory Watt had serious scientific interests and understood his father's business. Although in constant ill-health Gregory also had a taste for the libertine—at least as found in letters from his male correspondents. When on the Continent he bought an ample cross-section of books by the eighteenth-century French philosophes, the widely regarded leaders of the Enlightenment.[54] James Jr.'s political reading was also almost entirely radical or republican, and when not voraciously scientific, he added a goodly mixture of canonical texts cherished by the Enlightenment: Bacon, Locke, Hartley, Hume, Newtonian works, Voltaire, and Mirabeau. He did own a Bible and a Church of England prayer book.[55]

Somewhere between the Cornish coal mines and the successes of the Birmingham years, James Watt became an anti-Trinitarian, possibly under the influence of Joseph Priestley. When Priestley took up his clerical living in Birmingham among Dissenters originally of Presbyterian identity, his first sermon in 1781 sought to rationalize all religiosity around the unitary Godhead. He had made the secularizing odyssey from Calvinism to Unitarianism but there he resolutely stopped. We do not even know if Watt attended Priestley's first sermon. Some years later just after the pro-king and church riots of 1791 in Birmingham that threatened the life of Priestley and his family and destroyed their home, Watt claimed in a letter to Geneva that he had never been in a meetinghouse in Birmingham. But because of the French Revolution these were dangerous times for those suspected of being its sympathizers, and the Watts had to protect themselves. We may never know for sure where the Watts worshipped—if they did.

In general Watt's letters of a Sunday (his letter writing day) make no mention of sermons heard or pieties felt. He and Annie never hesitated to use Erasmus Darwin as physician to the family and his reputation for irreligion was well established. The will of their son Gregory recorded hastily just before his death in 1804 left nothing to church or chapel; Watt himself did the same in

1819.[56] James Jr., Watt's son by his first marriage, became so thorough a Jacobin with democratic tendencies that his letters barely deserve scrutiny for religious sentiments. None so far have been found, and the radicalism of his circle was a scandal in its time.[57] Other circles of early industrialists such as the Strutts in Derbyshire evince a similar disregard for formal religion accompanied by a taste for radical politics.[58]

Although never as radical as his errant son and certainly no Jacobin, somewhere along the way Watt became more a man of the Enlightenment than simply a non-Anglican Protestant. In that journey he precisely resembles Benjamin Franklin. The secular face of early capitalism Weber somehow missed. Neither Franklin nor Watt were original thinkers in matters religious as they can be found in various intellectual circles crisscrossing the Atlantic. We can imagine neither man editing his own Unitarian Bible as did Thomas Jefferson. They simply had values that were expressive of the way they saw the world and clearly, little about it invited supernatural explanations. But there were limits to Watt's heterodoxy. In the 1790s the Lunar Society made famous by Priestley, Watt, William Small (who taught the young Jefferson at the College of William and Mary in Virginia), Josiah Wedgwood, and Erasmus Darwin was not as radically involved as the literary and philosophical society headed by Darwin in Derby or the one in Sheffield. Nor did it become the Constitutional Society in Manchester, which through the good offices of James Jr., presented an address to the Club des Jacobins in Paris. Although disturbed by the political turbulence of the decade the Lunar Society would never have suited, or invited, Robespierre, but it could have welcomed Condorcet just as the president of the Royal Society, Sir Joseph Banks, welcomed the abbé Gregoire.[59] In all cases Burke would have been horrified.

The point about this search of Watt's soul is to try to assess what the progressive and universalist spirit of the Enlightenment might have contributed to the mentality of early industrialists on both sides of the channel. The emphasis placed here on the secular should correct an excessive reliance on Protestantism, or religion in general, as the single cultural well-spring of the industrial spirit. Practical, applied, utilitarian, innovative scientific culture, plus more and better science, became a credo in enlightened circles to which industrialists like Boulton and Watt belonged. This ethos, the moral economy of applied science, gave a social and cultural status to scientific and industrial practitioners from Boulton and Watt to their French imitators, the Periers. They also got more than status from science. They learned industrially valuable knowledge. Watt also believed that only scientific practice taught the method and regularity essential for industry and application, and that only men of science were worth associating with for that reason.[60]

Enlightened industrialists may have been monopolists to their competitors, or exploiters to their workers (for whom they had little but contempt), but among themselves early industrialists were also thoroughly secular, modern men. The Enlightenment permitted them to imagine that their industry had universal meaning. It validated them as improvers and progressives; being enlightened put an essential veneer on their unrelenting self-interest. Just as much if not more than their Protestantism, enlightened values inspired and legitimated their striving. In time, mechanized industry and the culture that spawned it would indeed come to be seen throughout the Western world as vehicles for progress, as forces dependent on deeply secular values that could be universally propagated. The moderate Enlightenment found throughout the Northern and Western Hemisphere from the 1720s onward belongs in the cultural history of the Industrial Revolution, both in England and on the Continent. The Watts as entrepreneurs and scientific people lived the Enlightenment just as much as any French philosophe lived its rather more abstract version.

7

Scientific Education and Industrialization in Continental Europe

In the previous chapters much attention was paid to the assimilation of scientific knowledge, particularly of a mechanical sort. Who knew what and when they learned it, the circumstances of the encounter with science, and the values of the clergy take on importance when we realize that science was never simply an abstract set of laws to be memorized. Scientific knowledge came enfolded within a "package" of beliefs, attitudes, and values that differed enormously depending on who taught it and under what circumstances. In Spain, for example, science belonged in the curriculum of all universities but Copernicanism either was not taught well into the eighteenth century, or when mentioned it was taught as a hypothesis, not as the foundation for an entirely mechanical understanding of nature. In Dutch universities of the eighteenth century, such as Utrecht, we can find Newtonian physics being taught throughout the century. Every thesis, or dissertation, done in the various science faculties whether in Leiden, Hardewijk, Utrecht, or Groningen, explicated its theorems mathematically and not by recourse to mechanical devices or machines.[1] Did the method make a difference? It did if you were trying to train civil engineers who needed to understand applications or if the goal was to offer a general familiarity with basic science to laymen with little mathematical background.

The kind of science learned and the timing of its introduction differed from country to country in western Europe. The differences influenced, but did not determine, which countries industrialized and when they did so. People cannot do that which they cannot understand, and mechanization required a particular understanding of nature that came out of the sources of scientific knowledge: textbooks, lectures, and classroom demonstrations that emphasized mechanics. In this chapter we want to look at certain key countries, France, the Netherlands both north and south, and briefly at Germany and Italy, to understand what kind of scientific education prevailed in those places. As we are about to see, the differences between scientific cultures in eighteenth-century France and the Dutch Republic or Britain are knowable, and once known, they inevitably force us to draw conclusions relevant to today's world.

If scientific education was a key variable in the Western experience and the timing and nature of this education differed significantly country to country, as we are about to see, then we need to revise the traditional model still offered to countries struggling to achieve technological development. The model of the semiliterate tinkerer as the key figure in industrial success, a model that downgrades formal or informal scientific education, is a prescription for success that may doom its followers to failure.[2] Put another way, the model permits the World Bank and other investors and creditors to ignore the educational infrastructure of a country while calling for growth and development. But they do so while operating with an incomplete history of eighteenth-century Western development. Our contemporary prescriptions to remedy economic backwardness lack the cultural dimension present in the Western history of industrialization, and this gap reinforces a willingness to ignore culture in late twentieth-century societies.

The ability to think mechanically—that is, scientifically, in the modern meaning of that word—permeated Western societies only selectively in the course of the eighteenth century. In less literate segments of the western European population, and in certain areas of eastern Europe, the penetration occurred only in the nineteenth and twentieth centuries. Knowledge has consequences. It can empower; if absent, it can impoverish and circumstances can be harder to understand or control. In 1787 one of the first flights ever undertaken by a large balloon occurred some 12 miles outside Paris. When the balloon came to rest, frightened peasants mistook it for the moon falling; they attacked the object and badly damaged it.[3] In the late eighteenth century the Russian government attempted to import many of the mechanical devices recently developed in the West. British engineers enlisted in the work of canal building brought with them models of mechanical devices, among them the steam engine. When in 1780 these models were shown to older members of the military engineering corps—that body supposedly most educated in mechanical principles—some of them simply did not understand how such a machine could operate.[4]

Relatively sophisticated mechanical knowledge had to be a part of one's mental world before such mechanical devices could be invented and, more to the point, effectively exploited. If you were a worker having to work in relation to a machine, understanding it meant coming closer to understanding how your employer might view all of nature, yourself included. Where mechanical knowledge was widespread and institutionalized in the educational system, and where capital, natural resources, and exploitable labor were also present, the results of that coincidence transformed both nature and society, creating in its wake the modern industrial world.

Wherever there were active industrialists, almost without exception, they had access to sophisticated mechanical knowledge if they wanted it. If entrepreneurs themselves could not build machines they could talk to those who could. Neither the entrepreneurs nor the engineers were on the whole tinkerers. The tinkerer model is historically inadequate for a host of reasons. First and foremost it presumes a distinction between the "scientist" and all others that simply did not exist in the late eighteenth century when industrialization began, first in England and then on the Continent.

Of the hundred or more leading British scientists from 1700 to 1800, for example, nearly half would have to be classified as "devotees" (to avoid the anachronistic term "amateurs"), and of that hundred, 45 percent made their living as doctors, technicians, or churchmen.[5] Getting at the diffusion of scientific knowledge can be tricky when there were so few people who hung out a sign that said "scientist." Indeed the term had not even been invented yet. They would have said "natural philosopher" or "engineer." In addition, everything we know about European history from the crisis of the 1680s onward tells us that in general science was "on the agenda" of Western elites. It had become the stuff of journals and textbooks. How then do we see differences between what the French knew and the Dutch didn't, or between styles of scientific inquiry?

There were significant and knowable differences in the educational systems around science found in each country. The remaining records—and quantification for the eighteenth century is impossible—suggest that in widespread scientific education of a mechanical sort, the British were at least a generation ahead of their European counterparts. As we saw in chapter 6, the generation from the 1760s to 1800 was critical in giving Britain its leap forward industrially. Men like the Watts gave the British a head start, nothing more, nothing less.[6]

In chapter 5 we noted in passing the relative slowness with which certain areas, particularly in Catholic Europe, picked up on Newtonian mechanics. Here we want to deepen and nuance the picture and examine the kind of scientific knowledge that was being disseminated within the rectangle made by Paris, Amsterdam, Berlin, and Turin. The purpose of this survey is to debunk the myth about how important inventions in the early stages of industrial revolution had nothing to do with knowledge systems. The tinkerer myth ranks with the belief that in Continental Europe governmental interference, pure and simple, caused technological backwardness.

But in Protestant Geneva lectures in natural philosophy were paid for by the government, free to students, and divided into the theoretical and the experimental. Governmental support did not extend to machines for the local academy. Its academy possessed none and in 1787 the experimental lectures had to be given privately by the local professor.[7] English entrepreneurs sent their sons abroad for such lectures and thought it money well spent on the acquisition of Continental ways and airs. In the narrow lens focused by our search for industrial culture, we might wonder why the Watts and Wedgwoods bothered to travel.

France

Despite the force of the Enlightenment in select French circles, industrialization on any significant scale did not occur until the early nineteenth century. Of course within the French scientific community, particularly but not exclusively when influenced by Newtonianism, the implications of applied mechanics were readily perceived. The French mechanist Jacques Vaucanson had attempted in the 1740s to establish factory production of a mechanical sort in the silk industry. He did so decades before Richard Arkwright founded his cotton spinning mill in Derbyshire.[8] There were French chemists of the early eighteenth century who also knew that their fledgling science should be applied and who wanted the state to intervene and assist in that process. The vision of such men was quite simply industrial. It included the training of workers whose skills would facilitate the entrepreneurs, who would in turn profit from chemical applications.[9]

The scientific lecturer of the mid-eighteenth century, the abbé Nollet (1700–1770), was probably the most important itinerant French promoter of the new science, complete with mechanical applications, on the Continent. He learned his techniques of demonstration in the 1730s from s'Gravesande and the Dutch Newtonians. Thereafter he opened his *cours de physique* in Paris, a lecture series that he eventually took on the road, to the French provinces, the Low Countries, and Italy. This series was the most popular ever given on the Continent, and Nollet's fame came to rest partly on his electrical experiments, which astonished and delighted his audiences. Popular enthusiasm for electrical effects cannot be ignored as one of the stimulants that drew the public to the new science. Experimenters thought that electricity possessed medicinal value and could cure everything from tumors to the gout. Given the state of medical practice at the time, it is little wonder that people flocked to see electricity in action.

Nollet's course of physics was grounded firmly in the practical uses of the new science. Like his British counterparts, he had to know the interests and limitations of his audience. He eschewed complicated mathematical applications and provided a glossary of terms for his readers. In general he avoided metaphysical or physico-theological questions in favor of practical examples to illustrate the "mechanism of the universe." In this last aspect Nollet's lectures reflected the general turn away from constant attention to religious questions, a shift clearly visible in scientific lectures given from the 1720s on both sides of the channel. In concentrating on the useful, Nollet claimed to be catering to public taste and machines were meant to facilitate their learning.[10] First Nollet concentrated on basic chemistry: how to dissolve metals, such as gold coins; how to use glues in porcelain making; how to use nitric acid to dissolve iron filings; the techniques of dying cloth and paper—in short, the chemistry useful in trade and hand manufacturing.[11] The general laws of physics, such as inertia and resistance, were then explicated verbally as well as illustrated by the impact of moving balls of lesser and greater size. Once these general principles had been established, the mechanical lectures embarked on explanations of how

The abbé Nollet from an engraving attached to the title page of his lectures. (Courtesy of Van Pelt Library, University of Pennsylvania.)

the laws may be employed "to the greatest advantage."[12] Nollet made much of windmills for grinding, or pumps that raise water "for our use or for the decoration of our gardens," or vehicles for transportation, and levers and pulleys for architecture and navigation—all to be constructed not by simple "machinists" but by true mechanical philosophers. He assured his listeners that sophisticated machines can replace human labor and consequently save money. The approach taken by Nollet in his lectures may be described as proto-industrial, rather than directly industrial, in that little is made of the actual uses of mechanical devices in coal mining, water engineering, or manufacturing.

The lectures of Nollet and the other French popularizers of the new science provided the French elite with an alternative to the relative scientific backwardness of the colleges and the University of Paris. It took up Cartesianism only in the 1690s, although even then Descartes's science remained controversial in the eyes of the church (and the state) well into the 1720s. The first Newtonian lectures at the University of Paris were in the 1740s, and Nollet himself received recognition from its officialdom only in the 1750s.[13] If we contrast the French pattern with natural philosophical teaching in British or Dutch universities, or even in the provincial Dissenting academies in England by the 1740s, it is clear that a generation or more of French students in over 400 colleges did not have access to knowledge directly useful to the process of industrialization.

Particularly in the colleges they controlled, the Jesuits fought the introduction of Newtonianism into the 1740s and beyond. By then the failure of Cartesian explanations was too obvious to be successfully ignored. Where the formal, clerically controlled educational institutions resisted or ignored Newtonian mechanics, the diffusion of industrially useful knowledge generally occurred a full generation or more after its British acceptance. It became available for young men educated after 1760 rather than before 1740. There is no point pretending that in Catholic Europe the scholastic clergy were teaching the new science with any dedication before 1750.

Put another way, it was possible to learn more about applied mechanics at a London coffee house lecture series than it was in any French *collège de plein exercise* prior to the late 1740s. Only then did the curriculum of the nearly 400 French colleges begin to shift decisively away from Cartesian metaphysics toward both a theoretical and applied Newtonianism. Focusing on the most backward of the colleges, the historian who has studied the curriculum of all of them, L. Brockliss, concludes that "if Newton finally triumphed in France it was probably over the corpse of the Jesuit Order."[14] The Jesuits were expelled in 1762. In the 1790s, despite the reforming efforts of the French revolutionaries, only thirty-one of the approximately 105 new French central schools (for students aged 15 and older) possessed significant collections of scientific equipment. Of course in an average year before 1789 only about 5,000 youths over the age of eighteen took a course in physics.[15] After 1789 the figure rose quickly, possibly to as high as 25,000. It is not surprising that in all of Continental Europe, including France, by 1790 there were fewer civil engineers in the private employment of mechanically knowledgeable entrepreneurs than was the case in Britain.

But France did have many active scientific academies. They made scientific knowledge available on an unprecedented scale. Yet aristocratic domination in the provincial societies and academies hardly permitted the kind of gentlemanly zeal for practical science that we see in late eighteenth-century Derbyshire or Birmingham. But aristocratic dominance did create a favorable environment for innovative and original science found throughout the French academies. Within the ethos created by noble interest, men did not sit in ordered rows, focused on an experimenter or lecturer. Rather they conversed spontaneously as equals within an elite, arranged around a large table where there was "much discourse without order." To be sure experiments and instruments were observed, but English travelers said that the quality of instrumentation, even in the Royal Observatory, was inferior to what they had seen at home.[16]

In 1793, at the height of the French Revolution, the radical Jacobin Convention (or parliament) abolished the French scientific academies inherited from the old order, both in Paris and in the provinces. Many of their leaders were executed. Two years later, the Paris Academy originally founded by Colbert in the 1660s was revitalized, reformed, and renamed; but its personnel was now quite different, with many scientists having perished in the Terror. We may well ask how and why a revolutionary government, however brutally or wrong-headedly, sought to abolish academies we might associate with enlightened progress and certainly with scientific innovation.

Answering the question requires that we take a close look at how scientific culture worked in eighteenth-century France. From the 1660s and Colbert onward, the French monarchical government showed a marked interest in science and its application. In the 1750s this interest focused on steam-powered boats, largely for military use; in the 1770s and 1780s encouragement went to the invention of mechanical devices for agricultural application.[17] The efforts to introduce "scientific farming" were very extensive and reflected the highest ideals of enlightened absolutism as found in the decades prior to the French Revolution.[18] Of course, the linkage between the scientific academies and the interests of the crown doomed their members in the eyes of the radical Jacobins.

The idealism behind the efforts of the French academicians derived partly from Baconian doctrines and partly from the secular idealism so commonplace among the educated elites influenced by the Enlightenment. Both they and the royal government supported scientific inquiry chartered or licensed by the crown. One of the major philosophes of the 1770s justified the linkage between absolutism and scientific inquiry in language that harks back to Italian debates of the early seventeenth century on the role of science within the state. In urging the Spanish monarchy to institute an academy in its scientifically backward country, Condorcet, a leading French philosopher of empiricism, explained that academies are "an advantage for the monarchical state." His reasoning was as follows: "In a republic all citizens have the right to meddle in public affairs . . . but it is not the same in a monarchy. Those whom the prince appoints have the sole right to meddle." But for men who have a need to agitate and who cannot abide the inactivity forced on them by the nature of the monarchical state "the study of science can only represent . . . an immense vocation with enough glory to content their pride and enough usefulness to give satisfaction to their spirit."[19] For such men academies of science are needed, or so Condorcet's argument went.

Other arguments of a less overtly political nature routinely came from the enthusiastic supporters of the state-sponsored French academies. In 1781 the secretary of the Paris Academy expressed both his nationalism and his enlightened liberalism when he presumed that the other European academies "owe almost all their existence to the noble emulation and mass of enlightenment that the work of the Paris Academy of Science has spread throughout Europe."[20] Had he just said "France" there might have been considerable truth to it. The Paris Academy permitted only Parisians to join and it excluded members of religious orders, such as the Jesuits. The academy maintained a very high standard of original scientific inquiry throughout the century,[21] and many provincial academies sought to imitate it. Their membership was overwhelmingly dominated by nobles, lawyers (many of whom worked with the nobility "of the robe," who were judges), and high clerics. All met together in the decades prior to 1789 "in search of prestige . . . believing that progress [would result] from their collective reflection on new ideas."[22] They did everything from sponsoring public lectures to becoming increasingly interested in technology, agriculture, and commerce.

Yet in 1793 the revolutionary government took its vengeance on the academies, not on their ideals or on science per se, but on their personnel. The Paris Academy of Sciences lost nearly one half of its members; the provincial nobility were equally detested, if not persecuted.[23] As we shall see in the next chapter, resentment against the scientific academies had smoldered for decades. Their Parisian bias against the provinces, their haughty dismissal of projects deemed to be not sufficiently scientific, made enemies among inventors, projectors, and would-be industrialists. The purging of the academies was not aimed at their science but at the attitudes, political and otherwise of their leaders.

Prior to the Revolution the academies had monopolized science and the degree and extent of public interest could not be accommodated by the elite academies. A new populist science with mystical overtones rushed in to fill the void and mesmerism, as it was called, captured the attention of men and women from high society to the poor. Some devotees dabbled in electrical cures performed by magician-like healers. They claimed to be searching for medical improvements that would benefit all of society. In that search we can see a profound disillusionment with establishment science, with the austere and rationalistic academicians and their private pursuit of scientific inquiry. The leader of the movement was one Franz Anton Mesmer, a Viennese doctor with masonic connections, more quick-witted than profound. Mesmerism attracted men and women in large numbers, and as one woman saw it, the progress she had made in her own health augured a general cure for the ills of society.[24] In the 1780s French social tensions engulfed science and pitted mesmerist reformers against the entrenched academicians. Their habit of exclusion doomed the academies from developing a populist constituency or from inspiring confidence amid the larger society. It may also be the case that a general absence of scientific education made mesmerism more plausible.

At the Revolution, the science that triumphed more closely resembled engineering than it did magic or mesmerism.[25] *L'Ecole polytechnique* founded in 1794 embodied the ideals of a revolutionary vision of science, of its "power to change the world."[26] Its founders were all men of the revolution and they wanted nothing less than a school for the science of the revolution.[27] They ignored the universities, which they regarded as moribund; they closed the academies and sought instead to reeducate teachers and hence the young. As we shall see at the end of the next chapter, they embraced an essentially industrial vision of the power of science to transform society and nature. A generation after his English counterpart of the 1760s and 1770s, the French civil engineer came into his own—not to displace his military counterpart (science in this period never abandoned the war-making needs of the state) but to complement him in the new national state created by the Revolution.

In this abrupt turn toward industrialization one aspect of Enlightenment ideals inherited from the old order now took preeminence over all others. Among the Parisian philosophes, especially those of bourgeois origins, there had been for some time a marked interest in applied mechanics of the sort popularized by Desaguliers and Nollet. The greatest project of the Enlightenment—in terms of scope, size, and personnel—had been Diderot's *Encyclopèdie*, volumes of which began to appear in 1751. Probably 25,000 copies circulated before 1789 and the outbreak of the Revolution. Its pages are filled with drawings and descriptions of mechanical inventions and devices. Its inspiration was Baconian; Diderot and his collaborators adored the new science and the promise it held to transform the human estate. As he put it, "Men struggle against nature, their common mother and their indefatigable enemy." In a utopian work intended to inspire the Russian monarch to establish the most modern of universities, Diderot urged that mechanics be the first science to be studied because it is "la science de première utilité."[26] Decades later the revolutionary instructors at *L'Ecole polytechnique* would have heartily agreed.

Never in this account of French scientific education do I mean to suggest that prior to the French Revolution there had been a massive backwardness in mechanical knowledge among all segments of the French elite. By far the most scientifically literate of the earlier period were, however, the military engineers.[28] The preponderance of the state and the army in the area of technical and mechanical education naturally meant that their interests would be served before that of society's. New mechanical knowledge was most systematically exploited in the service of state-run projects, in the making of war, also in agricultural improvement.[30] State control of engineering stifled the development of civil engineering relative to its progress in Britain. The tendency to bend science in the service of the state was made worse by the exclusivity of the engineering schools. Prior to the French Revolution they consistently chose men of aristocratic birth for places in their classes.[31] In them, incidentally, the abbé Nollet's lectures were the standard text. Throughout the eighteenth century French technician/scientists sought government patronage and the prestige that went with it.

In any survey of the social relations of eighteenth-century European science two patterns seem most prominent: the French, where scientists in the first instance serve the state, and the British, where they service the needs of entrepreneurs. The absence of a large standing army in mid-eighteenth century Britain and of the concomitant necessity of channeling mechanical knowledge and talent in its service may have been significant for the development there of a cadre of civil engineers and scientific lecturers eager to find employment in whatever capacity. They disseminated scientific knowledge on a wide scale in contrast to the less commonplace character of that knowledge even in the most highly literate areas of western Europe, especially in the Netherlands but also in France. Yet nowhere in the eighteenth-century, except in circles of radical reformers, does the application of science to the needs or interests of the majority of the people figure prominently.

Nowhere in eighteenth century Europe was universal education even for boys a seriously espoused ideal. It would emerge only as a result of the democratic revolutions late in the century and even then only decades into the nineteenth century did the ideal become a reality in much of the West. When we examine eighteenth-century educational curricula for what they did with science, we begin with the assumption that all such education was aimed at the already literate. But in France the male student had to be exceptionally literate as well as numerate if he was going to be scientific.

One of the earliest textbook explications of Newton's system in French, Sigorgne's *Institutions Newtoniennes* (1747), relied entirely on mathematical explanations and never mentioned machines or illustrated local motion mechanically. A few years earlier Madame du Châtelet had presented a sophisticated discussion of Newton and contemporary disputes about aspects of his physics. In *Institutions de physique* (1740) she also sought a grand synthesis of current science and metaphysics. She laid little emphasis on mechanics and its applications and the text would have been challenging to all but the most highly educated. She is important for the history of women and science, but it has to be remembered that she was also a participant in a particular style of scientific explication that had more to do with estate or class than it did with gender.

The emphasis on mechanics would come in the generation after Madame du Châtelet. As the student notebooks of the DuPont family confirm,[32] many French colleges of the 1770s and 1780s were indeed teaching applied mechanics. But the discipline had been available a full generation earlier in British universities and academies, especially in public lectures and philosophical societies. In the 1780s when the French academician, Coulomb, explained the Newcomen engine to his colleagues he referred back over forty years to the popular writings of the English mechanist and Huguenot refugee, Desaguliers. He then went on—for the first time this had been done in French—to explain the nature of Watt's improvement.[33]

Textbook knowledge of Newtonian mechanics explicated with an eye to industry, although important, was not sufficient. Scientifically and mechanically trained engineers from aristocratic background and destined for the military went on generally to become military servants of the state. After the French reforms of the 1740s, which aimed to improve engineering education, the statist aspirations of the graduates were further reenforced. The result of their training was to make them rigid when dealing with the citizenry who often regarded them as suspect because they were seen as representatives of the central government.[34] As we shall see in greater detail in the next chapter, French military engineers possessed abundant mechanical knowledge, sometimes learned from the same books available to Smeaton or Jessop, and very occasionally they had worked directly with fire and steam machines. The difference lay in their military effect and their sociology, which was complemented and affirmed by their more mathematical and theoretical understanding of science and in their real, as well as perceived, relationship to the state. All inhibited the industrially successful deployment of their knowledge. By contrast the British "civil engineer"—a category of professional first named by John Smeaton—had a different, more subservient, relationship to the entrepreneurs and local magistrates who employed him than did his French military counterpart. When French engineers visited Britain in the 1780s they were shocked and impressed by the egalitarian approach taken by civilians toward engineers.[35] The French engineer's self-imaging had included service to state and society, but not direction from, or employment by, the king's subjects.

Thus when we invoke a cultural setting in eighteenth-century Europe we must include the symbols of birth and authority—the political culture and value system of the *ancien régime*—just as we need to understand knowledge systems made available in formal and informal institutions of learning. So pervasive were the military mores of French engineers that when they migrated they seldom became civil engineers in private employment; they sought out other governments, state or local.[36] When they embarked on civil projects, canals, harbors, the drying of marshes, their first considerations were the military needs of the state; commerce came second. Not in every instance, but in general and because of their educational system, they tended "to scorn the instruments of the naissantly industrial."[37]

Science and the Decline of the Dutch Republic

When historians talk about Continental Europe in the eighteenth century and look for a country to which Britain should be compared they inevitably turn to the Dutch Republic. In the late seventeenth century it was the country to be emulated, and both the French and the Germans attempted to do just that.[38] We expect that when a Watt engine was installed near Padua in Italy in the early 1790s it "puzzled the Engineers here, not one of whom can comprehend it."[39] But our prejudices in favor of the commercially rich Dutch—because they were independent and relatively free by comparison to states where the Inquisition still had a role to play—may have the effect of our expecting more from them in the eighteenth century than their society and culture could achieve.

It should be evident from the discussion in Part I, that the Netherlands had also been one of the most scientifically advanced areas in seventeenth-century Europe. The Dutch scientists Beeckman and Huygens, among others, ranked with the leading mechanists of their respective generations. The Dutch universities responded first to Cartesianism and then to Newtonianism in advance of other Continental centers of higher learning. While this may not be as striking in the case of Cartesianism, because its penetration can also be observed in the Spanish Netherlands by the 1670s, it is unequivocally striking for the rapidity with which Newtonianism was accepted in the Dutch Republic. By comparison, the leading Belgian university just across the border at Leuven (Louvain—first under Spanish, then under Austrian, domination) enthroned Descartes in the 1670s, only to leave the statue untarnished, never mind untoppled, until well into the eighteenth century.[40] Similarly, Dutch lens grinding and superior optical work created the artisanal milieu wherein Anton Leeuwenhoek invented the microscope, and Leiden excelled in the early modern period as a center for medical education. No Continental country possessed a freer press or easier access to scientific treatises than the Dutch Republic.

The most important Continental Newtonian prior to 1750 was the Dutch scientist and professor of physics at Leiden, s'Gravesande. He excelled as a popularizer and promoter of applied mechanics. When the Dutch educational system in science declined by the 1750s, it did so from a position that had once been almost unrivaled. Unique among scientists in Continental Europe, Dutch scientists—such as Boerhaave, s'Gravesande, and Petrus van Musschenbroek (1692–1761)—learned Newton's revolutionary modification of the mechanical philosophy directly from the master himself or from his immediate associates and followers, such as Samuel Clarke or Archibald Pitcairne, who in 1693 was professor of medicine at Leiden. The Dutch Newtonians sought in turn to displace Cartesianism once and for all from the university curriculum. Writing to the aged Newton, Musschenbroek put his admiration and efforts on behalf of his science succinctly:

> Being an admirer of your wisdom and philosophical teaching, of which I had experience while in Britain in familiar conversation with yourself, I thought it no error to follow in your footsteps (though far behind), in embracing and propagating the Newtonian philosophy. I began to do so in two universities where the triflings of Cartesianism flourished, and met with success, so that there is hope that the Newtonian philosophy will be seen as true in the greater part of Holland, with praise of yourself. It would flourish even more but for the resistance of certain prejudiced and casuistical theologians. I have prepared a compendium for beginners with which, if it does not displease you greatly, I shall be well satisfied. I shall always endeavour to serve the wisest man to whom this earth has yet given birth.[41]

Musschenbroek had been in London in 1719, and he had proceeded on his return to the Netherlands to teach Newton's system at Duisberg and Utrecht.[42] His fellow Newtonian, s'Gravesande, like Boerhaave before him, embraced a similar project at Leiden, having also learned the new mechanical philosophy from its master. In 1718 s'Gravesande, another adoring admirer, wrote to Newton about how difficult it was to teach the *Principia* and of his efforts to use mechanical devices to get students interested. He too was worried by the opposition of the theologians:

> I begin to hope that the way of philosophizing that one finds in this book will be more and more followed in this country, at least I flatter myself that I have had some success in giving a taste of your philosophy in this university; as I talk to people who have made very little progress in mathematics I have been obliged to have several machines constructed to convey the force of propositions whose demonstrations they had not understood. By experiment I give a direct proof of the nature of compounded motions, oblique forces and the principal propositions respecting central forces.[43]

Like his British counterparts, s'Gravesande had encountered mathematical ignorance among his countrymen and students, many of whom came from abroad, and as a good teacher he responded with illustrations that relied on machines and devices. His practices were similar to those of his intimate associate

Desaguliers. He also gave his mechanical lectures in the Dutch Republic (probably in French), where they were then translated into Dutch and published.[44] s'Gravesande shared Desaguliers's enthusiasm for the industrial application of machines and his interest in the early steam engine.[45] Indeed part of s'Gravesande's obligations as professor of natural philosophy at Leiden—a position secured for him through Newton's intervention—included the surveying and improvement of water transportation in the Republic.[46] s'Gravesande was on his way to becoming a civil engineer.

In addition s'Gravesande belonged to a circle of publishers and journalists, many of them French Huguenot refugees, who were singularly important in transmitting the Newtonian philosophy through their French-language journals. They also became among the few citizens or residents of the Dutch Republic to be made Fellows of the Royal Society.[47] s'Gravesande's circle in Leiden and The Hague may now be counted as the first anywhere in Continental Europe to accept Newtonian science wholeheartedly and to promote it aggressively. In distant outposts of the Dutch empire such as Surinam, the propagandizing efforts of this circle, constituted as a private literary society with masonic overtones, were felt as early as 1723.[48] And most important, the propagandizing efforts were in French, the language of most literate elites in eighteenth-century Europe, as well as among the Dutch.

Out of s'Gravesande's Leiden classroom came the next generation of Dutch Newtonians, who took this mechanically explicated science to other Dutch colleges and universities, to Franeker and Harderwijk, for example, as well as to Amsterdam. The theses in physics were, however, entirely mathematical and none appear to have been so original as to warrant their being translated from the academic Latin in which they were written into French or Dutch. Some public outreach was attempted. The public lectures given in Amsterdam in 1718 were by Fahrenheit (made famous by his system for measuring heat), and he had worked closely in mechanics and the use of mechanical devices with s'Gravesande.[49] The master's influence lasted to the end of the century in the main Dutch scientific society at Haarlem and in the scientific thought of the Newtonian and revolutionary reformer, J. H. van Swinden. In addition the French philosophe Voltaire admitted that he learned a great deal from s'Gravesande's published explication of Newton's system, as did the most important French public lecturer of the first half of the century, the abbé Nollet.

Now we might ask, what happened? After this extraordinary head start, Dutch science seemed to stall. By mid-century the Republic evinced no widespread program of popular scientific education aimed at adolescents, merchants or elite audiences, nothing comparable to efforts visible in Britain at precisely the same time. The lethargy in public science also appeared within the universities. By 1750 the University of Leiden fell from its pinnacle of international prestige and the number of foreign students dwindled significantly. The Dutch elite, both landed and mercantile, and the ubiquitous theology students continued to attend, but the excitement of the previous generation had disappeared. Little original science appears to have been undertaken. The reasons for this change are complex and need to be addressed. They pertain to the history of Dutch science, but also to the question of how to explain the extraordinary retardation present by 1800 within the Republic. Using steam as but one indicator, by 1800 there were sixty-seven engines in Belgium, almost all in coal mines, and five or less in the Dutch Republic.[50] In 1816 by the government's count there were forty-eight engines at work in France. By 1850 there were 2,000 in Belgium and about 300 in the Netherlands.

Throughout the eighteenth century, the Dutch Republic rivaled, if not exceeded, England and Scotland in literacy and urbanization. Its transportation and manufacturing were on the whole efficient. Nonmechanized factories, often horse or wind powered, very occasionally coal burning, some employing over 200 workers (at least one was a salt refinery run by women), could be seen throughout the towns and cities of the 1790s.[51] But those factories of which there were over 1,100 and other, later ones did not, by and large, bring in the new mechanization, or employ steam until well into the nineteenth century. When the French invaded in 1795, they were impressed by the quality of Dutch workmanship and the canal system. By figuring out Dutch techniques the French sought to improve their own factories. Their engineers also noted, however, the relative lack of steam engines and discussed the problems with wind power for draining the polders.

By mid-century the scientific education offered in Dutch colleges reflected a profound lack of interest in applied mechanics not only by the professors, but also on the part of Dutch elites.[52] At mid-century a traveling instrument maker from Paris tried to make a living in the Republic but eventually settled in Liège.[53] When in 1790 James Watt was invited to lecture to the scientific society in Rotterdam, his friend and host, and the leading importer of the steam engine into the Republic, J. van Liender, advised him to "give as much explanation as possible and a great deal more even as you did to that of the Batavian Society's Engine because everyone there shall understand so little of the matter."[54] In one of the main philosophical societies of the period, when all the necessary technical knowledge lay in accessible published texts, there had not been sufficient interest among the members of the society to warrant their trying to master the new technology.

But the Rotterdam response to Watt can also be misleading. His engine was by far the most sophisticated of its day, and a failure to understand it does not necessarily mean an absence of interest in mechanics. Indeed in the Dutch Republic by the last two decades of the century (as in France) interest in applied mechanical learning markedly increased. The first scientific society for women anywhere in Europe, located in Middelburg in the Dutch province of Zeeland, took on the abbé Nollet's text in applied mechanics as its first effort in self-education. The lessons came from the Voltairean, Daniel Radermacher, and insofar as they can be reconstructed, his lecturers resembled the emphasis laid by the British educator, Margaret Bryan (p. 111) on piety and physico-theology. In about the same period the mayor of Middelburg was also trying to make himself knowledgeable in mechanics so as to better understand what engineers were saying about the town's harbor. Yet, here too, when the economic vitality of the town critically depended on being able to dredge its waterways and keep them from silting over, the local *regents* never brought in foreign consultants nor, as far as we can tell, ever contemplated using engines to help in the dredging.[55] They pleaded cost as the key factor. As we shall see in chapter 9, it is not clear that the Dutch knew very much about the available modern techniques, which were being planned in Bristol at precisely the same moment.

In general by mid-century the wealthy and mercantile Dutch elite valued astronomy for navigation, but not applied mechanics for manufacturing industries.[56] Also by mid-century the libraries of some Dutch technical colleges were noticeably deficient in applied mechanics. In Middelburg, for instance, the local equivalent to a university had not hired an instructor in mechanics as late as the 1750s; there were, of course, faculty in anatomy, history, and the classics.[57] In the library of the academy of Harderwijk, where the new science is very much in evidence during the second half of the seventeenth century, emphasis in the eighteenth century appears to have been legal, medical, and theological rather than scientific or mechanical, with the notable exception of works by s'Gravesande and Musschenbroek.[58] But there book acquisitions in mechanics and physics stopped. Only very late in the eighteenth century do we begin to see evidence in the province of Gelderland for the existence of public scientific lecturing intended for commerce, trade, and industry, and these efforts were sponsored predictably by the local scientific academy and the freemasons.[59] A similar lack of interest in science also plagued the academy at Deventer, and progressive parents in turn sent their children elsewhere on the Continent or to Amsterdam, where by the 1760s public agitation for reform in scientific education began in earnest.[60]

Many groups with vested interests in the *status quo* thwarted the development of a vibrant scientific culture. The traditional elite made its money from international trade and commerce and their wealth was such that little else interested them. Their power lay in the towns and cities, and there was no strong central government to offer a counterweight to their influence. The traditional Calvinist clergy had been receptive to philosophical positions that undermined Catholic doctrines and hence scholasticism, but this is where their interests stopped. By the 1730s the clergy led the reaction that set in against foreign influences, a reaction fueled by the increasingly obvious sense of decline and stagnation. In Deventer the local Calvinist clergy appear to have been particularly powerful at the academy and to have maintained a curriculum that had been innovative in the seventeenth century but was anachronistic by the mid-eighteenth. While Calvinism in the seventeenth century may have produced scientific rationalists such as Beeckman, by the eighteenth century the orthodox clergy had grown fearful of heresy among the laity. In addition the power of Calvinist orthodoxy produced widespread public opposition to aspects of the new science, for example, smallpox inoculation.[61] By mid-century only a handful of Dutch reformers were aware that something had gone wrong in the quality and quantity of Dutch educational efforts in science both pure and applied.

It is possible to illustrate the problem with a look at s'Gravesande's successor. By the 1740s the new chair of physics in Leiden was J. S. Allamand.

He, however, paid little attention to either its library or his laboratory.[62] There is no record of his having trained even one student who made a significant contribution to any aspect of Dutch science. But sometimes students ill-served by a university unwittingly take their revenge. A young and minor Dutch nobleman from Friesland and a student of Allamand bequeathed a private diary to his family. There the historian finds a depressing, if amusing, portrait of just how lackluster science at Leiden after s'Gravesande had become. On Allamand, Hessel van Claerbergen wrote that he had a long time to observe his character and counted him as a friend.

> [Allamand] is a *savant* distinguished by a great memory, but he has a knowledge more general than profound in many of the sciences. Metaphysics more than philosophy is his principal study, where he has made great progress aided by the mathematics of s'Gravesande . . . he uses the machines of s'Gravesande for teaching. He has many unique ideas about religion and deduces all metaphysics in a way that will accommodate with Scripture. He is very lively and loves the company and diversions of spirited people. If he has one bad habit it is politics.

Both Van Claerbergen and his sister did experiments with Allamand, who was a regular household visitor. But clearly the professor's cultural universe, when not absorbed in high society, was taken up with university and general politics and there is not the slightest suggestion that he did serious science or that anyone expected him to do it. Decades after s'Gravesande bought them, Allamand is still using the same mechanical devices. You may recall the contemporary fictional figure of Dr. Pangloss in Voltaire's *Candide*. He practiced a kind of metaphysics and taught about how this was the best of all possible worlds. In Allamand we may have found his real life analogue. As for Van Claerbergen (when not being treated for venereal disease), his view was that science should be a part of what an educated Dutch nobleman was supposed to know about, but not do. His diary records not the slightest interest in industry or scientific application; land and rents are wealth, and so too especially are governmental offices.[63] He regarded a practical scientific instrument maker as "a peasant."

Historians have sometimes just assumed that the Dutch must have been up on the latest technology and science.[64] But if its commercial elite had little interest in fostering application, who did? If the strong French state gets blamed in the traditional accounts of industrial retardation on the Continent for being too intrusive, a weak state coupled with an indifferent elite did little to improve Dutch economic or cultural development. Decline has a cultural component that encompasses political culture as well as educational systems. It is a point worth keeping in mind as advanced industrial societies at the end of the twentieth century, such as our own, seek to navigate in a more technologically complex and competitive world. The Dutch universities fell behind out of indifference and myopia as well as because there was no agency of church or state to prod them into competition with their British and European rivals. Whether by the end of the eighteenth or in the twentieth century, an absence of interest in science and technology boded ill for any society.

In their decline Dutch universities jealously guarded their monopolies on learning. Leiden opposed the establishment of any competitors and partly as a result of its opposition, Dutch scientific societies began to be formed only after 1752. In that year the first Dutch scientific society, De Hollandsche Maatschappij der Wetenschappen (the Holland Society for Science) sprang up in Haarlem.[65] Unlike almost all other scientific academies on the Continent, it (like the Royal Society in London) was a private body without an official relationship to the government, and certainly without one to the king—the Dutch stadholderate could hardly be described as an absolute monarchy similar to those found, for example, in France, Spain, Prussia, or Russia. The Haarlem society was supported by the dues of its members, and as such it reflected their immediate interests more closely than did other academies officially licensed by the crown.

A survey of the Dutch Society's proceedings during the first few decades of its existence reflects its interests and, incidentally, reveals that its largely clerical, commercial, aristocratic, legal, and medical members favored certain kinds of scientific inquiry over others. Christian natural religion, or physico-theology, was commonplace in their discussions, as were sophisticated astronomy and the latest medical problems. Some applied mechanics were explicated in the manner of s'Gravesande or Desaguliers, but this was a minor aspect of the society's transactions. Predictably interest abounded in canals and dikes, as well as, most notably, in navigation, although little mention was made of foreign innovations in hydraulics and hydrostatics. Like most European scientific societies or academies, the society posed annual questions for which prizes were given; yet significantly only in 1787 did it turn its attention to the question of the relationship between industry and commerce. In that year of revolution, however, no essay answers were submitted.[66] Industry was not what concerned Dutch society at that politically turbulent moment. Only because we are trying to answer larger questions in the history of European development are we justified in commenting on the relative myopia of the leading Dutch scientific society and the public at large.

So if the leading scientific society, unlike its British equivalent in London, would not sponsor technological innovation, perhaps maverick individuals would. Remarkably the Dutch equivalent of the traveling lecturer making his living from the fees charged his audience was rare prior to the 1760s. The difficulty appears to have lain precisely in the absence of significant interest in scientific and mechanical education among the old commercial elite or the merchant community, particularly outside Amsterdam. When the itinerant lecturers do make their appearance it was generally after 1760 and largely in Amsterdam. By then they had become strident in denouncing the lack of innovation they attributed to the Republic. Before turning to their efforts, we need to nuance what up to now has probably been a too bleak and monochrome picture of the Dutch situation and its perceived decline.

In the late 1740s radical critics of the existing political and social order based in Amsterdam condemned the corruption and apathy they attributed to the ruling elite, the so-called *regenten*, who monopolized wealth as well as governmental offices in the towns and cities. The radicals laid blame on an entire class for what contemporaries were beginning to describe as the century of decline—decline relative, of course, to the brilliant prosperity and inventiveness characteristic of the republic in the seventeenth century, in its so-called *Gouden Eeuw* (golden century).

The decline of the Netherlands can, however, be seen entirely in impersonal economic terms. It may be argued to have been nothing other than the inability of such a small country (population less than 2 million) to compete in an increasingly consumer-oriented Western economy. Its larger and more unified rivals, in particular Great Britain and France, enjoyed sufficiently large domestic markets. They did not have to keep up a vast international commerce just to compete. But economic arguments need not preclude or dismiss cultural ones. With regard to The Netherlands in the eighteenth century the concept of decline as a cultural phenomenon is difficult to dismiss, not least because the charges leveled by contemporary critics appear to be borne out by research in at least one area—namely, the absence of elite interest in scientific education useful for industry. The radicals of the late 1740s pointed precisely at intellectual lethargy in the sciences, as well as to the decline in manufacturing.[67] Indeed in 1751 the newly restored stadholder, William IV, was moved enough to establish a commission to study the decline of commercial and industrial activity, but nothing came of the inquiry.[68]

The radicals of Amsterdam had wanted to restore the stadholder in 1748 because they saw him as a counterweight to the influence of the corrupt regents. Before long the radicals grew disillusioned even with the stadholder, whom they judged (rightly) to be ineffective. After an initial courtship they were also little impressed by the reformers who belonged to his entourage in The Hague. William IV's principal adviser, Willem Bentinck, wanted a complete overhaul of the Republic's institutions, the establishment of a strong central government on the British model, and he took an interest in cultural and university life. He saw to it, for example, that liberal theologians were appointed at Leiden and he was a man of learning with wide intellectual interests. He too worried about decline and wanted to do something about it.

Within the stadholder's entourage scientific learning enjoyed a fashionable place. The enlightened aristocracy at The Hague led by the Bentincks, friends of both Diderot and Rousseau, attended scientific lectures of considerable sophistication, and these early public lectures in the Republic provide a welcome opportunity for the historian to compare what a lecturer there believed to be of interest with similar lectures routinely given in England or France.

In the lectures of Samuel Koenig, an associate of Madame du Châtelet and Voltaire, the new science from Copernicus through Galileo, Kepler, Descartes, Newton, and Leibniz, as well as the electrical experiments of Benjamin Franklin, were explicated as the singular achievement of European civilization. Koenig praised Descartes. At the same time he carefully recounted the achievements of Newton, although he accused some of his followers of attempting to reintroduce the occult qualities so carefully avoided by the mechanists of the seventeenth century. Indeed Koenig argued that the true physicist is neither

Cartesian nor Newtonian and he asserted that all true science displays God as the sole master of nature. Physico-theology and metaphysics were skillfully woven with the assertion that science and mathematics are useful for trade and commerce. Koenig gave pride of place to chemistry, to Boyle's law, and to the phenomena of density and porosity in bodies. Among the few instruments presented, the microscope was demonstrated. Clearly to please the audience, biological theories were discussed. In the section on dynamics where universal gravitation was explained, among other basic Newtonian principles, Koenig presented fairly sophisticated mathematics. In fact, for our pruposes the most significant aspects of the lectures consisted in the quality of the mathematical illustrations freely used and in the absence of any mechanical devices.

In Koenig's lectures we have a superb example of a sophisticated set of scientific lectures, more advanced and catholic than what was routinely available to a British audience of the period. Koenig offered his listeners what he believed they could absorb and what interested them.[69] Practical, industrial application mattered little to the aristocrats or government officials of The Hague. They entertained the importance of applications of the new science in metallurgy—for example, in the methods of weighing precious metals—or in the development of mathematical skills useful for trade. As a result there was far more "pure" and sophisticated contemporary science to be learned at Koenig's lectures than at the courses given by Desaguliers and his many British followers.

Koenig's cosmopolitan audience came away better versed in the state of European scientific knowledge at the mid-century than their counterparts across the channel. What they were not taught, however, were the many practical applications in mining and manufacturing to be extracted from simple mechanics. The value of mathematics and applied science for commercial transactions had long been recognized by the Dutch elite, indeed even the young son of the stadholder was educated in the late 1750s in mathematics directly useful for business.[70] In the transition, however, from commercial to industrial capitalism more was needed than mathematics for trade or astronomy for navigation or physico-theology for the inculcation of piety.

Not surprisingly, a specifically commercial spirit dominated the physico-theological literature pervasive in the Dutch Enlightenment, a genre of literature that also, to be sure, appealed to a broad European audience. J. F. Martinet's *Catechism of Nature* (1777) went through a multitude of editions in Dutch and then in English. It summarized a commercially focused piety that simply did not see the need to address the question of manufacturing through mechanical applications. In that sensibility, all of nature is arranged hierarchically and intended for human exploitation. The beauty of the heavens complements the order in the animal and vegetative world. Commerce and navigation stand alone as the keys to prosperity and to the exploitation of nature's riches. "The whole world is a grand storehouse for man"—the gold from Africa (slaves go discreetly unmentioned) as well as the tobacco from America are but examples of its yield. The child or adult instructed by this catechism should know that even if he were not a merchant he (there is no appeal to women) should know what is available and exploitable by means of commerce and navigation. This was a piety that self-consciously harkened back to the extremely popular Protestant physico-theology of the early eighteenth century, an intellectual invention that was primarily English in origin. It had been reinforced by an independent Dutch literature, much of it in turn translated into English.[71]

Physico-theology embraced the world of imperial commerce and sought to Christianize it. It never addressed the possibility of industrial development, and it was intended to ensure political stability and economic progress of a commercial sort. A brief survey of the textbooks used in Dutch schools up to the reforms of the early 1800s reveals the overwhelming influence of physico-theology in almost every chapter. Also in a remarkable diary written by an adolescent boy of the 1790s the evidence of physico-theological reading is everywhere apparent.[72] Otto van Eck went to scientific lectures with his father and he read Martinet's *Catechism* avidly. There he read a passage about the effects of sunlight. The sun rose for both good people and notorious villains. This injustice will be righted in the hereafter, as Otto put it, "though God at times allows the wicked to prosper in this world, yet he is just, and surely after death they will suffer the fate they deserve." Martinet's technical exposition of the nature of sunlight is simply left out. The scientific sensibility common in the eighteenth-century Dutch Republic belonged with the Protestant version of the Enlightenment, not with its industrial impulses.

Not everyone approved of the closed and smug nature of Dutch society and aristocratic culture, and with good reason. From almost any point of view the Dutch aristocracy of the eighteenth century—however mercantile its origins—was among the most entrenched in Europe. Some two hundred families, many based in Amsterdam, monopolized the senior offices in that city and many others.[73] Beginning at mid-century domestic stability was threatened by a widespread discontent among reformers and progressives within the Dutch Republic. By the 1760s many lesser merchants in Amsterdam were openly hostile to both the regents and the stadholder. As they explained to an English visitor, "their greatest grievance was to see their country enslaved by their own countrymen—by the very representatives who were chosen to protect their liberties and privileges."[74] The dissidents told their visitor that "the principal people in Amsterdam formed an association to shake off every connection with the rest of the provinces and they did not doubt but it would soon come to this."

The Amsterdam association with separatist leanings may have been nothing more than one of the new literary and philosophical societies where talk about the problems of the fatherland was commonplace by the 1760s.[75] The noticeable interest in science, learning, and reform found in the Amsterdam societies stands in marked contrast to the eating and drinking clubs of the wealthy regents, to the display of wealth that seemed "to lie in heaps" in Amsterdam, as another visitor put it.[76]

By mid-century in small pockets of Dutch society interest in industrial development was real, even crusading. In the absence of governmental intervention intended to improve technological capacity, two elements seem to have been necessary in the Dutch situation: the presence of entrepreneuring scientists with a distinct interest in applied mechanics and a large enough audience willing to pay for their knowledge. It was very hard to develop the first without the second, and only gradually did the critical mass needed to promote application come into being. For example in 1751 when an entrepreneuring Rotterdam clockmaker sought to install a steam engine he had to go to England to make his inquiries. Although the Rotterdam engine ultimately failed because of a weak and overelaborate mechanical arrangement for connecting its water pumps, the effort did lead to the establishment in 1769 of a scientific society in Rotterdam. In the 1780s it brought Watt's steam engine to the Netherlands.[77]

After 1750 patriot circles sprang up in Amsterdam where naturalists such as J. van Swinden and Benjamin Bosma attacked the widespread indifference to scientific matters and sought to redress it. These critics of the *ancien régime* in the Republic advocated that merchants learn the science and mechanics (*werktuigkunde*) once taught by Desaguliers and the few scientific lecturers, such as Benjamin Bosma, who had continued the tradition he began.[78] They advocated that a new scientific society be established in Amsterdam, similar to the one in nearby Haarlem, only this one should cater to merchants and their interests. They advocated a revival of manufacturing along English lines and asked pointedly, "Why do the English prosper more than we do in art and science?"[79] The nationalistic proponents of mechanical science intended for industry styled themselves as early as the 1770s "patriots," as the leaders of the Dutch Revolution of 1787 would be known. They participated in an international republican conversation that began in the 1770s. It was inspired by the revolt in the American colonies and included English radicals such Priestley, Price, and their friends. One vital piece in the conversation concerned industrial development through the application of science.

The leading literary-philosophical society of Amsterdam took up the fight for industrial development.[80] There lectures were given on French innovations in porcelain manufacturing as well as on the new porcelain techniques of Wedgwood, one of the pioneers of the British Industrial Revolution.[81] The nonelite founders of Felix Meritis, as that society was called, demonstrated a marked interest in all sorts of mechanical techniques intended for industry.[82] Similarly one of the other learned societies of the city, Concordia et Libertate, was also reformist and critical of the existing order. Bosma belonged to it, and his scientific lectures were among the most practically mechanical to be found in Dutch during the period.[83] They were very much in the manner of the lectures that were commonplace in Britain. Not surprisingly, Bosma passionately advocated mechanics as a way of facilitating human labor, and he inveighed against a life of leisure. He also attacked the moribund condition of Dutch science and pointed to Germany, France, and England, where "300 men have excelled in mathematics" while "I can count no more than ten in The Netherlands."[84] The one exception, he claimed, could be found in Amsterdam, where it was possible to find merchants with a genuine interest in the new mechanical science.

Sentiments like Bosma's about the need for merchants to learn natural philosophy can occasionally be found earlier in the century.[85] But in the late eighteenth century the frequency of such pronouncements vastly increased. Again Amsterdam led the way. Public lectures and courses offered at the Amsterdam private societies and its one advanced college, the Athenaeum, multiplied. A specifically industrial focus appears in the new lecturing, with the Dutch Newtonian scientist J. H. van Swinden (1746–1823) as a key figure. He lectured on the steam engine and porcelain making, as well as on the traditional subjects intended for trade and navigation, in particular astronomy.[86]

At the Athenaeum (which in 1877 became the University of Amsterdam) we find faculty during the last quarter of the eighteenth century who offered a distinct blending of industrially focused educational reform with a reformist political agenda.[87] This Dutch college began to look like an English Dissenting academy of the same period. The applied and theoretical scientist, Van Swinden, who became a patriot and an active participant in the Dutch Revolution of the late eighteenth century, exemplified the reformist trend. He advocated industrial science as part of a larger program partly inspired by British examples. The Amsterdam apothecary Willem van Barneveld (1747–1826), also lectured at the Athenaeum on applied science and in turn became an ardent patriot. Indeed a number of Amsterdam scientists as well as their mechanically minded scientific friends in Rotterdam became active first in the new scientific education and then as revolutionaries and reformers.

The faculty at the Athenaeum, like all scientific lecturers of the period, responded to the interests of their audience as much as they guided and refined those interests. Lectures at the Athenaeum were paid for by a subscription of about 30 guilders a year for a course and were even, on occasion, given purposefully at midday hours, when the Amsterdam stock market was closed. They were remarkably sophisticated in mathematics and astronomy, as notes taken by students at the time confirm. They also emphasized industrial applications. From the perspective of widespread scientific education the Dutch Revolution of 1787, like its French counterpart, reversed a trend of relative backwardness that had plagued the Republic in the middle decades of the century. To that extent the Dutch revolution, again like the French, paved the way educationally for the Dutch industrialization largely a generation or more later.

Perhaps of greatest significance with regard to Dutch industrialization was the number of societies that became commonplace after the revolution, all advocating utility and application. From the early 1800s onward a new generation of schoolbooks also emphasized basic scientific education for both boys and girls.[88] In the Dutch Republic the sciences needed to promote commercial capitalism, in particular astronomy and meteorology, were very gradually joined by applied mechanics (and also chemistry) necessary for industrialization. This transformation began only very late in the eighteenth century, and those who effected it looked self-consciously at the British example, in both political and intellectual matters. To achieve an industrial vision, Dutch scientific reformers required a revolutionary displacement of the old elite such as they sought to effect in the 1790s and beyond. In 1800 the new revolutionary government made a vast survey of the condition of industry in the Republic, which revealed a rather appalling decline.[89] But as a result of war and invasion, little was achieved prior to, or immediately after 1815 in reversing that decline. Scientific education aimed at industrial application could not in itself make an industrial revolution; but without it, systematic and sustained industrial development seemed unlikely.

Amid the political turmoil of 1787–1788 in the Republic, Watt's close friend J. van Liender told him that "were public circumstances in another turn, than they now are, the Steam Engine would undoubtedly take footing in this Country; but by being a work of Patriots it is quite condemned and abhorred."[90] Van Liender laid the blame for Dutch retardation solely on the shoulders of recalcitrant Orangists, followers of the stadholder, and the old regents. One of the causes of the Dutch Revolution lay in disillusionment with the possibility of reform initiated either by stadholder or regents. The condition of industry was part of that disillusionment.

By the 1770s reformers had had their fill of the rentier and place seekers of their world. Just as in France, scientific reformers of an industrial sort gradually came to see political revolution as a necessary step in attaining their goals. The Dutch reformers sought to imitate British industrialists such as Josiah Wedgwood, who had used their knowledge and their capital to improve—we would say to industrialize—the manufacturing process. The insights of the Dutch patriots, coupled with other evidence, adds another dimension to one of the thorniest problems frequently discussed by historians of Western industrialization—namely, why did the once-rich Dutch Republic, of all places, fail to industrialize in the late eighteenth century?

In 1778 a Dutch newspaper defined the country as a nation of "rentiers and beggars," and however exaggerated, the phrase suggests a great deal.[91] The term "rentier" denotes those who, like the Van Claerbergens, lived off their rents or investments, bought by the profits of commercial transactions, rather than those who generated capital through productive entrepreneurial activity. Indeed one of the earliest uses in Dutch of the word "capitalist" occurred precisely in this period, and it was used negatively to describe such people as rentiers.[92] Beggary was also common, especially by mid-century when the clothing industry declined as a result of foreign competition. An impoverished class existed in parts of the Republic that could have been proletarianized, as happened in Britain and the southern Netherlands. And there was certainly no shortage of capital in what had once been the richest nation, per capita, in all of Europe. What appears to have been missing in significant numbers were entrepreneuring capitalists interested in the industrial process.

Amid the complexity of factors that contributed to the absence of entrepreneurs with industrial interests culture as embodied in education needs to be added. Only after 1800 did Dutch educational reforms bring science and mathematics into the essential learning of both boys and girls. But by then political instability was by far the most pressing concern facing the Dutch. The Dutch Republic experienced a period of deep political turmoil from 1787 until 1815. First revolution, then Prussian invasion, and finally French occupation after 1795 brought an unprecedented instability. This was foreign domination unknown since the early seventeenth century when the Dutch had successfully revolted from Spain. By 1815, despite the growing attention to basic science and mathematics in the school and college curricula, Dutch retardation was marked by comparison to the southern provinces. The new united Kingdom of the Netherlands created by the Congress of Vienna incorporated into the old Republic the existing areas of advanced mechanized production in Flanders and favored their further development. This policy meant that in 1830 when the Belgian revolution succeeded and severed the southern provinces from the former Republic, it further exacerbated industrial retardation in the northern Netherlands.[93]

The Austrian Netherlands (Belgium)

The Austrian Netherlands, mostly Catholic and less literate but just as highly urbanized as the Dutch Republic, actually moved in an industrial direction earlier than either the Republic or France. In the abundant Belgian coal fields limited evidence suggests an interaction of entrepreneurs and engineers on technical subjects comparable to what is seen in Britain. The Austrian form of absolutism, which found the main government in Vienna administering a foreign colony through its representatives in Brussels, never exercised the kind of control or expended the degree of human administrative resources that could be seen in Paris or Berlin. The Austrians maintained a small engineering corps largely for fortifications. They collected taxes and by mid-century put in place policies that favored any set of circumstances that would weaken the authority of the traditional aristocracy and clergy.

The historian searches in vain in Belgium for evidence of governmental involvement in local affairs, the kind that any French bureaucrat routinely practiced. Further complicating the story is the coal-rich Principality of Liège, which was not under Austrian control. It belonged to the bishop who as far as can be seen did little but collect taxes. Liège also possessed a distinctive and indigenous tradition of interaction between entrepreneurs and coal miners. Indeed a French reformer of the 1740s had even advocated that his country's coal entrepreneurs seek out the expertise of the Liègois coal miners.[94] How engineers and entrepreneurs worked together in Belgium can best be illustrated by some local examples. The examples prove that some technical knowledge existed among some entrepreneurs, even among artisans. They in turn found engineers who could work for them. The Belgian partnerships are comparable yet different from what can be seen in Britain. What we may never know in the Belgian case is how or where the knowledge was learned, given the control that the clergy enjoyed within the school system. In Belgium industrial progress owed more to enlightened secularism than it owed to religion.

The coal-rich areas around Mons and Maastricht (in the eighteenth century also administered by the Austrians but today part of The Netherlands) put civil engineers, owners, and entrepreneurs into close proximity. The owners of the land were frequently clerical or noble, or in some cases widows of the original owners. Their interest in the mines was active but confined almost entirely to profit taking. Without the assurance of long-term leases that these owners were reluctant to grant, the investment of local entrepreneurs could not be protected. Predictably entrepreneurs frequently held off introducing sophisticated and expensive engines. When seeking to innovate, the entrepreneurs made formal requests for government permission from Brussels to levy new taxes, raise the price of coal, or for assistance in the form of tax reductions. They also sometimes sought new guarantees from the landowners to protect their investment in an engine. With the absolutist Austrian state as a relatively passive partner in the sense that it had little direct technical assistance to offer—although it eagerly permitted new taxes and even granted subsidies—civil engineers had to be found and consulted and engines assessed by the investors.

In the case of the coal mine at Bois du Luc near Mons, the steam engine installed in 1780 became the central investment of the entrepreneurs, who renamed their company after it, giving the machine equal financial importance to the mine itself.[95] Their extensive meeting records document a decades-long process of discussion and consultation first with "worker-experts" skilled in the extraction of coals—who were nevertheless regarded as belonging to a separate and disorderly "estate." Then by the 1770s, consultation began with local professional engineers. At various times the members of the company visited other horse-drawn machines in the region, sought technical evidence about their practicality, surveyed the terrain, and finally in 1773 at a general convocation of the company, came to the conclusion that to ensure their profits, the directors needed to go deeper in the mine. They could not go on without putting up the money for a new conduit and a "fire engine" to extract the water. Local engineers, the brothers Dorzée, were hired to construct the standard engine of the day, probably modeled on Newcomen or Savery engines seen elsewhere. But the engineers managed, cunningly, to extract additional compensation by extending the work and by keeping horses on the site for an extra six months. They claimed that the time and expense and additional power were necessary for laying the piping. There is no evidence in the records that any one in the company possessed the necessary knowledge to challenge this claim or the ability to offer day-to-day supervision. But in the end engineers employed by the entrepreneurs installed a steam engine, which in the years that followed, markedly increased profits. What the engineers built was, however, so wondrous that its inauguration required the local priest to bless it, a large feast, as well as a new clock to supervise more closely the work habits of the miners.

The story in Mons can be matched by one near Maastricht in Brabant. There in 1772 the coal master and the director of the coal works at Klosterrade explained that they had made detailed observations over a two-year period, which proved to everyone's satisfaction that the hand-powered and horse-powered pumps currently employed were no longer adequate. The abbé that owned the mine and the surrounding lands petitioned the government for the right to raise local taxes to pay for an engine, to be powered by wind or water, which could be used to extract water from the mine. It will work "not only for the considerable profit of this abbey but also at the same time it will be a great utility to the countryside, and considering that there is a great shortage of wood in the region . . . two experts and the coal directors have examined the area and they have judged it necessary that a New Hydraulic Machine be erected."

But the abbé did not have the needed capital, and so he petitioned Brussels for the right to raise 6,000 écus or more. "Public interest" was enlisted as the rationale sent to the royal officials who were being asked to give permission for the erection of a hydraulic machine driven by water or wind. It was needed, it is claimed, because this was an area where "no manufacturing or factory existed" and where the needs of the poor for employment were clearly visible. The arguments employed by the Dutch-speaking coal master indicate that at least for the sake of the government a conceptual link had been imagined between mechanization and putting the poor to work.[96] By the 1740s in Britain that conceptual link had been abandoned and the labor costs reduced by technology were openly acknowledged.

The abbé also sought profit from his efforts and that motivation was clearly articulated. The machine to be powered by wind or water for which permission was granted, was not state of the art for the time; it was not steam but it may nevertheless have been adequate. The available reports indicate no consultation with a hydraulically trained engineer, and indeed the technological decision to use only wind or water power may also have been economically motivated. Such a pumping device could be installed with the aid of observation and with a partnership of sorts between skilled coal masters and owners with government assistance. Its involvement on a financial level was quite real and important, but—and this is a point to keep in mind when in the next chapter we analyze the French situation—this did not extend to actual technical assessments or assistance. The government in Brussels used military engineers for defense or public works only. It appears to have been content to receive detailed reports from men described as "experts" who were on site.[97] Both of these Belgian examples, although suggesting less involvement in technical matters by owners or exploiters, loosely conform to the pattern witnessed earlier in England where skilled engineers, or their pre-professional equivalent, negotiated directly with owners to utilize technology for profit. The interaction of entrepreneurs and engineers, or the entrepreneur turned engineer, seems to have been in both Britain and Belgium the key social relationship in the early mechanization of mining. In the absolutist states on the Continent government involvement inevitably occurred, but when we compare the French and Belgian settings we see that the nature of the involvement could differ considerably from country to country. In the Belgian case the introduction of power technology occurred almost entirely in mining, although machines were introduced for cotton spinning. Neither became widespread phenomena until the educational reforms of the late 1790s.

The pattern witnessed in Britain, whereby we see a link between sustained inquiry of a mechanical sort and early industrialization, would seem to hold for at least the important areas of industrialization in the southern Netherlands, for example, in the area around Charleroi. Just as it had in Mons and Maastricht, the Austrian government promoted manufacturing in the countryside where an impoverished peasantry, unprotected by guilds, could be proletarianized.[98]

Again the rich coal deposits tempted entrepreneurs and government assistance also helped. In addition traveling English mechanists or engineers made a significant contribution. Yet in none of the areas so far studied by historians do we see the kind of systematic, sustained partnerships between engineers and entrepreneurs that were commonplace in coal fields across the channel. The differences between Belgium and Britain are, however, of degree rather than of kind.

In the province of Liège under the control of its bishop, governmental assistance seems never to have occurred. Among entrepreneuring manufacturers in the province we can observe later in the century a sustained agitation for mechanical and technical education, and this *esprit* was, as in England and the Netherlands, also associated with enlightened reform. Predictably, freemasons were active in the scientific reform movement. The enlightened propaganda around the cause of industrialization equated economic development with the highest ideals of social utility.[99] But clearly there were quite traditional elements within the scientific circles in the principality. As we shall see in the next chapter, when the French took over the Belgian educational system after 1795 they found it necessary in Liège and elsewhere, to remove the "pure" science faculty and replace them with people more interested in application. By then the university as a whole may have been out of step with what reformers had been advocating for decades.

Wherever we see industrial activity in Flanders after 1770, we can observe the presence of a significant interest in mechanics, chemistry, and technology. The Flemish journal of the period, *Vlaemschen Indicateur*, reflects that interest on the part of literate elites who were also eager to promote the reforming policies of the Austrian monarchy. Its ministers consistently used the new science and the establishment of scientific academies as a stick with which to beat the clerically controlled universities.[100] The ideology of reform, and with it the promotion of industry through scientific inquiry, fitted well with the imperial need of the Austrians to overcome the localized interests of the indigenous Flemish aristocracy and clergy. The evidence shows officials of the Austrian government intimately involved in the industrial process, in particular the nascent chemical industry, where they fostered research, lent money to entrepreneurs, and licensed their factories. The government's day-to-day involvement in the industrial process rewarded the indigenous entrepreneur and attracted foreign projectors, who brought with them new scientific knowledge, frequently from England.[101]

The Austrians never really succeeded in imposing their imperial will when it came to the educational system. The power of the Belgian clergy was awesome to behold. As late as 1777 a plan for reforming education intended that the education of adolescents still be predominantly religious. Yet this clerically devised plan did recognize the need for improved mathematics and science education. The plan directed teachers of elementary science, mathematics, and geometry only to scholastic, Cartesian, or physico-theological texts. No mention is made of any Newtonian texts or of mechanical devices or illustrations.[102] The Austrian Netherlands did make strides toward the industrial during the period of its *ancien régime* but the formal educational system had contributed little to the effort. In the 1790s revolution and invasion brought new elites into positions of power, and many of them were coal developers or bankers. As in France after 1800, industrial development took hold; in Belgium it was particularly rapid and all-pervasive.

Germany

To look at scientific education in Germany we would have to examine all the German-speaking lands from the Rhine river to Vienna. Germany was not a unified nation state in the eighteenth century, nor until 1870. Such a survey is neither possible here nor even necessary. Given the rural character of much of the territory and the extraordinary power of the nobility over its peasantry, particularly in the eastern regions, we want to look only at certain cities where progressive policies were imposed by the local prince or were the work of educational reformers.

The ideas of German educational reformers were remarkably similar to what British, then Dutch and French mechanists and reformers, advocated or practiced. Their efforts were important but sporadic. Some textbooks late in the century reveal what was or was not taught. Certainly by the 1780s Newtonian mechanics were fully integrated in such texts.[103] "School programs" have also survived from many institutions and these show a widespread basic numeracy. In the higher grades, however, entries such as "naturlehre" or "die physicalische Classe," "die mechanische classe," "die optische classe," and "die manufactur classe" tell us very little about what was actually taught, except that such classes were generally for boys, girls being limited to religion, reading, geography, history, and sums. In one Berlin gymnasium of the 1730s the language used to describe the lessons tells us that Cartesian and Leibnizean science were still the dominant paradigms.[104] In one school for boys over fifteen, intended to prepare them for university, the Collegium Carolinum in Kassel (in Hessen), we see mention in 1771 of a Prof. Matsko's lectures on Newton's philosophy. A less elite *Realschule* in Berlin taught Newtonian astronomy in the 1760s. The students did experiments modeled on the work of the Dutch Newtonian, Musschenbroek. English lessons included a comparison of physico-theology with religion based solely on revelation. The public examinations required that machines be described in detail. By these decades, just as in France, applied mechanics had begun to penetrate at least the Berlin school system.[105]

Education in the German-speaking lands where first Lutheranism, then Calvinism and Pietism had promoted reasonably high rates of literacy was largely a matter of private schools set up by churches or educational reformers. The literacy promoted often centered on what was needed to read the Bible. These efforts date as far back as the Thirty Years War (1618–1648) when links also developed between German Lutherans and English Puritans. Yet despite the contact with English Puritanism, German Lutheran and Pietist reformers in general education display little or no interest in the scientific side of the Puritan reformist ideology.[106] The motive for educational initiatives late in the seventeenth century centered on instilling discipline throughout a society where poverty was rampant and begging endemic.

For much of the seventeenth century an atmosphere of political and economic crisis brought on by the war dominated reform efforts in whatever sphere. Only in the absolutist states like Prussia late in the seventeenth century did the state take an active role in education—then in a serious state of decay—and the founding in 1692 of the Lutheran university at Halle created what was to become the leading German university of the eighteenth century.[107] By 1700 policies were in place to promote commerce and manufacturing in imitation of the Dutch. The Berlin Academy of Science was established in the same year with the great mathematician and natural philosopher, Leibniz as the key figure and inspiration. Within the decade primary schools were also established in some rural villages and towns in the remoter parts of Prussia. Schools under Pietists emphasized the necessity for a practical education and children were exposed to technology and working models of machines. Not innovation but work training for artisans was the goal. The commercial policy of the Prussian kings and the curriculum of the schools maintained a neat fit.

The royal policies received a remarkable boost from an influx of French Huguenot refugees arriving in Berlin after 1685, driven from their homeland by religious persecution. They were often commercial and manufacturing in orientation and eager to serve in the king's army as officers. The books used in the Huguenot school on Fredericstadt as late as 1781 indicate, however, that these French Huguenots had continued to follow the scientific learning that came out of France and not the Newtonianism being promoted by their journals emanating from the Dutch Republic.[108] In the long run, however, the commercially enlightened policies of the absolutist Prussian kings, Frederick III(I), Frederick William I, and then Frederick the Great had far less success than their efforts to build up the military. Largely through massive taxation they created the largest and most efficient land army in the West. It would remain the key Prussian institution, an enormous fiscal burden until defeat at the hands of Napoleon brought major reform efforts within the state and civil society. To service the army, textiles for uniforms flourished and techniques of production were developed that look forward to the mass-production of a later era.

The reign of Frederick the Great (1740–1786) brought enlightened values into education without challenging its disciplinary focus. As early as the 1740s private educational reformers such as Johann J. Hecker, turn up who specifically addressed German secondary education and chided it for a laxity in scientific education. They cited the excessively academic and classic nature of German education and the need for an improvement of the mechanical arts intended for production and agriculture. In 1748 Hecker's schools in Berlin were instructing over 700 pupils and he sought to introduce boys to knowledge leading to professional lives as pharmacists, chemists, and architects. As they always had, girls received instruction in reading, mathematics and theology.[109] Hecker's new economical-mathematical *RealSchule* was also in business by 1747. It maintained the ideology of disciplined religiosity, obedience, and active participation in society with lessons in mechanics, geometry, architecture, and manufacturing. Machines were brought into classrooms that were populated largely by boys who were to become artisans as well as professionals. On occasion the students went into the field to observe working mills. Brewing and chemical applications were also stressed as was the link with experimentalism at Halle.[110] This one school taught theories and practices close to what entrepreneurs needed for industrial innovation and did so self-consciously. So too later in the century did King Frederick's Gymnasium in Berlin, where the school program recommended the use of machines for scientific education.[111] Just outside the Prussian border in Braunschweig near Hannover a new curriculum at the Collegium Carolinum also attempted to tailor education to professional needs.[112]

If, however, changes were to come on any large scale in German education they would have to come out of the Universities of Halle or Jena, which trained almost all the Lutheran pastors as well as teachers in Prussia between 1713 and 1740.[113] The education they received had been almost entirely evangelical, and in the 1720s they exhibited hostility toward the physico-theology and experimentalism of Christian Wolff, the main representative of the scientific version of the Enlightenment. In 1723 they drove him from the university and these attitudes only softened in the 1740s. Thus there is limited evidence that, again just as in France by the 1740s, state of the art mechanical science, Newtonian mechanics, was gradually becoming a part of the Prussian educational system for boys aged 15 and older.

Of course throughout the eighteenth century there had always been contact between German scientific circles and their French, Flemish, or English counterparts.[114] German-speaking academic culture in science resembled its French counterpart and scientific knowledge was also traded with the academy in Brussels.[115] These contacts can be traced back to the era when Leibniz (d. 1716) and Christian Wolff (d. 1754) emerged as leaders of their respective scientific generations. Both maintained active contact with their English and French counterparts. In the case of Leibniz academic culture and service to the state entailed an interest in the productive capacities of scientific knowledge. As a servant of the absolutist state, Leibniz wanted to make science into one branch of its efforts at economic development.[116] In his lifetime these included the employment of alchemists whose craft centered on the constant search for the miraculous production of gold. When not toying with the mystical arts or chemical spectacles at court, the natural philosophers were charged with economic tasks from developing coal mines to making lists for the prince of useful and recent inventions.

In the physics and mechanics of the next generation Wolff also cultivated empirical and experimental approaches and his influence briefly radiated from the University of Halle, and it helped to push German-speaking scientific practitioners in practical, applied directions.[117] He defined experience and theory as being of equal value, and in Baconian fashion said that the work of artisans was to be mastered so that it might be developed by scientifically sophisticated philosophers. The interaction between German academic culture and the needs of the absolutist state resembles what occurred in France throughout the century; only in Prussia, the key player in the Germanic lands, the state entirely predominated. Haltingly at first it tried to set policy in every area of culture and education. The best and the brightest students from the Prussian schools went on to the University of Halle, which became the largest German university. Theology remained the dominant subject, followed by medicine, and both were intended to produce healthy bodies and souls. Gradually after 1750 scientific education with a mechanical focus grew in importance within the curriculum.

In 1725 the medical professions, including pharmacy, became subject to state regulations and medical training was regulated to provide students with professional degrees guaranteed to impart a standardized level of knowledge. Chemistry benefitted from these reformers and became a part of university curricula. Civil servants were also now expected to have a minimal degree of technical competence.[118] The German lead in applied chemistry visible during much of the nineteenth century can be traced back to these eighteenth-century reforms in medical practice initiated by the state.[119] But none of these absolutist policies favored the mechanician/engineer or entrepreneur as such. In the 1720s and 1730s the entrepreneur was viewed as "a poor patriot" who followed his own interests and not those of the state. State-run enterprises intended to serve the needs of the military remained the major priority of the Prussia state well into the second half of the eighteenth century. In Prussia the state's extraordinary control over the economy was not matched by direct control over the classroom. There the clergy predominated and as a result if the Prussian state wanted to be innovative in scientific education intended for industry, it had to bring the clergy *en masse* along with it.

Throughout Germany attention to science and technology intended to promote economic and industrial development occurred randomly in the first half of the century, but increased during the reign of Frederick the Great, generally in the period after 1750. By the 1770s the base of academic technical knowledge had significantly improved. The university at Jena supported chemists of the caliber of Johann Göttingen (d. 1809) who knew Priestley and Boulton personally and who taught chemistry and technology to an entire generation of students. Application was central to his preoccupations and he published on industrial processes such as sugar extraction from beets. Late in the eighteenth century, if not earlier, faculty at Jena lectured extensively on Newtonian physics and mechanics and both the theoretical and the practical were effortlessly blended in the lectures of Prof. Voigt.[120] Within the Prussian bureaucracy an official like Friedrich Anton von Heynitz (d. 1801) initiated more efficient methods of mining and helped to found several institutions of technical education, the Bergakademie in Freiberg and the Berliner Bauakademie in civil and mechanical engineering.[121]

Yet, as Eric Brose has demonstrated when he picks up the story of Prussian industrialization after 1809, what went on in the reform-minded eighteenth-century universities and schools had on the whole not been translated into the larger society.[122] The earlier marriage of convenience between the dominant absolutist state and the Protestant clergy had produced little widespread, practically oriented scientific and technological education. The universities and the technical schools were exceptional in offering leadership that recognized at least in principle that "scholarship will help to increase trade."[123]

When Napoleon extended his influence and authority over western Europe as far as Berlin, his ministers solicited reports on the state of industry there. The word came back from German manufacturers interested in mechanization that little had been done by previous Prussian governments to improve fabrics or manufacturing.[124] Such statements certainly had the effect, if not the intention, of currying favor. They downgraded the sporadic industrial progress that had been made in Prussia and of course they ignored the proto-industrial development of textile manufacturing that had occurred in certain rural areas of Germany, in Silesia, Saxony, and the Rhineland.[125] In the last decade of the eighteenth century the manufacturer Johann Bruegelmann established spinning machinery and a water-powered spinning mill near Ratingen. He had illegally smuggled the machinery out of England.

The Napoleonic occupation of the Rhineland was at best a mixed blessing for the industrial development of the region. Evidence does suggest that the French administrators reformed education with economic necessities in mind.[126] But their efforts came to an end with the defeat of Napoleon. After 1815 and the return to home rule, the story of German industrial development in the Rhineland and Prussia begins in earnest. What is important for the story I have outlined here is that the scientific culture seen by the Prussian state as relevant to industry was mechanical and applied. The goal of industrial development had, however, to be negotiated politically with vested, often aristocratic, interests that opposed every innovation from steam to railroads. It took decades into the nineteenth century for Prussian industrial reformers to achieve their goals. By the middle of the nineteenth century German industry began to challenge British preponderance, but that is a separate story.

Italy

Where literacy was weak and the power of censors strong the dissemination of the new science was infinitely more sporadic than in France, the Low Countries, or Germany. In Italy, where once Galileo had captured the attention of both the elite and the censors, the new science of Gassendi, Descartes, and finally Newton held a tentative claim to allegiance among select circles in Rome, Naples, and Turin. In Rome, the city of the Inquisition, a circle of Gassendian atomists had met in the mid-seventeenth century, and for a brief time an academy there under the direction of Giovanni Ciampini dedicated itself to Galilean experimentalism and to the study of Cartesian metaphysics.[127]

In the late seventeenth century the intellectual crisis that afflicted much of western Europe—the shift toward the secular which we discussed in chapter 4—was also felt in Italy, and out of it came the linkage between science and heterodoxy. The search for philosophical liberty among scientifically minded Italian intellectuals in turn galvanized the Inquisition "against mathematics and physico-mathematics" because they were seen as pernicious "to the sincerity of religion."[128] Yet for all the dangers attached to the study of the new science, its penetration south of the Alps was real and durable. The writings of Hawksbee and s'Gravesande were known almost as soon as they were published. An Italian edition of Francis Hauksbee's London lectures, *Physico-Mechanical Experiments on Various Subjects* (1712)—one of the first of those series of public lectures so central to this process of dissemination—appeared in Florence in 1716. This exposition set the tone for a scientific empiricism that pitted Italian Newtonians against Cartesians and scholastics for much of the century.

The works of Robert Boyle also made their way south as did various visiting British Newtonians. By 1707 Newton's *Optics* and *Principia* were the subject of avid discourse, and the polemic against Cartesianism had begun. The link between Newtonianism and Galilean mechanics was readily perceived, and not surprisingly it was an Italian engineer of Naples, Celestino Galiani, who made a major contribution to the formation of a Newtonian school in Italy.[129] Interest in the rationalization of navigation and agriculture, rather than industrial application, characterized the scientific *esprit* of these Newtonian circles. Not surprisingly they were also in close contact with the first generation of Dutch Newtonians.

The Enlightenment in Italy made anticlerical polemics central to its concerns, and the Italian Newtonian Francesco Algarotti sought to enlist educated women in the enlightened camp. His *Newtonianism for the Ladies* (1737), published in an Italian edition emanating from Milan, became the most widely read and translated general explication of the new science in the century. It may be seen as a bold appeal to enlist women against both the church and the Inquisition. Throughout the eighteenth century various scientific lecturers—the abbé Nollet, and Benjamin Bosma, for example—reached for the support of that new segment of the literate population, not to offer women full membership in the scientific community but to enlist them as its supporters. There is some evidence from late in the century that women responded in their own discourse to that appeal, and they made use of natural philosophical arguments to criticize the inequity of their place in every European society.

Newtonianism permitted liberal Italian Catholics to formulate a moderate and enlightened religiosity, indebted in part to the early Boyle lectures of Clarke and others who offered a *via media* between the materialism of the radical Enlightenment and the scholasticism advocated by the official church. In the face of material and social conditions totally unsuited to the promotion of industry, Italian Newtonians such as Antonio Genovesi concentrated their energies on the reorganization of the schools and academies. In Naples, one of the centers of the Italian Enlightenment, they sought nothing less than the modernization of their society and culture. Genovesi attacked the near feudal conditions that prevailed in the countryside and sought through the new science of economics to address the problems of poverty and agricultural backwardness.[130] The integration of science into Italian society may have produced a more immediately humane response to social problems than that found in either the British or French models. Yet it should be noted that in every European society a scientific approach to agriculture gained acceptance in the course of the eighteenth century and contributed significantly to the elimination of food shortages in major areas of western Europe.

Perhaps one of the most remarkable examples of the enlightened implantation of science occurred in Turin in the northern province of Piedmont. There in 1757 its aristocratic ruler simply began a new scientific academy where none had existed before. It quickly moved to the vanguard of contemporary European science, both pure and applied. Its proceedings, and those of the laboratory established to serve the needs of the army, display a remarkable interest in applied mechanics of an industrial sort. In this Piedmontese enlightenment, fostered by an absolute ruler, aspects of the modern relationship between science and the state appear with an uncanny prescience. Reform and improvement through science, progress, and liberalism are inextricably linked with war and war-making. One out of fifty Piedmontese were involved in war or war-making; the laboratories belonged to the technician/scientists of the army. The scene conjured up in our imagination of those decades looks toward the state-sponsored industrialization of the nineteenth century, toward the military-industrial complex of the twentieth.[131] We should not view the past with such "Whiggish" spectacles; yet at moments it is difficult to remember that we are wearing them.

Despite the interests of select Italian intellectuals or the Piedmontese state, applied science for industrial purposes never took root before 1800. There was simply too much opposition and too much censorship affecting its fortunes. Italy and Spain before the later half of this century took economic directions vastly different from what occurred in the rest of western Europe. Their relative poverty has lifted only in the postwar era. The cultural life that did not build around science and technology in both places in the course of the eighteenth century is a part of the story of their industrial retardation.

8

French Industry and Engineers under Absolutism and Revolution

[When traveling in England] I saw with dismay that a revolution in the mechanical arts, the real precursor, the true and principal cause of political revolutions was developing in a manner frightening to the whole of Europe, and particularly to France, which would receive the severest blow from it.

—a French industrial spy writing to the ministry in Paris in 1794.[1]

The portrait of British scientific culture presented in earlier chapters depicted a deep and broad penetration of simple mathematical as well as mechanical knowledge. As we have just seen in the previous chapter, within the educational systems of other western European countries the penetration of mechanical knowledge came later. Wherever it spread, the common vocabulary offered by the culture of practical science built an unprecedented bridge between people with capital and men with mechanical knowledge. Where uninhibited by social or institutional barriers, this common vocabulary permitted a high degree of interaction between engineers and entrepreneurs. Their mutual conversation conducted at mine pits, harbors, canals, and factories was one of the keys to British industrial success, and some contemporary observers, such as the French spy quoted above, knew it. He was writing to tell the new revolutionary government about another, very different revolution that he had witnessed across the channel a full decade or more earlier. The British industrial revolution of which Le Turc spoke threatened to revolutionize the balance of power in Europe—in the context of his letter that is what he means by political revolution. The parts of the revolution his spying had uncovered consisted of new machines, engines, factories, and skills organized within a division of labor. Through direct, if covert observation, Le Turc, who was by training an engineer, had spent years describing the new machines in detail. His work had cost serious sums laid out by the treasury of the *ancien régime* largely in the hope of ultimately recruiting English workers and engineers or imitating their techniques. In short, from the 1730s onward French ministers of commerce nursed *une véritable obsession* about English competition.[2]

Right up to 1800, and well beyond, French governments had reason to be worried. Late in the eighteenth century the revolutionaries in particular became obsessively concerned with economic reform, and they were convinced that the policies of their prerevolutionary predecessors had failed. They brought fresh insight into the nature of the British competitive edge. Most important for our story industrial reformers after 1789 saw the political and economic dimensions as well as the cultural elements we have been describing. But with the passage of time their insight was forgotten by historians. Now historians of culture, science, and technology are in effect rediscovering what competitive French observers, actually from as early as the mid-eighteenth century onward, believed they were beginning to understand.[3]

In the 1790s the French revolutionary governments were so taken with their insight into the culture of British science that they sought to foster and replicate it. As we shall see, they self-consciously sought to make possible technical conversations between entrepreneurs and engineers. But first they had to invent the civil engineer—as opposed to the long established and highly professionalized military engineer—and then they had to educate and favor technically competent entrepreneurs—sometimes embodied in the same person. The French even perceived the specifically Newtonian and mechanical mindset in the ongoing technical conversation so necessary for the trial and error at the heart of successful technological invention. Describing British manufacturing prowess in 1807, worried Napoleonic ministers privately used this aptly Newtonian metaphor: "The absolute necessity to create and sustain French industry is a problem that England has resolved for itself in a very decisive manner. It is with this powerful lever that she sustains the enormous mass of products. Their weight produces an overwhelming gravitation which pulls everything else into its orbit."[4] As we saw in the discussion of Newton and the Newtonian Enlightenment, the French metaphor (perhaps unwittingly) pointed to one of the key cultural elements embedded in a new economic order.

The seeds of Anglo-French industrial rivalry had been planted in the previous century, in an era which by 1790 French revolutionaries had come to call their *ancien régime*. The label quickly took on the pejorative connotation of old (not just former) and backward, but it would have been meaningless to the pre-1789 men who made up the former royal administration. It had ruled France for centuries, and by the mid-eighteenth century there was nothing "old" about the regime in its goals or aspirations. In matters economic the administrators were often innovators who availed themselves of the best scientific talent as consultants and advisors. As we saw in chapter 2, the system of patronage for French science went back to the 1660s, to Colbert and the reign of Louis XIV. All knowledge, even Cartesianism, was meant to enhance the glory of the state and the grandeur of the king.[5]

By the middle of the eighteenth century French administrators aggressively sought to stimulate commerce, to further enhance through manufacturing, invention, and industry the king's wealth and the well-being of his subjects. In theory any invention or scheme would receive a fair hearing especially if it might rival the progress that reports said was being made in Britain. Its perceived and real technological advantage in textiles, metallurgy, and mechanical application in general, had grown with each decade. British industrial success fueled French ministerial initiatives and the ministries of commerce and the marine as well as the local intendants—the king's regional representatives—and local assemblies (or estates) were willing to give not only patents, but also subsidies and prizes for inventions and pensions to their inventors.[6] Yet from the 1790s until today the verdict on all this activity has been in, and seldom modified: by and large eighteenth-century France remained relatively backward in technological and industrial matters.

Retardation, it might be said, is solely in the eye of the beholder. Indeed the verdict is usually made after the fact; hindsight is always 20/20. Yet in the case of old regime France many contemporary observers had already made the observation and worried about its causes. From a cultural perspective the challenge becomes to figure out the context or circumstances of French inventiveness. We need to shed some light on how the absolutist system—the chain of command from Paris to the provinces, the system of recommendations, *mémoires*, and administrative hierarchy within a social universe of *corps*, estates, privileges, blood and birth—might have influenced industrial moments. The case will be made here that old regime political culture, precisely the universe of divided realms and fixed stations, played into economic and technological life in unexpected ways. Those who lived within the rules of its decorum and worked within the absolutist bureaucracies could not always see the dynamics we are about to trace, any more than we can get outside the parameters of our own social and cultural universe.

French Scientific Culture during the Ancien Régime

Perhaps all the industrially focused activity undertaken after 1800 has distorted the record achieved by the *ancien régime*. Certainly the French revolutionaries gave the old order a bad press for being obscurantist and interfering. Indeed one of the shibboleths in the historical literature about French industrialization in the eighteenth century is that everywhere "it was checked . . . by government interference. . . ."[7] It is true that the old regime government was a more directly involved player in economic development than was its British or even Austrian (and absolutist) counterpart. Indeed the pre-1789 central archives of the French state, the ministries concerned with commerce, industry, and the marine, as well as the provincial archives are rich with examples of efforts to drain harbors, develop the glass and chemical industries, and import British technology in silk and cotton manufacturing, as well as in steam. Yet with all this energy being expended, a marked gap, noticed at the time, had developed between the technical and industrial skills found in Britain and what the French had done by the 1780s.

Numerous projects, some developed by foreign engineers and entrepreneurs, were brought to the attention of the old regime government.[8] Rather than being seen as opponents of voluntarism and development, as the French revolutionaries characterized the previous generation of administrators, they should be seen as immensely interested in economic developments, in some cases eager to facilitate. Indeed the French state was even willing to provide subsidized monopolies for technological innovation, which many a traveling British inventor sought to acquire.[9] Not surprisingly, from the 1730s onward a steady stream of reports reached the government on a wide variety of mechanical subjects, from both home and foreign industries. By the 1780s any spy was operating on well-traversed paths.

The reports from French travelers sent to the state ranged widely, but never did they fail to note cultural elements. When in the 1780s the French engineer, Pierre-Charles Le Sage, saw the stock exchange in London, he drew its floor plan. There he placed each faction as he saw it: men clustered in groups, some arrayed by new commercial identities (as drapers or traders), yet others congregated by religion or nationality.[10]

We would be delighted to know how a Quaker draper negotiated his identity and chose where to stand: Did he go with the clique of Quakers or with the manufacturers? Did Anglicans and Lutherans on the floor feel religious antagonism as well as commercial rivalry? Did both feel superior to the Jews, who also stood among themselves? In effect did the modernity of the impersonal market coexist with traditional cultural values and beliefs? If men brought their religion to the floor of the exchange, might not they also bring other values to protoindustrial moments, sometimes having them work at odds with the impersonal competitive, market-oriented goals they so assiduously pursued?

French ministers of state even said that blind spots resulting from the educational system of the day necessitated their intervention. The academician Etienne Mignot de Montigny spent years observing textile manufacturing in Britain as well as industries in Switzerland and provincial France. It was good onsite job training. Eventually back home he was given the position of commissioner to the Department of Commerce. As inventors learned, one needed to stay on his good side. His job entailed the oversight of "all the discoveries, inventions, and machines that would prove useful to the arts which can interest the king's commerce." In a *mémoire* written shortly before his death in 1782 de Montigny described his many activities on behalf of stimulating French industry, and helpfully for our purposes, he explained why it had been necessary to establish his office in the first place.[11]

De Montigny's job in the old regime had been to determine the value and merit of innovations, and to prevent the government from being deceived by the false appearance of utility, or from buying or subsidizing pretended "secrets" that were already known. The historian might ask, why could not local officials or merchants closer to the industrial sites which de Montigny had once visited, make the reports? Why did the central government interfere? He explained: "The Magistrates are lacking instruction in the details of the arts and of commerce, machines, in knowledge of crafts, machines that work on metals and minerals." As a result they are often "duped by charlatans and allow the public to become engaged in ruinous enterprises by according Privileges which they abuse and with which they live at the expense of the authorities." Thus, on the testimony of the French commissioner most directly involved in promoting industry during the middle decades of the century, it was the underlying lack of scientific education, an absence of interest in machines, chemical procedures and inventions on the part of local elites and authorities that necessitated his intervention. In his *mémoire* de Montigny referred primarily to textiles and porcelain, as steam is never even mentioned. In his opinion local indifference or ignorance brought about an elaborate and countervailing system of formal academic evaluation. It sought to overcome educational blind spots partly due, as we have seen in the last chapter, to the clerically dominated school and *collège* system that had been slow to pick up on Newtonian mechanics. But as we are about to see, the privileged academicians, despite their good intentions, often also inhibited industrial development.

When in doubt old regime French officials like de Montigny routinely called in the academicians of science or the military engineers for consultation and assistance. Indeed French military engineers were of higher social origins and on the whole better trained and more professional than their British counterparts.[12] Projectors and mechanical engineers submitted proposals for evaluation and adjudication, and the archives of the Parisian Académie des Sciences are rich in utilitarian projects brought for assessment to the academicians. The historian, Robin Briggs carefully sorted through the voluminous literature found at the Académie. It demonstrated governmental interest in the application of science, especially in mechanics, and the relative backwardness of eighteenth-century French industry particularly in metallurgy. He further combed the archives of the Académie to illustrate its fitful, but real, commitment to utility. He concluded that "The Académie was party to far more innovative ideas than the French *ancien régime* economy could absorb, and the French scientists compare favourably with their English contemporaries, at least from the end of the seventeenth century onwards . . . if relative French backwardness does need to be explained, then the answers must be sought elsewhere."[13]

The elsewhere suggested by Briggs resides in other French archives that illuminate the actual nature of the countless interactions between the academicians, inventors, and entrepreneurs. As Briggs notes, the Académie "was also inclined to be ultra-cautious in giving its approbation to inventors," but he then wrongly concludes that this was "a natural reaction to being pestered by optimistic cranks." More was at work in these rejections. The French academicians held to a particular definition of scientific standards when they assessed the projects of would-be engineers and entrepreneurs. Their oftentimes theoretical approach brought to industry a social and cultural style best described as aristocratic and hierarchical. It was comparatively less egalitarian than the trial and error, even competitive, exchanges around scientific or technical knowledge that occurred between entrepreneurs and civil engineers within scientific societies and academies in Britain and more rarely, the Low Countries.

In pre-1789 France a different and much more hierarchical kind of division of mental labor prevailed. The division was social and cultural, operated between academicians and inventors, and it worked throughout the eighteenth century in ways that could frustrate or inhibit industrial applications of science and the trial and error development of technological innovation. This human side of the French division of elite scientific labor was played out in hundreds of small dramas enacted every time an inventor sought approval from the Académie. The formality of the application, the rigor of the evaluation, the yes or no quality of the Académie's judgments, rendered from afar, spoken down to inventors as masters speak to artisanal supplicants, worked against the development of the trial and error that lies at the heart of technological innovation. The pursuit of utility without recourse to the relatively egalitarian framework found in the scientific and philosophical societies in Derbyshire, or Birmingham, Manchester, or Rotterdam—while no absolute guarantee that industrial application would successfully occur (witness the retardation of industrialization in the Dutch Republic)—hampered French efforts to promote invention and innovation throughout the eighteenth century.

The Académie gave advice to government officials about the viability of specific proposals for new techniques and technologies. Without its approval, a privilege or subsidy would not be forthcoming. In short, a great deal was at stake from an inventor's or entrepreneur's perspective. Mechanists, both French and foreign born, had to convince the Parisian ministers and academicians that they were the best in the business, or that their devices would work. "Dr. Desaguliers who was my master, was the best engineer we ever had and has left ye Best Instructions of any I ever read, though there is as good French authors . . . ," wrote an English engineer of the 1750s in search of a contract to drain the marshes around Dunkirk with his Newcomen engine. He was eager that his English experience and the reputation of his Huguenot mentor precede him.[14] Little could he have known—as we now do—that a few decades later the French ministers would devise a kind of "revenge of the Huguenots" thesis to explain the superiority of British industry. In their private reports the ministers said that manufacturers in Manchester, Leeds, Halifax, and Birmingham excel and "nearly all [are] French refugees enjoying a very great liberty."[15] But the point of the English engineer's mention of Desaguliers was not to guilt-trip the French ministers. Rather, he wanted it known that he was trained by a practical scientist, and that his reputation as a man scientifically literate accompanied his proposals. These would then be sent by the ministers in charge of commerce and industry to the austere academicians of science.

Objections from the academicians, on whom the ministers relied if their own training was not sufficiently mechanical or scientific, doomed a proposal. Sometimes the academician wrote out a formal report with his findings; other times his objections were vague: "Camus of the Academy does not have a favorable opinion of this pump but he has not told me why,"[16] an official reported. On one occasion the academician said vaguely that only time would tell if such a machine would work.[17] The system of formal examination, based on the written account supplemented often with drawings and calculations, doubtless aided by connections and political affiliations, was hated by the supplicants; they saw the government ministers as being "intimidated by the Académie des sciences of Paris." Or they responded simply, but firmly, "It is not a question of science but of skill and mechanics, and in fact when it comes to the Marine men of the sea are able to judge soundly." Both Parisian and local academicians were seen as insensitive to the interests and merits of inventors and entrepreneurs. The inventors accused them of not having verified their negative reports by experiencing "the reality of their discoveries."[18] Even in industries such as silk that would apparently require little formal scientific training the scientists were involved and their opinions solicited. Entrepreneurs with little mechanical knowledge played an elaborate cat and mouse game with academicians, hiding from them a remarkably simple industrial "secret" while trying to win over their support. In their mind governmental ministers like de Montigny and intendants merged with academicians as men with power and authority who had to be wooed and placated.[19]

Indeed the academicians undoubtedly knew more formal science and mathematics than did the engineers and inventors whose proposals they were asked to evaluate. When asked to observe an industrial site at first hand, they routinely evaluated the level of mechanical expertise possessed by the entrepreneur. "Mr. Badger is little versed in mechanics," one inspector wrote to the minister who had subsidized Badger's silk calender factory in Lyon.[20] That this migrating English silk entrepreneur was particularly skilled in finishing silk, but not in the mechanics of metal machine and factory building hurt his business but also his reputation among the inspectors.

The skepticism and pessimism of the academicians was not simply the result of the crank proposals they occasionally saw—although doubtless there were enough of those—rather it grew out of the depth of their knowledge, their understanding of just how much science or mathematics they imagined that one needed to know to make successful interventions in nature. Indeed in the old regime the academicians' grasp of theoretical science was a mark of their stature. It formed the content of their exclusive schools, often intended only for the younger sons of the aristocracy. These equipped them to serve in an elaborate system of civilian and military bureaucracies, *les corps savants*. The *corps* were in turn restricted by law to graduates trained in the schools.[21] The royal engineering *corps* was a key player in the commercial and economic life of the state. It collected the road taxes; more important it fixed the roads, built the canals and bridges, and acted as the technical adviser on all projects where government money was involved. Let us look briefly at how late in the century, but before the fateful summer of 1789, this elite group thought of themselves and their role in society.

French engineering students at L'École des Ponts et Chaussées (founded in 1747) were expected to be highly literate as well as well-versed in mathematics, physics, and hydraulics. Their examinations at the end of five or six years of training also included essays in which they were asked to reflect on "the utility to the state and society" of their school. Answering that question in Paris in late April 1789 and in a high state of tension, the students told of their dedication to public service, *le bien public*, to the splendor of the state, and to making commerce and agriculture flourish through the public works that they built in every province as well as in the capital city.[22] They saw their work as enlightened, but they also knew that "the marine, the artillery, military engineering offer to the nobility the resources [for] men whose fathers are not very illustrious, or who cannot buy with money the privileges which merit alone should confer." In the spring of 1789 the school is under attack by unnamed "Reformers" who are trying to demonstrate its uselessness or to accuse the engineers "of having the cruelty of traversing their terrain in order to construct a route" or who, the engineers think, just dislike change. In the spring of 1789 it was still possible for enlightened and scientifically trained young aristocrats to imagine themselves as the prime engines of change.

The enemies of their privileges had, however, a quite different take on what was needed. One of the changes being proposed by the reformers was local schools of engineering in many parts of the country. To a man, the students think this to be a bad idea. Every essayist says that he would not be able to respect the teachers in such places and "it is always an advantage to reside at the center of good taste and the source of knowledge." At present engineering students receive a good education, grounded in mathematics and "their mores are known and irreproachable," indeed they are so studious that "there is not one student who does not want to be a professor" and all possess "a disinterested love of the sciences, and the arts." Universally they say that it is "for the good of the state" to have one school where the instruction is both theoretical and practical. "The state is composed of subjects and the state directs all public works." In the school are "young and well-born men who sacrifice a part of their fortune to acquire knowledge, and who require little salary from the state before being instructed in being truly useful. Their most noble ambition is for glory." As one soon-to-be engineer put it: "Frequently disinterested, [the engineer] limits his ambition to meriting the approval of those who employ him and he finds his compensation in the pleasure of being obliging." It is also useful for the provinces to find in the royal engineers men already sufficiently educated in their subject that they know what they are doing. Never once in these essays is interaction with local magistrates or private entrepreneurs mentioned. The vision of the state is absolutist; the vision of engineering as a profession is elitist, even noble. The old régime engineer's self-perception, as one historian puts it, was "a mélange of authoritarianism and abstract generosity."[23]

By contrast ten years after the French Revolution, in 1802, the graduating student engineers from the purged and reconstituted school have begun to talk about the complexity of their tasks—an engineer "must be a geometer, a physicist, naturalist, *commerçant*, and administrator." Uniformly they speak of the revolution as having been a good thing, and of the other new schools for science and engineering as welcome additions. These are new men, speaking to their teachers as did their predecessors (as all students everywhere do) with the intention of pleasing and getting a good grade. Most remarkably, in these postrevolutionary essays another figure, the entrepreneur—as well as a new self-definition of the engineer—appears. Both were totally absent from the pre-1789 essays. Now one essayist says: "The engineer must make estimates for projects . . . know the price of material . . . he must prevent the ruin of an entrepreneur who takes on a work for too low a price for which the estimate was made without proper consideration; or he must know how to force an ambitious and unembarrassed entrepreneur to be content with legitimate gain authorized by the government."[24] This moralizing vision of the engineer's service to the government would never have occurred to British engineers like Smeaton or Jessop, but they could have related to the financial interests of industrial entrepreneurs as they were now being described. Thus by 1802, as a result of reforms put in place during the mid- and late-1790s, French engineers had arrived in a different era, where a different story could be written about the cultural and social foundations of industrialization.

In the pre-1789 world the differences between the social systems and behavior of French and British engineers did not go unnoted by the French. When the graduates of the École des Ponts et Chaussées went to Britain in the 1780s and observed its navy, they noted how "the employees consider themselves to be civilians . . . and do not feel that they are inferior to the military . . . Perhaps in France, our customs, our prejudices . . . would make this impossible to hope for, although this way of thinking is certainly one of the reasons for the prosperity of the British navy."[25] Engineers such as Le Turc were keen to import British technology but they had no intention of actually becoming manufacturers or business men. As he put it to his paymasters in Paris, "You know my repugnance at being asked to head a business, however profitable." Le Turc supplied mechanized looms that had been requested and even put them in operation, but there his interest stopped: "That is all they can require of me and I feel I am capable of. It is impossible for me to repeat each day what I did the day before, and above all the detail of a manufacture than which nothing in the world is more boring."[26]

Ministers and even engineers with formal scientific training, of which by the 1770s there were many, and academicians approached technology in remarkably similar ways. Men like M. de Montigny wished to examine it formally, test it, and pass or reject it. The governmental ministers also badly wanted the most advanced technology, particularly in steam. But the gap between what they wanted and what their hierarchical social and administrative system could achieve was another matter. The problem with the French *ancien régime* lay not with intervention *per se*. If that had been the problem then there never would have been any industrial development in France because in the course of the Revolution and certainly in the Napoleonic period a new system of assistance and indirect intervention was put in place, a "tempered liberalism," which with some significant modifications remains to this day. The problem lay instead with the very structure of old regime society, with its divisions, barriers to knowledge exchange, inhibitions on technical trials, and errors. Had the ministers, academicians, entrepreneurs, and engineers all met more on the level, there is no telling what enlightened state intervention might have achieved. Once again, the cultural cannot be understood without attention to the social.

The Watt Engine in Old Regime France

The absolutist system of French administration also entangled the Birmingham entrepreneurs Boulton and Watt and their new engine. From early in their effort to market a steam engine with a separate condenser, Boulton, the manufacturer and entrepreneur, and Watt, the engineer and inventor, had had an eye on the Continental market. The French system in particular held out the prospect of government subsidies and monopoly rights. As early as the 1750s French government ministers and spies had surveyed and sought to import various British engines. Among the most favored projects entailed using the engines to supply Paris with water from the Seine.

Dealing directly with representatives of the French government, even one who was a London-based spy, Watt and Boulton first obtained a *privilège*, the exclusive right to set up the engine in France, and then they sought the opportunity to install and test their superior machine. As Watt put it, "our business in France is only in its infancy yet—that is we have obtained an arret [decree] of the king and Council for an exclusive privilege but cannot have the force of a patent until we have erected an engine and after trial made thereof it has been reported by certain commissarys appointed by the arret that our engine is superior to the com[petitors?] one which we hope will easily be proved."[27]

But the *privilège* did not guarantee that a competitor could not steal the design either by importing it from memory—an almost impossible task—or by posing as a friend, working on behalf of the inventor and then running off with the machine before Watt's property rights could be secured. The *privilège* had to be followed by trial demonstrations within a year, leading in turn to another *arrêt* (effectively a patent), which had to be registered in all the provincial Parlements of France. Because none of these steps had occurred Boulton and Watt did not want as yet to appoint anyone as an official agent in France. To make matters even more precarious, they did not exactly trust the French. As Watt put it to his friend and scientific collaborator the Glasgow professor, Joseph Black: "I had lately a letter from Mr Magellan at London who was our agent in the French business, is I believe a Carthusian or Benedictine monk, by profession a dealer in and retailer of philosophy, and perhaps a spy—however thus he has acted honestly and honourably by us—he made many enquirys about your latent heat, which I answered in so far as was expedient—he wants to know when you invented it. . . ."[28] Magellan was indeed a spy for France and clearly his instincts told him that Watt's engine bore watching, even importing, and the ever-calculating Watt had figured him out.

Newly discovered letters from the French archives supplemented by Watt's private papers tell a fascinating story about how Watt sought to protect himself and how the scientific and administrative culture of the old order thwarted the introduction of Watt's engine by at least ten years. In 1779 Watt wrote from Birmingham to explain to the French minister, the mechanically knowledgeable Count de Hérouville in Paris, that his machines worked better than any others. In support Joseph Jary, a visiting French inspector and *concessionaire* of mines in Nantes also wrote from England to explain that he too had just seen the new Watt steam engines pumping away successfully at various sites. The spy Magellan confirmed the trustworthiness of the descriptions.

The purpose of all these letters was to arrange a test demonstration of Watt's engine in a technologically appropriate setting that would show it to greatest advantage and at the same time protect Watt's interests. This required the presence of a skilled mechanist with hands-on knowledge of the steam engine. After initial misgivings, Watt had come to trust the character and mechanical competence of Jary: "We have agreed with a Mr Jary that the trial engine will be erected at a colliery he has near Nantes in Britany.—And Mr Jary who is a very ingenious man possest of the necessary knowledge has undertaken the care of the erection house . . . the finishing which will require the attendance of some person practically acquainted with putting our engines together, until this matter is finished our property in the invention in that kingdom is dubious."[29]

But none of these plans for a trial run or testimonies from skilled eyewitnesses satisfied the French ministers. Perhaps in consultation with the water company created in 1778 to supply Paris with a better system for domestic consumption—a company directed by Watt's potential French competitors, the brothers Perier—the ministers also wanted a working model of Watt's engine installed in Paris. True to the system of French scientific testing for inventions, the ministers wanted only their academicians to observe it firsthand.[30] And Parisian academicians did not travel for long stays to the provinces if they could avoid them.

Watt and the academicians held to very different notions of how a machine should be understood and tested. He wanted his engine observed only by comparison with an existing, older engine, in other words by trial and error in the presence of a mechanician like Jary who understood how to exploit the engine most efficiently. Jary also had an older, Newcomen engine at work near Nantes. Of course, Watt had other worries. He feared the machinations of his Parisian competitors, and he was suspicious of the Parisian judges who would inspect his engine, regarding them, in his words, as "literati [gentlemen of leisure and letters]." As all these obstacles presented themselves, poor Magellen wrote to say that he could barely conceive the engine making it into the Seine, doubting that the project would ever "be able to force water into *les louvetes, reservoirs & tuyaux des administrateurs.*" Magellen's water metaphor was apt: He imagined the project of importing Watt's engine as drowning in the man-made lakes, streams, and reservoirs of the bureaucracy.

The French opportunity to acquire Watt's engine through Jary had been doomed by different cultural styles of technological inquiry, by the power of the Parisian academicians and administrators, and by Watt's fear of Perier's machinations to bypass his patent. The cozy relationship of the Parisian administrators with the distrusted M. Perier rather than with Jary may also have caused Jary to be delayed in getting permission to import the new engine.[31] With his knowledge of applied mechanics, machines, and their use for drainage Jary was the right man to exploit Watt's engine, but in the late 1770s he was in the wrong place at the wrong time. His rivals, the Periers, were big-time entrepreneurs and skilled mechanists who had seen Watt's engine in England as early as 1777, and their hope was to finance their projects by monopoly rights to supply Paris with water. All they needed were Watt's superior engines complete with their cast-iron parts, which could at the time still only be made in England.

Credit for introducing Watt's engine would eventually go a full ten years later to the Periers. They had the skilled mechanical knowledge, capitalist backers, and plenty of government contacts.[32] After years of trying backhandedly to copy Watt's design they finally agreed to pay for an engine. In the late 1780s they set up a working Watt engine at their factory in Chaillot and in violation of Watt's arrêt, it became the model for turning out (slowly) about 100 French imitations. When Watt saw the Periers' engine factory in 1787 he was mightily impressed. One of its first engines was furnished to the old regime government. With assistance from the Académie, the ministers sent it to Saint-Domingue [Haiti], where it was used to irrigate the fields in this slave and sugar colony that was seen as economically vital to the state.[33]

The delay in bringing Watt's engine to France was not simply the result of Boulton and Watt's desire to strike a better deal with the brothers Perier or of Watt's well-placed fear that they would steal the design. Certainly there was plenty of self-interest and concern, but the administrators and academicians with their procedures that thwarted Watt and Jary also aided and abetted the delay.[34] Yet by 1790 Watt's superior version of his engine—the details passed on to French academicians by a Spanish industrial spy and mechanical engineer, Augustin de Bétancourt—had become an attraction for mechanically minded visitors to Paris and just as the revolutionary minister, N. L. François de Neufchâteau wanted, was being studied by student engineers.[35] The mechanical knowledge of men like Jary and Perier, aided in some cases by technological spying, facilitated this technology transfer. They were part of a new breed of French entrepreneurs/engineers educated in the 1770s and 1780s who, as we are about to see, would come into their own in the decades after 1789.

Jary learned enough about engines to go to England in search of the new technology. Yet he was a different kind of industrial entrepreneur than is classically conceived, more French than British. In good *ancien régime* fashion he had inherited his father's concession, a monopoly on coal mining in the region north of Nantes.[36] His privilege appears to have in no way inhibited his being one of the earliest French coal entrepreneurs to see the immense usefulness of Watt's invention. Despite the absence of a formal academy of science in

Nantes,[37] Jary's technical understanding, amply demonstrated in the Watt affair, arose from his intimate knowledge of engines at work in the marshland and mines around Nantes, his conversation with the visiting Birmingham manufacturer and engineer, William Wilkinson, as well as from his own interest in exploiting his coal mines.[38] By the last decades of the eighteenth century Jary could participate in what had become an international conversation based on mechanics and hydrostatics, fueled by the desire for profit and justified by notions of improvement and the public good. He resembles J. C. Perier who in a final memoir described himself as scientifically trained, a possessor of electrical instruments and pneumatic machines, a fashioner of centrifugal pumps, as well as a man who had always worked in the public interest. Indeed J. C. Perier had learned his science, not surprisingly, from the mechanist and imitator of Desaguliers, the abbé Nollet himself.[39] The industrial entrepreneur girded with skill in applied science had become an internationally recognizable figure; in time his mentality would come to be seen as characteristically Western.

But in the 1780s in France a different kind of scientific mentality and approach to industrial development still dominated. The delay that Jary experienced importing Watt's engine occurred despite the willingness of the academicians to approve many projects sight unseen, but in particular one for applying "la pompe à feu" to riverboats.[40] They did so on the basis of elaborate drawings and their expeditious response may have had a great deal to do with current government policy, which sought first and foremost to improve transportation. The academicians of the 1780s, like the engineers at Ponts et Chaussées, emphasized central government policy. This tension between the needs of the government and the interests of manufacturing and mining sometimes worked to enhance industrial development, but it could just as easily work to thwart entrepreneurs who, like Jary, did not have easy access to the tributaries of administrative power. They possessed the necessary mechanical knowledge; they were self-fashioned engineers/entrepreneurs themselves or they could talk to engineers. There was nothing "pure" or original about the scientific knowledge they possessed; it was intended to serve their interests just like the learning of the academicians facilitated their prestige as well as governmental policy.

By contrast to the Jarys of the world, *les grands* of the academies or the royal engineers who by the 1780s were beginning to show an interest in industrial capitalism could make little sense out of the entrepreneurial or management side of industry, having had little practical training in either.[41] On occasion they were also repelled and bored by it. Such men of applied science who worked on industrial applications for the king also found themselves thwarted by a system that overburdened its administrators or worked according to their personal priorities or interests. As the engineer Brunelle told Watt: "I could not get the intendant of finances to listen to me even for a quarter of an hour . . . I work for the king and I can never obtain an audience!"[42] Events in 1789 and beyond would abolish the system that thwarted Watt, Jary, and Brunelle forever.

The Impact of the French Revolution on Industrial Policy and Practices

In the wake of the political revolution in France a new ministerial elite came to power; they too, like their *ancien régime* predecessors, vowed to do something to combat the British advantage.[43] Men with industrial and scientific experience such as the chemist and Montpellier factory owner, J. A. Chaptal, now saw their opportunity; some of them even took over the Ministry of the Interior. They set up policies and institutions intended specifically to address the technology gap; indeed Chaptal can be seen as the key visionary in the revolutionary approach to industry.[44] Then under Napoleon when Chaptal became the head of the Ministry of the Interior he was the prime architect of industrial policy. The total effect of all these efforts from 1790 to 1810 was to alter French industrial policy forever. Indeed had the French Revolution never happened we might legitimately wonder when or how France would have moved into the industrial era.

Jean-Antoine Chaptal dressed in the imperial fashion of Napoleonic France. (Courtesy of the Bibliothèque Nationale, Paris.)

The revolutionaries were guided in their deliberations by an overarching explanation of precisely how the perceived retardation had occurred. Such contemporary explanations should be of more than passing interest to cultural historians of science, technology, and industry. We will take Chaptal as representative of the new thinking. His interest in mechanization was long-standing, going back to the 1770s. His scientific education had been Newtonian in method and vitalist in philosophy, principles that helped to organize his work in chemistry. He also knew about steam engines and had a direct interest in the application of mechanical techniques to save on labor costs.[45] Chaptal was also a major applier of chemical knowledge with a large factory that made nitric and sulfuric acids needed for the dyeing and bleaching of fabrics.[46] In the 1780s he was also on good terms with the chemist Berthollet and he in turn was in fairly regular communication with Watt, who had visited in Paris in 1787. Chaptal probably also read English scientific publications, but it is unclear what direct knowledge, if any, he had of English industrial practices.

It was probably as a result of his own industrial experience that Chaptal became convinced that French education and industrial policy needed drastic alteration. As early as 1790 the economic section of the patriotic club in Montpellier where Chaptal was a leader offered free, public lectures in mathematics. Within this revolutionary milieu and its reforming focus on industrial retardation, Chaptal developed complex arguments to explain Britain's success. He identified mechanical and chemical knowledge along with the division and tight control of labor, as the critical elements in British entrepreneurial prowess in factory production.[47]

Early on in his career and before his rise into revolutionary circles of government, Chaptal made the case for applied science. Indeed in 1790 his own *Catechism for Good Patriots* had argued that the *savant* finds his calling in the enlightenment of his country in matters agricultural and industrial; he enriches industry by inventions; it is impossible to isolate him from society. Others picked up on Chaptal's insights. By 1800 such sentiments turn up in French journals concerned with improving manufacturing, mining, and technology. They said that success depended not exclusively on competition between individuals but rather on "the aggregation of industry, skill [du génie] and the talents of a mass of manufacturers." Their long, intimate and combined knowledge of theory and practice gives them the edge. What is bad is the tendency to separate "l'homme savant [from] l'homme ouvrier,"[48] the knower from the worker.

Attention to applied mechanics and mechanical knowledge actually became revolutionary policy as early as 1791 when Antoine Blondel received an extensive report on the mechanical manufacturing of cotton and its obvious superiority.[49] In 1795 the Committee of Public Safety proposed the establishment in Paris of "The Factory for Improvement" where anyone could come and receive a fair idea of "la Mécanique en général" and at the same time learn "the rapport between theory and practice." Some of the items in this factory were to be machines confiscated by the revolutionary army as it moved north, which, as instructed, sent back to Paris for inspection any machinery captured in the Low Countries.

The new French approach put in place after 1789 stressed how particular social settings foster the ability to think mechanically. Those settings, so the claim went, would breed manufacturers with scientific training who then acquired personal experience in application. The accumulated weight of their numbers extended rationalizing thought processes throughout society, to the entire range of industrial activity from machines to the workers themselves.[50] Even in textiles where the self-taught tinkerer still commanded a place, it was the "exact notions of workers" such as could still only be seen in England that would give imitative French entrepreneurs the confidence to establish factories.[51] When asked about setting up an industry to produce agricultural tools made of iron, Chaptal as minister of the interior told the government official to master the theory and then to integrate the theory with the practice. The rest depended on individual enterprise and "the natural progress" that comes from following one's interest.[52] James Watt or Matthew Boulton could not have said it better.

The revolutionary French ministers laid ideological emphasis on the actual contextual factors, the milieu wherein manufacturing occurs. Chaptal's distinctive contribution was simply to supply a sophisticated synthesis about the interplay, the context composed of social, cultural, political, and economic factors. His analysis pointed to English protective tariffs, strong home-based consumption, and superior coal. Yet it also paid serious attention to the cultural component. Chaptal was convinced that British industrial success owed a major debt to mechanical knowledge, which he thought had been conveyed to entrepreneurs through the educational system. In France education had been retarded, he believed, because of clerical control during the old regime. Since the Revolution improvements had occurred that brought enlightened authors into the curriculum, but more needed to be done, especially in the training of workers. He saw young French workers as being badly prepared and not given the mental resources through which they could "exercise a mechanical profession."[53] As a remedy he wanted to institute an elaborate system of trade schools, which by their description, look as if they would have been remarkably similar to the mechanics institutes that appear in Britain during the second decade of the nineteenth century. There the schools worked to extend mechanical knowledge and work discipline even deeper into the artisanal population.

Chaptal wanted a specific kind of scientific education particularly for boys. He recommended that actual machines be carried into the classroom because, given the relationship between words and things, machines once absorbed by eye and touch give "the advantage . . . to fix or form one uniform language for the arts which is all the more necessary because now the variety found from town to town makes difficult the transmission of the discoveries in mechanics."[54] One of Chaptal's predecessors in the ministry of the interior, François de Neufchâteau, was also concerned in the 1790s with introducing machines into the instruction of young engineers at the École des Ponts et Chaussées.[55] As we saw when we looked at the mentality revealed in the pre-1789 essays of graduating royal engineers of the eighteenth century, François de Neufchâteau made an important change.

Decades later in one of the first histories of the Industrial Revolution ever written, Chaptal claimed victory for the revolutionaries' vision: "In this epoque . . . the study of the sciences has become so general, and the relationship between *les savants et les artistes* so intimate that they have arrived at an extraordinary degree of perfection in their art . . . the Industrial presumes on the part of the artist an extended knowledge of mechanics, various *notions de calcul*, a great dexterity in work, and enlightenment in the principles of the arts."[56] What Chaptal published in 1819 about the benefits derived from the marriage of theory and practice could have been written years earlier by the new revolutionary ministers, or for that matter by someone in the Lunar Society of Birmingham or the Manchester Literary and Philosophical Society. By 1810 the French had put in place the elements of a new scientific culture that glorified application as much as it championed the entrepreneur conceived mechanically, as self-propelled when acted on by material interests.

Brought into government by the French Revolution but implemented largely under Napoleon, the new industrial vision of Chaptal and company had also been imperial. It was exported wherever the conquering army installed French administrators and their local allies.[57] Under Chaptal's brief ministerial reign, and that of his successors, even the schools and universities in the Low Countries experienced educational reforms introduced through a cadre of Napoleonic administrators.[58] Operating under conditions characterized by war abroad and hostility from local peoples who now suddenly found themselves subsumed into the expanded French Republic, the imperial administrators nevertheless stuck to a vision of reform that brought with it educational changes intended to create industrial development among a new class of citizen-entrepreneurs.

In Belgium these reforms accelerated the movement toward technical education that had developed earlier but informally in various centers of industrial development. Education in the Austrian Netherlands had long been the province of the Catholic clergy who in the reforms of the 1770s, as we saw in the previous chapter, managed to introduce in place of Aristotle, not Newtonianism but Cartesianism. The administrators of 1798 were undoubtedly biased against the faculties and schools that they found particularly in the Catholic areas of the Low Countries and on the left bank of the Rhine. But we have no reason to doubt their inventories of books. Even in Maastricht, so near to Liège and the coal deposits of the region, the inventory revealed many titles in theology and few "in the sciences and the arts." The new French-instituted École centrale elevated the sciences and appointed a distinguished scientist to its faculty; he labored for years to get the necessary books and instruments. Yet French policy always favored French education first. Much to the detriment of the locals, the French army besieging Maastricht was under orders to send back to France any mechanical items that looked distinctive.[59] Prior to the 1790s and the impact of the French Revolutionary vision, the Newtonian science to be found in the Dutch and Belgian universities had simply not penetrated into the school system, although the retardation was far greater in the Austrian territories than in the Dutch Republic. Similarly in the

German universities and schools to the east (but west of the Rhine), theology was downgraded by the French and the sciences systematically favored.[60]

The French took abroad educational policies intended to make citizens who would contribute to the war effort and also become economically viable. Between ages 14 and 16 they were now to do very little other than science and mathematics, and finding faculty for this educational experiment (which proved to be short-lived) was no easy matter. In Liège a struggle broke out between the "party of pure mathematics" and the appliers; at least during the period of French occupation the faculty friendly toward application won. In Brussels the science faculty had to be organized so quickly that it was never very diligent or particularly adept.[61] North of the border books and instruments were even scarcer than in France. In Amsterdam the French inspectors found the university library to need books in physics; they also betrayed their other interests in wanting it to have more books in French.[62] For a brief period the educational visionaries had their way. In the Belgian coal mining town of Mons a Napoleonic engineer believed that he could safely educate his children locally because the government had put in a school for design and courses in physics, chemistry, and mathematics.[63] French imperialism may have subordinated the peripheries to the center, but in the process an industrial vision was also exported throughout western Europe.[64]

What was educationally valued abroad was also to be instituted at home. For example, French schooling for girls prior to the Revolution leaves not a shred of evidence of serious mathematical training; after the Revolution there is limited evidence that professors of experimental physics taught all pupils over the age of 14—in at least one province—and books in simple science begin to appear addressed to both girls and boys.[65] A similar expansion of educational opportunity for girls occurred to the north as evinced in French and Flemish (i.e., Dutch) texts aimed at Belgian and Dutch students. The Dutch shifted emphasis from the fashionable physico-theology of an earlier era toward application and mechanics.[66] Overall French scientific education after 1789 remained markedly more theoretical than what can be seen across the channel, especially for engineers, but it also became much more commonplace than it had been prior to the Revolution.[67]

Industrial competition primarily with Britain fueled the educational reforms at home and abroad; so too did anticlericalism and the revolutionary ideal of creating the virtuous citizen who would be made rational not by education in religion but in science. Professors of physics wrote from the French provinces to say that they were trying to create "enlightened citizens," with "the characteristics of republicans," and they also complained that verbal descriptions of machines were insufficient and no substitute for hands-on manipulation of the devices.[68] They began to instruct dyers in the new chemistry and believed that "already many workers have abandoned their old prejudices for following the new procedures." At least one of these teachers also became an inventor of textile machinery. Throughout the new school system the professors of mechanics and physics begged for money for machines, for the opportunity to wed "theory to practice."[69] In 1793, after his release from prison where he had been put by the Jacobins, Chaptal led the way and gave a public course in physics and chemistry in Montpellier.[70]

Beginning in 1798 public expositions were also held in Paris to transmit the researches and wisdom of "a class of artists . . . between *les savans et les fabricans*, . . . a mechanician, a simple *contre-maître* (foreman), a skilled worker [endowed] with the spirit of an observer [who] by fortuitous discoveries elevates manufacturing all at once to an even higher degree of prosperity."[71] In 1798 the first exposition at the Champ-de-Mars, organized by François de Neufchâteau, featured clockmakers, craftsmen, and some craftswomen in porcelain, leather, bookbinding, cabinetry, metallurgy, scientific instruments, and planetary orrery makers, as well as cotton manufacturing by machines.[72] On that occasion the highly skilled artisans, whose luxurious and finely wrought goods were destined for the elite market, separated from the *fabricants*, the manufacturers of machine-made goods.[73] Gradually in 1801 and 1802, the former as well as engineers and larger manufacturers became more prominent, and so did the machines. At the 1819 exposition the first prize went to a French student-engineer for his steam engine, and heavy machinery dominated the event.

The overall effect of these policies, expositions, reorganizations, and educational reforms, which included a brutal purge of the old academies of science during the Terror, turned France decisively in an industrial direction. They may also have contributed to the rapid industrial direction well under way by 1810 in the French-speaking parts of Belgium. The turn toward an industrial mentality cannot be separated from the rise of democratic and republican ideals, however thwarted they may at times have been. The application of mechanical and chemical knowledge to manufacturing of everything from fabrics to engines became a doctrine built into the ideological framework of French republicanism both at home and abroad.[74] After 1789 the departure from old regime practices and beliefs had been both real and abrupt.

Hierarchically Arranged Knowledge and Industrial Retardation

Certain stories drawn from the era prior to the Revolution point to conclusions about the nature of eighteenth-century French industry, engineers, and the problems of both. When asked in the 1770s to drain the harbor at Nantes, the military engineers and professors of hydraulics sent by the government because they were skilled in "la théorie hydrodraulique" found themselves described and treated as "experts" by the local merchants. From the extant records in Nantes there is not a shred of evidence to suggest that the engineers sent to the site by the government consulted with local entrepreneurs about their transportation interests or needs. Nor were they asked to consult. Merchants and industrialists were busy at the time quarreling with the monarchy over what local authority they would be permitted against the claims of other powerful groups in the third estate.[75] In that drama the condition of the harbor was one of many grievances. Amid the political maneuvering entrepreneurs had little interest in the engineers, except insofar as they represented state authority. Engineers were representatives of the king, and the application of their knowledge concerned only public works. Before 1789 that assumption operated wherever the more than 400 royal engineers could be found.

Take the region around Montpellier, Chaptal's home base. When royal engineers consulted with the local magistrates through commissions set up by the Estates of Languedoc—a body that had quite a significant interest in fostering regional industrial development in textiles and coal—never once prior to the 1780s are the engineers consulted in manufacturing or industrial matters.[76] Conversations with engineers on the subjects of canals, roads, ports, and especially the financing of these projects, could be quite extensive, but never in the records we have examined were the royal engineers called in to consult on matters to do with manufacturing. The Montpellier entrepreneur (like Chaptal) was basically on his own prior to 1789, although limited evidence does point to occasional collaboration between local coal entrepreneurs and engineers from the small Royal School of Mines in Paris.[77] With the Revolution the mining school was revived and its graduates put into state service to exploit iron and coal mines servicing the sometimes conflicting needs of both industry and warmaking.[78]

On the whole prior to 1789 French entrepreneurs and engineers occupied separate universes, did not possess a common technical vocabulary, and the details of public works were a matter solely for the engineers. The entrepreneurs' languid response to the details of technology—however simple the device—surfaced when the English silk finisher from Lyons, John Badger, arrived in Nîmes. At the government's request he brought the "secret" of his *calandre anglaise* to the city, but the local Protestant silk merchants and Catholic magistrates were busy in confessionally motivated political quarrels. Badger left, no one sought to imitate his closely guarded secret process for enhancing the shine on silk, and the project fell by the wayside.[79] Badger remained the expert, his technical knowledge having been seen as something separate from the political process, unrelated to the forms of power most avidly sought or disputed by manufacturers and merchants. His use of heavy weights, rollers, cardboard, and cold water went with him.

Compare Badger's experience with that of many British mechanists. The complaints against the interference of entrepreneurs made by engineers like Smeaton and Watt allow us to conclude: what they would not have done to be left alone by the magistrates and entrepreneurs, to be awarded the title "expert" so freely given to their French counterparts in royal service across the channel. The British civil engineers could have done their calculations without technical interference from their beneficiaries or employers. In France the merchants' and magistrates' interest in the advice provided by the experts was largely political, and not intellectual or applied. There is no substantial evidence (at least from Nantes or Nîmes or Montpellier) for the close interaction between merchants and engineers or skilled craftsmen, which can be observed in the British archives. Just as Watt's report to the Scottish magistrates interested in building a canal showed, the next chapter will also reveal the critically important interaction between entrepreneurs, engineers, and magistrates as it occurred in Britain from at least the mid-eighteenth century onward. We see the interaction at the repairing of the Bristol harbor, at the testimonies given by canal companies before Parliament, and at the installation of steam engines in Derbyshire. By contrast, the hierarchical social and governmental structure of the old regime, and hence of old regime French science, inhibited the establishment of the common scientific culture out of which technological innovation of an industrial sort emerged. By 1800, after the destruction of the traditional academies, a new scientific culture arose in France and "the industrial take-off was then controlled very directly by the top echelon of the scientific elite."[80] In Britain, by contrast, the distinctive cultural matrix of voluntaristic ideology and applied Newtonian science available to practical-minded scientific elites arose out of the political and intellectual revolution of the seventeenth century. It did not export easily. Perhaps in a European context before industrial culture took root or became dominant it would require the presence of new men and new institutions, the post-1789 Chaptals of the world.[81]

Bringing culture into the story of western industrialization goes part of the way toward answering the question that historians are now beginning to ask: Why is it that in some societies talent can be unleashed around technical problems, whereas in others, as Joel Mokyr puts it, "this kind of talent is either repressed or directed elsewhere."[82] Having a common cultural legacy—a common technical but utilitarian language—provided by applied scientific education, formal and informal, buttressed by a voluntarism found more in civil society, in associations, societies, and clubs than in the individual singularly conceived—or than in the formal state institutions of an aristocratic or rigidly oligarchic regime—may go part of the way toward explaining the particular unleashing of industrially focused talent that occurred in eighteenth-century Britain.

In effect, when seen comparatively, the French evidence suggests that the emergence of *relatively* democratic patterns of social interaction, the new public culture of the eighteenth century, may have been far more important than has been previously imagined in making Western economic development of an industrial sort happen. French military engineers could not evolve into civil engineers because their understanding of their role in the state and their social place prevented such an evolution. Their status and position blocked the emergence of civil engineering on any widespread scale.

Recent scholarship has dwelt on the creation and expansion of the public sphere in eighteenth-century Europe. When seen comparatively and thought about industrially, the relatively greater expansion of societies, public education, and conversation across the classes in Britain gave it an advantage. The evidence suggests that true *anciens régimes* deserve their posthumously acquired meaning; they inevitably become old and backward. By and large they do not industrialize easily because they cannot. Vast inequities, hierarchies, and privileges when operating on the microlevel of industrial interaction inhibit the transmission and development of technical knowledge. The knowledge in itself is no guarantee of success; on the other hand, you can have all the coal or capital or cheap labor you can possibly need, as France did, and still not have entrepreneurs and engineers working together to harness these elements with technological innovation. Similarly in Prussia industrialists of the early nineteenth century had to do battle with traditional elites just to establish "even the desirability of industry itself."[83] If the examples presented here have led to a historically accurate conclusion—and *if* history holds lessons—imagine what might be possible in this society or any other, were democracy to be extended, broadened, and deepened, not for the purpose simply of mastery and profit—although these are not motives that can ever be ignored or discounted or downplayed—but for the long-term improvement of the human estate.

9

How Science Worked in Industrial Moments: Case Studies from Great Britain

In the first place, when the law of acquisition is established, it is impossible to restrict the amount. That must be left to a variety of personal qualities . . . each person is but an engine in the great mechanism of circulation, [and] . . . the general plan, upon which wealth is distributed, is also the most salutary.

—From an anonymous writer, *A Dialogue between a Gentleman and a Mechanic*, c. 1798

Nature mechanized provided early industrialists with an arsenal of new knowledge, as well as with new metaphors of self-justification. All could be applied in the service of their economic interests. The image of the individual as an "engine in the great mechanism of circulation" rendered the disparities between rich and poor as simply the social consequences of the same mechanical laws at work in the natural order applied to the social order. The entire process of industrialization could therefore be seen—then and now—as the coming together of impersonal factors, necessities arising out of the application of surplus capital to raw materials, or the appearance, willy-nilly, of new technical inventions called forth in an attempt to overcome low profit margins. Once begun the innovations became "cumulative" in number and effect, literally "self-sustaining," or alternatively, the effect simply of a neat interaction among a variety of factors: rising demand at home for consumer goods, higher wages, and the impossible burden faced by the entrepreneur of trying to render technical innovations unnecessary when the supply of skilled labor was always so precarious.[1] The standard economic histories depend on mechanical laws for their credibility. In this chapter we want to turn to decision making once seen as relying entirely on economic considerations and actually watch scientific culture at work.

Historical explanations that exhibit simplicity and monocausality as their distinguishing characteristics are not only flawed, they are also frequently boring and ultimately irrelevant. Consequently, to emphasize in this chapter the importance of scientific thinking in the decision making of entrepreneurs, and then to assert the primacy of these mental operations over economic considerations or material circumstances, would be to err in the same monocausal way, only using a different set of simplistic assumptions about historical change, as do the many economic and social histories that emphasize material and impersonal forces as the critical, and indeed as the only, factors in early industrialization. A balanced account, one true to the complexity of human nature and hence history, would seek to show that certain moments of decision making—where the decision to proceed or not directly assisted or inhibited industrial development—were shaped by the availability of scientific knowledge, among other beliefs and judgments. Here we can do no more than present a few concrete examples to illustrate the generalization that to an extraordinary extent scientific knowledge had penetrated the thinking of literate Britons by the late eighteenth and early nineteenth centuries. That penetration contributed directly to the process of industrialization, to creating the world in which we now live.

The Application of Mechanical Knowledge

A variety of projects, long regarded as central to the historical process of British industrialization, in particular canal building and the use of steam power, illustrate how scientific knowledge could affect profit and productivity. The contention here is that the long process of cultural integration through which science passed in the seventeenth and eighteenth centuries created by the second half of the eighteenth century a new kind of Western European. Such a person possessed a signal characteristic: access to, and understanding of, the mechanical aspects of the new scientific learning. He is found most readily in Britain at the vanguard of industrial and commercial activity, but he (seldom she) might also be a landed gentleman using his land in capital-intensive ways and profoundly interested in agricultural improvement. He is, of course, to be found in many western European countries by this period, although not always with access in those places to the political power necessary to effect the changes he desired. That he can be located so conspicuously and so early in Britain—in the west country, in Derbyshire, and of course, in Scotland, as well as in the houses of Parliament—must be acknowledged as one important piece in the puzzle of why England industrialized first.

Such a person understood nature mechanically and wished to use that knowledge for financial profit and sometimes also for the general improvement of society. He (occasionally she) approached natural obstacles inhibiting transportation or manufacturing or mining with this mechanical knowledge of nature at the forefront of his thinking. Such knowledgeable capitalists did not think less about labor costs, or compete less ruthlessly with other entrepreneurs, or treat their workers more or less harshly. But they did engage in economic activity armed with a new and compelling kind of knowledge. Early industrialists understood the ways things worked in the natural world, and that knowledge provided a consciously felt power, which sometimes also resulted in personal economic advancement in excess of that achieved by other less knowledgeable competitors. At the least this knowledge translated into a new self-confidence. The entrepreneur could apply the mechanical model of nature to society and assume, therefore, that profit based on the wage labor of others and on the manipulation of markets was simply in the nature of things, that order existed beneath the apparent disorder of self-interest and market forces.

Frequently, scientifically educated gentlemen considered themselves to be as mechanically knowledgeable as the professional mechanists. For practicing engineers such as John Smeaton and his successors—William Jessop (whom he trained), Benjamin Outram, and John Rennie, whose canals and bridges revolutionized transportation in Britain—the scientific learning of the entrepreneurs offered immense opportunity for employment and even for business collaboration. Their learning was also the source of much irritation and conflict. Writing in the 1760s Smeaton complained of his entrepreneurial employers, that is, the members of the canal companies who paid his fees to draw up plans and then to explain and defend them in Parliament. He said that they frequently interfered with the execution of the plans. Writing in the first instance of the foremen who supervised the on-site work of the navvies or canal diggers and who then reported back to the directors of the canal companies, Smeaton angrily observed: "Not only all of the inferior departments are ambitious to be practical engineers," but "even members of the company *have a propensity that way* too; by which means *all becoming masters* . . . the parties interfering *suppose themselves competent* to become Chief Engineers." He ruefully argued that "they cannot have great length of experience [like himself] in conducting public works" and do not possess the degree of theoretical knowledge they imagine themselves as commanding.[2]

As must have been evident to Smeaton and his contemporaries, people are not born with the ability to conceive of nature mathematically and mechanically, nor with the ability to invent mechanical objects of anything but the most rudimentary simplicity. That seems so obvious, yet when considered in relation to the late eighteenth century it acquires a startling historical significance. In the world in which we live such deficits at birth are quickly remedied through daily experience of mechanical devices or their effects, and then, of course, through universal formal education in basic science. As a result it is extremely difficult, indeed it requires a leap of historical imagination, to conceive of the time when the mechanical understanding of nature was new and anything but commonplace, when its assumptions violated centuries-old explanations about nature that rested on nonmechanical beliefs.

Imagine being in the unique situation of Smeaton and his employers, of being a member of a small elite conversant with science—someone who had attended scientific lectures, who may even have been taught Newtonian science by a particularly advanced schoolmaster, or who had possibly become so passionately interested in scientific learning as to qualify, as did Smeaton, as a fellow of the Royal Society. You would have been, and may have conceived of yourself as being, in possession of a new and powerful wisdom. Such scientifically knowledgeable men and some women could be found in every town of any significance in late eighteenth-century Britain. They frequented literary and philosophical societies, attended scientific lectures, read scientific books, promoted new transportation schemes, joined agricultural societies, or even installed steam engines in their factories, often at considerable capital risk. When they thought about the natural world they saw it as measureable, as a set of push-pull interactions that release power that could be maximized by its application to machinery. They thought of water, wind, hills, and valleys as places where canals might be built or steam engines used, provided the terrain could first be measured and the correct principles of leverage and pressure applied to regulate the flow of water or the power of an engine. And they thought of these man-made objects as beautiful in themselves, as aesthetically pleasing as well as useful and profitable.

Such mechanically minded men also went to Parliament to represent their towns and shires; first and foremost they represented the interests of people like themselves. Having them in Parliament proved critically important as key moments in the process of mechanizing occurred. In 1775 the House of Commons set up a committee to investigate Watt's claim that his engine was markedly different from all competitors. It had to be for his patent to be worthy of an exclusive privilege. It would allow him alone to make the engine but only if the exclusive right would not affect the making of other, common engines. To establish the difference between Watt's innovation and the other engines meant that members of the committee had to be able to talk to ordinary mechanics as well as to men like Matthew Boulton who was in wealth much more like themselves.

For the hearings on Watt's engine the members of Parliament called in one Joseph Harrison, a lowly smith at Boulton's factory in Soho near Birmingham. From him they wanted the exact measurements of the cylinder, the height of the column of water raised by the engine, and the diameter of the pump used. The smith knew his engines; indeed twenty years later he could have qualified for membership in the third "class" of skilled artisans instituted in the Society of Civil Engineers as Boulton, Watt, and others set it up in 1795.

In the Birmingham company of Boulton, who was about to go into business with Watt, Harrison and Boulton "erected [a Watt engine] for Experiments to prove what could be done." They tested Watt's engine against what Mr. Harrison knew other engines could do. The results were "5 Times as much work with the same Quantity of Coals." This engine was fitted with a rotating wheel, which when "steam was let into it . . . the wheel weighed upwards of a Ton." The rotating device would become critically important when fitting the engine to other moving machines particularly in the new cotton factories. The smith explained that all the work of Watt's engine was being done by "the steam itself, being a greater power than the Common Atmosphere." The MPs wanted a precise figure: "How much more water will be thrown the same heighth with Mr. Watt's Engine, than the old one—the dimensions being the same?" Mr. Harrison replied with equal precision, "12 to 7." Not quite satisfied, the MPs want to know "how many pounds to the square inch does the common engine work?" Mr. Harrison answered in kind while Mr. Boulton supplied the key information about the extra force actually saved on coal: "1/4th of coals in one engine will produce the same quantity of water and to the same heighth." The MPs wanted to know how precisely this saving was accomplished. Boulton explained in detail how the separate condenser for steam means that no cold water need be injected into the cylinder, "the vacuum is not injured," the metal stays perfectly hot, and hence no loss of energy occurs as with other engines. If a business were to lay out 1,000 pounds on a Watt engine and the same on a common engine Watt's would do as much work with a third or fourth as much coal.

One of the MPs then held forth on the "imperfections of the common fire engine," which he had examined himself. Another civil engineer, Robert Mylne, came forth and also was interrogated. He commented on the vibrating balance beam and pump gear, on Newcomen's and Savery's engines and the differences among and between them and Watt's engine, on the cost of common engines, "not an opinion or an estimate but from experience within these 3 or 4 years." Mylne explained how Watt's will cost twice as much and do double the quantity of work, and finally that "a machine consists not only of the power raised, but of the business done." Watt's engine had to be considered as both a mechanical and an economic entity.[3] From all the experts on mechanics present, enough was learned and understood to convince the MPs that Watt's engine was worth protecting. The rest would be up to the ingenuity of Boulton and Watt.

When entrepreneurs of transportation and industry put up the capital for projects, their foremen in turn went out and hired unskilled or semiskilled workers to dig the trenches and tunnels for the canals or to feed coal into the burners of the engines. The human element in the early Industrial Revolution was the least subject to the mechanical laws explicated by the scientific lecturers. As one engineer complained: "Stone, wood, and iron, are wrought and put together by mechanical methods; but the greatest work is to keep right the animal part of the machinery," that is, the workers.[4] We know from other sources that these unmechanized men, who frequently lost their lives in the digging of tunnels or the mining of coal, would consult the position of the stars before they began their arduous and sometimes dangerous work.[5] For them astrological calculations and magical beliefs were appropriate to the uncertainty of their livelihood, even of their lives. As such explanations had once been for the seventeenth-century Watt family, astrological predictions for the laborers were more meaningful perhaps than any other form of natural explanation. And certainly if a worker were illiterate or could not afford scientific books, and the majority of workers in this period could not, he also had no access to an alternative rendering of the relationship between the natural order and everyday events, to the new mechanical learning.

When the promoters of transportation and industrial improvements, or members of Parliament, consulted their communities to gain their support or financial backing for a new canal or engine, predictably they did not consult either the workers or the stars. Rather they appealed to men with some capital who were also literate and hopefully knowledgeable in mechanical matters so that they could understand the merits, if not the actual engineering details, of the proposed plans. It is possible to distinguish levels of scientific learning among such middling sorts of men and to compare their understanding of the world with that possessed by engineers or fellows of the Royal Society. This microscopic examination of individuals poised at the beginning of a world where the signs of industrialization—the factories, canals, harbors, bridges, and steam engines—were now apparent should reveal the multiplicity of natural systems of explanation still prevalent at that moment. The examination will be conducted in Bristol, the second largest British city by 1750, and it should illustrate the singular relevance of mechanical explanations for the promoters of commerce and industry.

Bristol: An Early Example of the Application of Mechanical Science

The city of Bristol in the west of England, the commercial metropolis for the west country and center of the Atlantic trade, provides the historical laboratory for examining the application of mechanical science.[6] In this city (population 60,000 by the 1760s) many of the preconditions of industrialization were already present. The west country was rich in mineral deposits, mining was commonplace, advanced forms of iron production were already widespread, while surplus capital needed for investment came to Bristol's merchants from the Atlantic trade—especially in slaves, tobacco, and sugar.[7] Yet despite these signs of early or proto-industrialization, Bristol and the west country in general would ultimately lose ground to its northern rival, the growing port city of Liverpool and its environs. Indeed the response of Bristol's merchant community to that threat provides an occasion for discovering the degree of mechanical knowledge to be found among its commercial elite—the leading merchants who often invested in industrial ventures and who exercised an inordinate amount of authority over day-to-day political life and town government.

The New Science in Bristol

Bristol is all the more interesting for illustrating the widespread availability and use of scientific knowledge because it possessed no scientific society of its own. Indeed one of the leading scientific minds among the city's urban gentry, Richard Bright, had to use the services of the distant Manchester Literary and Philosophical Society as he struggled to convince his mercantile colleagues of the necessity for a new floating harbor to be built with the best engineering advice of the day. The evidence for the varieties of scientific learning found in Bristol and its environs comes not therefore from a single source, such as exists for Manchester, Derby, or Spalding, but from a variety of sources, and as a result it is all the more fascinating.

For instance, when the prominent scientific lecturer of the 1760s and 1770s, James Ferguson, F.R.S., gave his course on mechanics, hydrostatics, and hydraulics in Bristol, he dined with a member of his audience, a locally prominent accountant, part-time doctor, and student of natural philosophy, William Dyer.[8] Dyer was also an electrical practitioner, well read in the latest experimentation, who used electrical shocks, as was common at the time, in his medical practice. He applied the technique to a wide range of diseases and ailments—from rheumatism, the gout, and "lombago," where it was apparently helpful, to consumption and deafness, where it seems to have been quite useless. This man of the new science, on the very evening he dined with Ferguson, also visited his close spiritual friend, one Rachael Tucker, a prophetess who "possessed the intimate worship of God."[9] Dyer was an intensely religious man who also believed in witches and diabolical possession. He became swept up in supporting the veracity of accusations of witchcraft made against a local woman, a sensational episode that captured the attention of Bristol's citizens in 1762. He also corrected learned scientific treatises by his electrical friends while at the same time being drawn to Methodism, where at one meeting house he was able to view a new electrical machine. Although well versed in Newtonian mechanics as taught by Ferguson, Dyer distrusted aspects of the Newtonian tradition and described William Whiston, the early Newtonian, as a "Deist."[10] By his own admission Dyer was largely uninterested in business matters, and he displays no interest in industrial developments or in the mercantile world of his city, despite the fact that so many of its merchants were Dissenters or Methodists, as was Dyer himself. In him we see the assimilation of scientific knowledge by a practical man for whom religion remained the central preoccupation of his life. We can compare him with a contemporary Bristol schoolmaster, John White, also devout as his diary reveals, who gave his students "A Train of Definitions according to the Newtonian Philosophy." White apparently possessed no mystical tendencies, and one of his major intellectual interests, if we can rely on his diary, was in the new science. His Newtonian definitions closely followed the outline of any number of the lecture courses described earlier, and in his classes he dwelt at length on gravity, pulleys and levers, the laws of motion, hydrostatics, and electricity, in that order.[11] By the mid-eighteenth century such classes were commonplace not only in Bristol's grammar school (secondary or high school), but also in a variety of technical and mathematical schools intended to equip boys for practical careers.

Indeed contemporary clerical opponents of Newtonian science, and Bristol and its environs were particularly rich in one such group called the Hutchinsonians,[12] believed by 1774 that the new science, which they regarded as a threat to Christianity, had permeated the minds of the landed gentry in deepest Somerset.[13] In that year among the books proposed for purchase by the Bristol Library were Benjamin Franklin's *Letters on Electricity*, the *Philosophical Transactions of the Royal Society*, s'Gravesande's and Voltaire's works on Newtonian philosophy, and a materialistic essay from the French Enlightenment, Helvetius's *De l'esprit* (1748). In addition James Ferguson also donated technical works to the Bristol collection.[14]

The Problem of the Bristol Port

One wonders if Richard Bright, or any of his fellow merchants in the Society of Merchant Venturers, requested those books, or had been taught by White, or had listened to Ferguson or debated with John Hutchinson.[15] If they had been taught by the Newtonian schoolmaster, they learned well—and none too soon. In the last decades of the century they were forced to bring their scientific knowledge to bear on a complex engineering problem centering on Bristol harbor, a matter critical to the maintenance of their prosperity and ultimately to the commercial as well as industrial future of the city. As we have seen a multiplicity of explanations of natural phenomena existed simultaneously amid Bristol's population—astrology, witchcraft, divine intervention, electrical principles, mechanical and Newtonian models—yet only the last would be brought to bear by engineers and merchants attempting to come to terms with the problems presented by the extreme tides of the River Avon and their effects on Bristol's harbor.

By the late 1750s the growth of commercial life and material consumption, which provided the money for transportation improvements and industrial development, began to overwhelm Bristol's harbor and rivers. The number of coastal and river craft sailing in and out of Bristol rose from an average of 900 a year in the 1750s to well over 1,700 a year in the 1770s.[16] But the exceptional tides—often over 40 feet—meant that when ships in the harbor unloaded their cargoes at low tide they literally sat on the mud banks, presenting a bizarre sight described by Alexander Pope as "a long street, full of ships in the middle and houses on both sides, [looking] like a dream."[17] But the dream quickly became a nightmare when ships listed over, causing loss of the cargo or damage to the vessel, or when the absence of water at low tide made possible the spread of serious fires leaping from docks to ships or vice versa. In addition the River Avon, the main access route to the harbor, was treacherous in places, and large ships had to be towed in by rowboats. In theory the corporation, that is, the town government of Bristol, was responsible for the maintenance and improvement of the river and harbor; in practice the council had delegated its responsibility to the Society of Merchant Venturers.

The society was a most elite body, composed of only the wealthiest merchants in the city and its environs, with an average annual membership throughout the century of fifty to sixty men. Among them were landed gentlemen, some with aristocratic titles; at a meeting in 1776, fifteen of the assembled possessed the title of baronet or above.[18] Great merchants and landed gentry had intermarried and secured common interests in England from at least the sixteenth century. In this Bristol society and the attempt it made to come to terms with its economic interests, and hence with the need to improve the city's transportation system, we can test the mechanical knowledge available to the mercantile and landed elites who controlled British political life in many places at the time of the Industrial Revolution.

The decision to proceed with plans for the improvement of Bristol's harbor was taken so slowly by the Society of Merchant Venturers that by the time the work was actually undertaken, in the first decade of the nineteenth century, the commercial lead had passed to Liverpool with its excellent system of new canals giving access to the industrializing Midlands. The explanation for the delay seems ultimately to lie in the extreme wealth of the society's members, who felt no economic need to compete at that moment with their distant northern rivals. In their relative inertia they might be compared to the great merchant families of Middelburg in Zeeland. There too, as we saw in chapter 7, the harbor needed serious work. Although the problem there was silting and not as technically complex as Bristol's, not enough was done soon enough.

Bristol's ruling merchants may also have feared that improvements to the port would enhance the wealth of local small industrialists who operated outside the city. They were competitors, not to be assisted. Not least, the "middling sorts" of citizens saw the tax levies needed for these schemes as just another instance of their having to pay out of their own pockets for the profits enjoyed by Bristol's mercantile elite.[19] Indeed Bristol was a socially troubled city in the eighteenth century. Many of the hand workers in the city and environs had been proletarianized decades before industrialization began on a large scale in the north and in the Midlands.[20] There were riots in the 1750s in the coal mines around Bristol; and in the 1790s magistrates ordered the troops to fire on rioting citizens. In this highly commercialized city only a small proportion of the population enjoyed the surplus capital derived from the flourishing Atlantic trade, and the extremes between rich and poor were particularly visible.

Amid these social and economic tensions the Society of Merchant Venturers took up the matter of transportation improvement. But how were mercantile gentlemen to proceed with matters as complex as tidal currents, sluices, new canals and dams, the possible installation of steam engines for drainage and the pumping of cleaner, less saline water to the city, and the additional problem of sanitation if the water, once trapped in the harbor for the benefit of the ships, should become stagnant and polluted by waste from the surrounding city? Any one of these problems might have been commonplace enough. Indeed elsewhere in the 1760s canals on reasonably flat terrain were being built by engineers, such as James Brindley, who possessed no sophisticated mechanical knowledge. But taken in total, Bristol's harbor and rivers posed one of the most difficult engineering problems of the century.[21]

Fortunately the rich archives of the city provide unique evidence of the extraordinary number of plans or schemes that were put forward from the 1760s onward by engineers and other natural philosophers. More important, the Bristol records allow us to follow the gentlemen merchants as they make their way through these extremely sophisticated (for their time) mechanical discussions and, most fascinating, offer their own opinions or even corrections of engineering plans. Like the contemporary members of Parliament who had interviewed the engineers concerned with Watt's engine, they took their mechanical learning for granted. But we should not.

The society possessed sufficient familiarity with the latest mechanical techniques to seek the services first of John Smeaton and then of William Jessop, probably the best civil engineers of their day. But for these merchants technical matters did not end there. In our own highly specialized world, where in the first instance all scientific knowledge has become the province of highly trained scientists or technicians, such a specialist, once chosen for his skill by industry or government, would be allowed to get on with the task of designing plans and executing them, provided these activities were carefully reported and always assessed in relation to cost feasibility. In the eighteenth century cost was certainly an important factor in all of the society's discussions of engineering proposals; but so too were the engineering plans themselves. The Society of Merchant Venturers became the arbiter of mechanical knowledge, with rival engineers and natural philosophers competing for approval of their plans, appearing before the subcommittee of the society concerned with these mechanical matters and never, as far as the records indicate, adopting a non-specialist vocabulary for the society, although one was certainly used when addressing the general public and trying to convince it of the accuracy and economy of a particular plan.[22]

In 1765 John Smeaton presented the society with "Proposals for laying the Ships at the Quay of Bristol constantly afloat, and for enlarging this part of the Harbour by a new Canal through Cannon's Marsh." To get a sense of this proposal and its degree of complexity, we must read a portion of it along with the society, and so I quote from Smeaton at some length:

> First: It is proposed to keep the water in the quay and new canal to the constant height of the 15 foot Mark upon the lowermost marked staff upon the quay next the Avon and by clearing away 2 or 3 feet of mud there laying to make from 17 to 18 feet of water. N.B. the 15 feet mark is about 6 feet below the top of the quay and about 4 feet below the spring tide high water mark of 24th and 25th January 1765 which, though not the largest, were never the less accounted considerable tides.

And that was just the beginning; the document goes on:

> Secondly: It is proposed to dig the new canal as far as the sluices so deep as to make 18 feet water therein at the said proposed level, and to make the same at least 100 feet wide in the clear.
>
> Thirdly: to drop the tail of the new canal into the River Avon at the bottom of Cannon's Marsh just above the Glass house.
>
> Fourthly: to construct two separate sluices, one as near as conveniently can be to the River Avon upon the tail of the canal, the other at the distance of 400 feet from the former further within the canal, both these sluices to be furnished with two pairs of pointing gates, one pair in each sluice pointed to the land, the other to seaward. The width of the chamber of space intercepted between the two sluices

> to be 60 feet and the width of the sluices to be capable of taking in the largest ships that use the port which I suppose will be done by an opening of 30 feet wide.
>
> Fifthly: The thresholds of the upper sluice to be laid at the depth of 18 feet below the constant water, that is even with the bottom of the canal but the floor of the chamber between the two sluices, as well as the threshold of the lower sluice, to be laid as low as the bottom of the river in the shallowest place below the tail of the canal.
>
> Sixthly: These things being executed . . . the present mouth or opening of the River Froome into the Avon to be stopped up by a solid dam of earth; furnished, however, with such draw hatches, as may be necessary to assist the hatches in the gates of the sluices in discharging the freshes of the River Froome in rainy seasons; but yet as so to make a communication for all kinds of carriages from the back and quay down along side of the new canal between the same and the river.
>
> Seventhly: The whole of the new work to be wharfed with stone . . .
>
> Eighthly: To erect draw hatches at the new bridge at the head of the quay, capable of retaining the water behind them, when the water in the new canal and quay is let off.[23]

And so it went on; the above being by no means the entire contents of the proposal. These proposals were accompanied by drawn plans, making visible the proposed changes, and they were followed by an explanation of how the entire system would operate. Its success depended on correct estimates of the volume and therefore the weight of the water, and the force and pressure of the tides to be admitted or closed out of the canal chambers, permitting the vessels to enter and leave them safely.

Building a canal and sluices on relatively flat terrain was not in itself extraordinary and there were time-worn procedures for doing it. What is important in these Bristol proposals is the sheer size and complexity of the problem of controlling two rivers and the tides in such a manner as to keep the harbor constantly filled with water. The cost estimated by Smeaton for the implementation of his plans was 25,000 pounds, and "at a numerous meeting of the merchants" he was unanimously thanked for his proposal. But the matter of Bristol harbor only began there.

Another mechanist, William Champion, who was a successful local industrialist in part because he was the first person to develop a chemical procedure for "making" brass (a compound of zinc), put forward his own set of complex plans for the harbor. He proposed not only a dam for the River Froome, but also one for the River Avon, and he added a further proposal for erecting a steam engine "to serve the city with water at a less expense than the present great water wheel aqueducts."[24] In so doing he introduced a subject that would plague these schemes for Bristol's harbor in the years to come, and added yet another scientific question that would require expert assistance and on which the society would also have to pass judgment. If the water in Bristol's harbor went too low or was trapped by these new dams, the result would be stagnation and pollution from the sewers that dumped waste into the harbor, and with that would come disease. The sewer system in Bristol was already poor; these schemes for a wet harbor would, it was argued, make it worse.

In appealing for the assistance of engineers the society opened itself to a flood of conflicting proposals from engineers, mechanists, or natural philosophers, who, as one projector put it, did not "write this with any hope or intention of getting a job. I do not profess engineering, but having been many years a teacher of experimental philosophy, my experience in hydrostatics gives me full confidence in the above effects."[25] The conflicting babble of scientific tongues played into the hands of those merchants who opposed any further improvement in the harbor, who found the existing situation sufficiently comfortable as to warrant their doing nothing. The matter dragged on for many years, and then came the American Revolution. The concomitant trade embargo imposed against the colonies intervened to depress Bristol's economy and to delay the issue of the harbor into the late 1780s.

By then, however, the Society of Merchant Venturers had within its own leadership a man of the new science, Richard Bright, F.R.S., who had been taught chemistry by Watt's friend, Priestley, as well as the latest mechanical philosophy at the Dissenting academy at Warrington.[26] Bright was a merchant capitalist and a landed gentleman who was worth 70,000 pounds in personal and landed wealth in 1777. In other words, Bright was a member of the urban gentry and oligarchy whose Whiggery complemented his scientifically grounded faith in the necessity of progress and improvement. He made the project of promoting Bristol's harbor his own personal crusade, and he enlisted in this struggle all his natural philosophical talents and contacts, as well as his political influence.

Bright saw more clearly than many of his contemporaries in the society, of which he was for a time secretary, that its profits hinged on the improvement of the harbor, that this was essential for Bristol to compete effectively with Liverpool.[27] He sent copies of new engineering proposals to his friends in the Manchester Literary and Philosophical Society for their approval, and in so doing he demonstrates to us his identification with a scientific society at the forefront of the industrial application of science.[28] He also sought out optimistic medical opinion that would contradict the opinion of medical experts who had judged various plans for the harbor to be dangerous to the sanitation system of the city. At the end of one such favorable medical report, offered by a Dr. Falconer, F.R.S., Bright penned in his own expectation "that nothing will impede the prospect of our improvement."[29]

Once again, as in the 1760s, the society was deluged by contradictory engineering plans, but in the 1780s new factors were evident. Not only were the plans more complex and more expensive to implement, but conflicting medical evidence had also become a public issue. The professionalization of scientific knowledge was everywhere evident; as one fellow of the Royal Society commented to Bright: "I will decline giving any answers to the queries you sent relating to the proposed undertaking at Bristol as none but Physicians are proper judges of many of them and the Engineers they [the Society] have consulted are much better judges of the remainder."[30] Once again the best engineers were enlisted, only now the society could no longer put off its decision.

In the course of the ensuing debate within the society its secretary, now another local merchant, Jeremiah Osborne, was instructed to query the actual theoretical principles on which Smeaton's hand-picked successor, William Jessop, based his engineering plans. Obviously, the discussions within the society had ranged to the very philosophy of nature that lay at the basis of these complex proposals. Jessop's response was a short lecture on mechanics, just the sort of basic scientific information that was disseminated up and down the country by the itinerant lecturers; only Jessop confesses that he has forgotten some of the fine points of the new mechanics:

> To make you completely acquainted with the Principles on which the Calculations are founded, respecting the discharge of Water over cascades or through orifices would take me much time and some study; for having in the earlier part of my time endeavoured to make myself acquainted with these Principles, and having been once satisfied with the result, I have, as most practical men do, discharged my memory in some measure from the Theory, and contented myself with referring to certain practical rules, which have been deduced therefrom, and corrected by experience and observation. But I can in a few *Weeks* inform you in the general principle on which these calculations are grounded. It has been found by experiment that a heavy body falling from rest will descend about 16 feet in a second of time; and that the velocity acquired at the end of that second is such, as proceeding on at the same rate, or without any acceleration, would carry it on in equal time through a sphere of double the height which it fell from, or 32 feet in a second; that bodies falling from different height acquiring velocities in proportion to the square root of these heights. And that water in going through orifices runs with the same velocity as would be acquired by a heavy body in falling through a space equal to the height of the waters surface above the virtual center of the orifice. So that while a height of 16 feet would produce a velocity of 32 feet in a second, a height of 9 feet would produce a velocity of 24 feet in a second or as 4 the square root of 16 is to 3 the square root of 9. But as this is the greatest possible velocity that can be acquired; it is found in practice to vary from this rule, conformable to a variety of circumstances, such as the shape of the orifice, the manner in which the water is introduced into the mouth of it, the friction in passing through it, etc. and so that while in some cases it might not discharge about 2/3 of the full quantity, in others, it fluctuates between that and the full quantity, in degrees which only experience and nice observation can nearly ascertain. So in flushing over cascades it is found that the velocity is something less than would be due to a height equal to half the thickness of the sheet of water, for instance, in a sheet of 18 inches thick, the velocity would be such as would be produced from falling a height of about eight inches or about a fifth part of what would be due to 16 feet. If these hints will threw light upon your enquiries it will give pleasure to your most obedient servant, W. Jessop.[31]

Jessop could have been copying from a text by Galileo; in fact he was remembering what he had learned from a standard Newtonian textbook of the eighteenth century.

The basic principles of the mechanical philosophy, coupled with observation and experimentation—as described by Jessop—along with the professionalization of mechanical science and its application by practicing engineers such as Jessop, had come to be accepted by entrepreneurs and merchants alike. These are the elements that went into making what we may reasonably describe as the scientific culture possessed by the industrial mind. Armed with this mechanical understanding of nature, and willing to give credibility to the superior knowledge claimed by professional mechanists, merchants, entrepreneurs, and industrialists could, and did, make decisions that formed an essential part of the history of the early Industrial Revolution.

By the 1790s the merchants of Bristol were finding it necessary to comprehend engineering plans that described nearly 200,000 pounds worth of alterations and land purchases, which were now needed to improve the harbor. Jessop's technical drawings were accompanied by detailed descriptions:

> AB.—is a cylinder of 5 feet diameter open at the bottom or base on which it stands, closed at the top, and perforated by four large apertures at the sides, or in the circumference. The cylinder, C, is suspended to a beam which moves on a Center D, and at the other end of this beam is suspended a cast iron bucket E, which moves up and down in a well. When the water rises above its common height, it will flow through a pipe F, and fill the Bucket, causing it to preponderate and raise the cylinder C. . . . I must observe that as the pressure of the water counteracts itself on all sides of the cylinder C, it will move under any head of water without much friction. (I believe this cylinder was the invention of the ingenious Mr Westgarth, and applied, on a small scale to an engine for raising water from mines i.e., a steam engine and has been used by Mr Smeaton for a similar purpose.[32])

We might be tempted to imagine that in the face of all this technical verbiage the merchants who favored the harbor, led by Bright, simply consigned themselves into the hands of Jessop, the most famous and accomplished engineer they could find, a man whose prestige could sway public opinion and stand up well under parliamentary cross-examination. For in the first instance, Parliament would have to legislate on the basis of the plans presented to it by the society; an act was required whenever private land was to be requisitioned and purchased, or businesses were threatened by the diverting of local water supplies, or money was to be raised by the selling of shares to the public. But minutes of the society, as well as Bright's private notebooks, show the merchants themselves discussing engineering plans in some considerable detail, sitting through complex sessions with engineers—discussing water levels in the harbor, or weighing ecological objections based on health considerations and sewage control, or deciding the merits of placing steam engines on various docks.[33] The merchants became the final arbiters of scientific knowledge and its application. They were capable of assuming this role because they had learned enough about the mechanical philosophy, through reading or lectures or personal experimentation, that they were able to apply it at this level of sophistication.

The point can be nicely illustrated. In 1792 Jeremiah Osborne had to arrange with a London engraver for the printing of Jessop's hand-drawn plans. But he advised the engraver to hold off drawing one section and noted his disagreement, based on direct observation, with Jessop's figures:

> One of the observations I have made on the sections and which I wish cleared up before that part of the Plate is begun to be engraved, is that the 15 foot level on the quay appears to be rather above the 14 foot mark on Hillhouse Dock gauge in the section of the Avon; but I believe that the bottom of that gauge should stand six feet higher than it does in the drawing. . . .[34]

As it turned out Osborne's confidence in his knowledge was well founded; he had caught an error in the engineering plans of William Jessop.[35] The society also took the advice of one of the medical experts who wanted two sluices on each side of a dam to prevent flooding.[36]

Bristol in the Nineteenth Century

Finally, in 1804, work on the floating harbor began; the improvers had won out. The problems of class resentment so commonplace in Bristol's history continued, and predictably the company formed to execute Jessop's plans became an object of resentment. It alienated many local citizens by its secretive and high-handed manner. One irate citizen warned the directors "you may chance to get a ducking, in that stagnant lake, to which your own enlightened conceptions have given birth."[37] Bristol's lead in industrialization, based on the iron foundry business of Abraham Darby and its local brass and glass industries, nevertheless gave way to its northern rivals.[38] Its merchants had made their decisions on the basis of technical knowledge, but they had done so too slowly. Industrial dynamism now came in particular from Birmingham, where the application of steam power to the glass and brass industries was coupled with a better canal system, and hence glassware and "toys"—that is, all small metal objects—could be made more cheaply and transported more efficiently. Yet it should not be imagined that Bristol's elite lost out entirely; the city continued as a banking and commercial center of importance throughout the nineteenth century. It remains so to this day. By 1825 it possessed its own literary and philosophical society, where lectures on natural philosophy, very similar to those that abounded earlier in other eighteenth-century provincial cities, were commonplace and offered the latest scientific knowledge on electricity and magnetism.[39]

The Politics of Mechanical Application

As we can now see, a multitude of factors went into successful industrial and commercial decision making: the ability to comprehend increasingly complex technical knowledge through a mastery of basic mechanics, the presence of entrepreneurs willing and able to push a particular project through Parliament, MPs learned enough to understand the technical details, sufficient surplus capital from large and small investors to be invested in shareholding companies, and never least, the availability of laborers to dig the canals or feed the furnaces. Scientific knowledge was only part of the story in Bristol and elsewhere, but it was a vital part.

The industrial application of scientific knowledge constitutes historically the single most important use to which Western science has ever been put, and this occurred first in England in the second half of the eighteenth century. Obviously there were many moments in the early Industrial Revolution when scientific knowledge, particularly in the area of transportation, mattered not at all.[40] Indeed the basic techniques of simple canal building and water control had been used in ancient China as well as in seventeenth-century France and Holland. But in the late eighteenth century in England these techniques first became commonplace and eventually mechanically sophisticated beyond recognition.

Feeding into the energy of early industrialists was their political vision. As early as the 1760s reformers seized on canal building as the fulfillment of public interest at the expense of "self-interested motives and local views."[41] The pioneers of early canal development, without which the coal needed by the industrial centers would never have been made available cheaply or in sufficient quantity, saw themselves as reformers, in opposition to monopoly interests. The projectors and developers, whether involved in transportation or industry, could employ the political language of the opposition when their interests were thwarted. They could talk like "the country" when they were angry. When placated they largely identified with the Whig oligarchy, and they sought out and on the whole received its support. This point needs to be stressed as we survey the uses to which science was put in this period; namely, that its successful application required support from the landed classes that controlled both houses of Parliament. From the 1760s into the 1790s all evidence suggests that the Whigs (and not the rival Tories) were identified with industrial innovation and canal building.[42] Once again we find secular-minded elites drawn to science as a way of increasing their wealth and power; only in late eighteenth-century England, political stability and centralization permitted this application to become national in character, with consequences that would soon become global. Once again, we are reminded of the far-reaching effects of the Enlightenment.

Canals

By the 1790s we can see how technically sophisticated the revolution in transportation had become by surveying the records of canal companies from the period. Once again, those records illustrate the depth of mechanical knowledge to be found among gentlemen of land, industry, and commerce. The example of the Bristol merchants can be multiplied across the country, although few towns, counties, or cities, to be sure, faced the complexity of engineering problems presented by Bristol's harbor. Sometimes the records also clearly indicate that technical knowledge, which could have been used to great advantage, was simply absent. The results were often disastrous, as lives were lost and money wasted on canal projects that were badly designed.[43]

As the mania for canal building swept the country in the last decade of the eighteenth century, hundreds of canal companies were formed. In partnership with them engineers like Jessop grew to be wealthy men. In the north of England the same engineers who designed the canals frequently invested in the companies or became industrialists in their own right.[44] Indeed there the linkage between industrial development and canal engineering was perceived almost immediately. In case Parliament or local interest groups needed reminding of the necessity for new systems of transportation, natural philosophers and radical Whigs like Erasmus Darwin stood ready to admonish them and to urge on the projectors.[45]

Not every industrialist of this early period supported canal building or bothered to understand the mechanical principles used by canal engineers in making their plans or by mechanists such as James Watt and Matthew Boulton in designing and installing their steam engines. Even renowned industrialists—Richard Arkwright, for example, famed in part for his self-taught mechanical ability—opposed improvements that might threaten their profits.[46] And, of course, there was competition from older monopolies where earlier improvements had served to entrench their wealth and position. But where a neat fit existed between profit and improvement, we once again see merchants, landed gentlemen, industrialists, engineers, and natural philosophers allied in common self-interest, abetted by the mechanical knowledge they held in common.

The landed capitalists and factory owners who hired an engineer, as for example Philip Gell, and the promoters of the Cromford Canal in Derbyshire, who hired William Jessop in 1788 and asked him to draw up plans for it, sought always to get the best mechanical knowledge that money could buy. They knew that whomever they hired would have to go before parliamentary committees, where lords could be found who were "for teaching anybody [their business], the Bishops', religion; the Chair, the law; and now Jessop as an engineer."[47] In one case a mere "teacher of Mathematics and Philosophy," of whom no one had ever heard, was brought before the House committee to testify against the canal. When he managed to get his calculations close to Jessop's, the promoters were genuinely surprised.[48] They had come to accept the professionalization of scientific knowledge of a mechanical sort, to rely solely on engineers, preferably famous ones—if they could be found. The promoters sat through parliamentary cross-examinations of experts, following in detail their estimates of the weight of water lost through the diversion of river water into a canal. In some cases the projectors understood more about the mechanics involved than did the lawyers who were doing the questioning.[49]

The knowledge was necessary for lobbyists if their case was to be won against those who felt they stood to lose water power for their factories as a result of diversion by the canal and who brought in their own mechanical experts to argue against the proposed canal bill. At moments in the hearings less knowledgeable witnesses on both sides damaged their client's case. We can watch the self-satisfied response of canal promoters who happened to possess greater mechanical knowledge and used it more effectively before these parliamentary committees. The promoters and engineers alike recognized the necessity of making actual "experiments" to be able to present the most accurate information before Parliament.[50] On those occasions lords on the committee could be seen "constantly taking notes and making good observations and asking really very pertinent questions."[51] Others, however, admitted that they simply "did not understand it." Mechanical knowledge among the English aristocracy was widespread but not universal.

The best guide to the depth of knowledge and its application lies in the parliamentary records of committee hearings. These document moments of decision making in the early Industrial Revolution that reveal the key role played by central government, particularly in the transportation side of that revolution. Without the canals, harbors, and turnpikes approved by parliamentary bills, the revolution would have been stillborn. Many factors went into those parliamentary decisions: political pressure brought by local interests, the reputation of the engineers, and outright bribery; but not least was the mechanical knowledge of the committee members and their enlightened faith in the value of improvement.

In the hearings for the Cromford Canal, for example, the committee pressed the matter of the effect of the lost water pressure on the profits of some factory owners and the concomitant danger of increasing unemployment in the district. Time pieces were fitted on water wheels to supply that evidence, while expert witnesses were called in to testify on the relationship between water pressure and the power of turning wheels, to explicate as one natural philosopher put it, "by known hydrostatick Principles agreed to by all authors,"[52] or to give evidence based on conversations "with many who are scientific, and I have read most books upon the subject."[53]

The questions and answers, the flow of discussion on these occasions tells us much about the use of mechanical knowledge in the period and, most important, about the ease with which it was assimilated and used. Witness this cross-examination of one Richard Roc, "a surveyor and teacher of mathematics," questioned by the House of Lords committee:

HOL: You are a surveyor and teacher of mathematics.
RR: Yes.
HOL: Suppose two shutters of a Mill of 4 feet each are elevated 17[inch] with 4 feet over them. What quantity of water flows in a minute?
RR: 278 Tons per minute.
HOL: That is when the water is 4 feet high from the bed of the river?
RR: Yes.
HOL: Do you speak from observation or calculation?
RR: I calculated it from the Dimensions given by Mr Snape.
HOL: How do you ascertain it?
RR: By known Hydrostatick Principles. . . .
HOL: Then can you say that with a given velocity a certain quantity of water flows in a minute?
RR: Yes.
HOL: Do you make that Calculation upon a supposition that the water is not retarded by water contiguous to the wheel?
RR: Certainly.

HOL: Then in point of fact, supposing the water to be retarded, would not the quantity be less?
RR: Yes, very much less.
HOL: Have you ever measured the River Derwent at Cromford Bridge?
RR: Yes.

The questions proceeded from general theoretical principles to Cromford's proposed canal in particular and ended, once again, on the theoretical:

HOL: What is the rule by which you stake your calculations?
RR: From the height a body falls in a second of time, it is said to fall 16,'1[inch] in a second, and then acquires a velocity which will take it through twice that distance, I then proportion it by the square root of the height.[54]

When the lords turned their attention to the vexed question of the interests of factory owners dependent on water power for their profit and afraid that the canal might decrease that power, engineers in favor of the project presented elaborate mechanical arguments to refute those objections.[55] There are poignant moments in these deliberations—for example, when a mill foreman whose livelihood had already been adversely affected by the canal, is asked:

HOL: Could you increase the power of these wheels still more if you tried?

The foreman's answer reveals that he simply does not understand the mechanical principles that have already been applied to the detriment of his water supply:

Fore: Upon my word, I do not know, for the power of a wheel is what I do not understand.[56]

Doubtless there were many mill owners who did not understand the power of a wheel in mechanical terms. Indeed one of the myths about the Industrial Revolution is that few, if any, of its tinkering progenitors understood the science most commonly disseminated throughout the eighteenth century. We know of a few such early industrialists who apparently possessed no theoretical knowledge, yet it is also possible to find historical evidence that significantly contradicts the myth.

Steam Engines

If we go to the lead mines of Derbyshire, to the industrial center of economic development in the late eighteenth century, there too we find evidence of mechanical knowledge applied by mine owners who possessed no known formal or academic scientific education. The business of deciding to install a steam engine was tricky; bankruptcy quickly followed if the wrong engine was installed in the wrong place. In 1794 a steel firm failed immediately after installing an engine; it was "too heavy [too expensive] a concern."[57] Yet it was obvious from as early as the 1720s—as Desaguliers, Martin Clare, and the scientific lecturers pointed out—that the steam engine had enormous potential, particularly in mining, where its power could be used to extract water from the subterranean tunnels, which were always prone to flooding. By the 1770s innovations in coal mining, at least in the Derbyshire area, were being introduced by men with experience in mining who also now possessed geological knowledge as well as an understanding of the Newcomen engine and soon the significant improvements made on it by Watt.[58] These were lead merchants, such as Benjamin Wyatt and John Barker, who had a wide knowledge of their industry and of the principles of commercial life, but also, at least in some cases, of the technical and theoretical aspects of the mechanical philosophy.

The historical literature about the early Industrial Revolution tends to describe the use of steam power in production as if its application had been an automatic process. The benefits of the engine were perceived immediately; if capital could be found by owners they simply called in engineers who installed the necessary engine. The engineers knew mechanical science, the owners did not—or so the argument goes. They made the decision to install solely on the basis of economic considerations, cost of fuel, labor, and so forth, and their own relationship to the machine and its power was largely passive and inarticulate. But many of these owners were smarter than historians have allowed them to be. They knew that there were many variables and so much at stake that it behooved them to think well on these machines, to understand what they could and could not do. Lead mine owners recorded their anxiety about what a steam engine could do: "How this may answer no one can say; so much depending on accidents."[59]

Where we can find evidence of the process of consultation about steam engines between owners and engineers, there we find the "experts" speaking in considerable technical detail to their employers, complete with mechanical drawings.[60] Sometimes encouragement to proceed with installation came from Sir Joseph Banks himself. As president of the Royal Society he had an interest in the application of the mechanical science that so many of its Fellows had been instrumental in disseminating.[61] He had also invested in mineral veins from which he sought to profit. Aside from taking advice from Banks or others, mine owners went about the countryside observing steam engines at work and then told the engineers what they wanted. They sought estimates and again, like those Bristol merchants, had to choose among designs.[62] They had to comprehend the technical data laid before them and, as did William Wyatt who inherited Benjamin Wyatt's successful lead mining business, they knew in the end what they wanted:

> I have through the hand of mr Snyed received your estimate for a steam engine of sixty horse power but before any further steps are taken I will thank you to furnish me with the following particulars, viz., the diameter of the cylinder, the construction and size of the boilers, the length of the beam, the weight of the ply wheel, the number of strokes to work per minute, the quantity of water lifted 240 yards deep at each stroke, the diameter of the working panels and the quantity of coal

> which will be consumed in every 24 hours. Perhaps we can use a plunger to advantage as our pitt will be a great depth. Please write me by return of post as I am anxious to determine about an engine soon as possible. Are there any very superior pumping engines to be seen at work in your county, if so I should like to see them—I mean engines that do a deal of work with a little fuel.[63]

Levers, beams, pulleys, and weights—the stuff of the experiments by which the new mechanical philosophy was illustrated in books and lectures—when combined with the profit motive, cheap fuel, and access to transportation for coal and materials as well as consumer goods—transformed the means of production first in England, then in western Europe. When we ask ourselves why this happened first in Britain, we must remember the English Revolution and the relationship it forged between its landed and commercial beneficiaries and the new science, both as ideology and congenial practice. By the late eighteenth century other Western elites aided by progressive intellectuals in various Continental countries aspired, as we have seen in the case of the French revolutionaries, to the industrial application of mechanical science. But political factors—perhaps more critical than the presence of coal or surplus capital or labor—would delay that process in France and The Netherlands until the nineteenth century. By then the English industrial model existed in reality, no longer solely in the dreams of the natural philosophers. It was in part the consequence of a series of discrete decisions made by entrepreneurs who aspired to scientific knowledge because self-interest and the enlightened ideology of improvement demanded that they have it.

The historical development of scientific culture from Copernicus to the steam engine illustrates that the widest possible dissemination of scientific knowledge, the democratization of learning, will do more to foster an indigenous creativity in matters of application or innovation than will the importation of foreign experts or the maintenance solely of elite cadres. The language of science must be capable of absorption by thought processes also expressive of other commonplace elements of a culture or society. If an aristocratic and closed system dominates in the academies, or if clergymen more concerned with orthodoxy than with material progress control the schools and universities, then theory may dominate over application, or science in general will receive less attention. If oligarchs grown rich through commerce control local education and sponsor academies, as was the case in the Dutch Republic, then innovators with industrial ambitions may have no place to implant their values or, as happened in the eighteenth century, to foster mechanics. If religious beliefs promote suspicion toward foundational ideas in science, then its cultivation will be sporadic, confined to select academies such as we find in Turin. The open, public science commonplace in eighteenth-century Britain may not be highly original science—although that is by no means foreclosed—but it can be innovative in application, widely adaptable to profit seeking. The framing of nature cannot be divorced from other experience. In that sense, the language and practices of science are also socially anchored, and true creativity, relevant to its time and place, is rooted in social experience as transformed by ingenuity.

Notes

Introduction

1. See Samuel Y. Edgerton, Jr., *The Heritage of Giotto's Geometry. Art and Science on the Eve of the Scientific Revolution*, Ithaca, N.Y., Cornell University Press, 1991.

2. Here I wish to pay tribute to a text that has stimulated my thinking when it was a dissertation and now as a book, Richard Biernacki, *The Fabrication of Labor. Germany and Britain, 1640–1914*, Berkeley, University of California Press, 1995.

3. Bridget Hill, *Women, Work, and Sexual Politics in Eighteenth-Century England*, New York, Basil Blackwell, 1989, pp. 63–68.

4. For a very helpful discussion of the different meanings of all these terms see Ronald Kline, "Construing 'Technology' as 'Applied Science': Public Rhetoric of Scientists and Engineers in the United States, 1880–1945," *Isis*, 86 (June 1995): 194–204.

5. For how France was being swamped by British textiles by 1789, and for the inferiority of French techniques in iron and steel, see Jean-François de Tolozan, *Mémoire sur le commerce de la France et de ses colonies*, Paris, Moutard, 1789; Bibliothèque Nationale, microfiche V.17731. For memoirs making similar points and for the mixing of cultural and economic arguments see the vast collections of the Archives nationales, Paris, in particular F12 677; F12 661 and examples of the somewhat desperate search for cultural explanations: "Reponse du Sieur Clicquot Blervanche, April 1778," where it is thought that French Protestant refugees are the key to British success; F12 647–48, 1768 memoir by Dubroeuil, where Jews are blamed for trouble in the Lyon textile industry. For the economics of English coal and its superiority, see F12 724, memoir of 28 August 1789.

6. Patrick O'Brien and Roland Quinault, *The Industrial Revolution and British Society*, Cambridge, Cambridge University Press, 1993, pp. 13–14.

7. For an intelligent statement of what the British Industrial Revolution means, see David S. Landes, "The Fable of the Dead Horse; or, The Industrial Revolution Revisited," in Joel Mokyr, ed., *The British Industrial Revolution*, Westview, Conn., Westview Press, 1993, pp. 132–70. A French minister in 1789 estimated that two-thirds of the cost of manufacturing cotton in France was in labor costs; AN, microfiche V.17731, mémoire by Tolozan. By this date France is a net importer of cotton, much of it British.

8. Archives Nationales, Paris, F12 661 "Mémoire du Sieur Holker fils sur les fabriques d'Aumalle, Amiens et Abbeville," no date, but from the 1780s. A fascinating attempt to calculate per capita income: Holker thinks there are 26,673,000 people in France with a per capita income of 115 livres 7s 8d, of which 21 l. comes from industry.

9. Thomas Young, *A Course of Lectures on Natural Philosophy and the Mechanical Arts*, 2 vols, 1807; vol. 1, p. 250.

10. See Timothy Claxton, *Memoir of a Mechanic*, Boston, 1839. I owe this reference to Joyce Appleby.

11. See Margaret Bryan, *Lectures on Natural Philosophy: The Result of Many Years' Practical Experience of the Facts Elucidated*, London, 1806; and James A. Epstein, *Radical Expression. Political Language, Ritual, and Symbol in England, 1790–1850*, New York, Oxford University Press, 1994. The novel is Elizabeth Gaskell's *Mary Barton*, the opening of chap. V, and I owe the reference to Ruth Perry.

12. But by the 1660s experimenters were interested in the application of steam; see Richard L. Hills, *Power from Steam. A History of the Stationary Steam Engine*, Cambridge, Cambridge University Press, 1989, chap. 2.

13. A wonderful description of Birmingham can be found in the diary of Chrétien G. Malesherbes, "Voyage en Angleterre, 1785," Philadelphia, American Philosophical Society, MS B/M 291. Cf. Gordon E. Cherry, *Birmingham. A Study in Geography, History and Planning*, New York, John Wiley & Sons, 1994, chap. 3. For a guide to manufacturing activity in the town see K. J. Smith, ed., *Warwickshire Apprentices and Their Masters 1710–1760*, Oxford, Dugdale Society, 1975.

14. See Richard Margolis, "Matthew Boulton's French Ventures of 1791 and 1792; Tokens for the Monneron Frères of Paris and Isle de France," *British Numismatic Journal*, 58 (1989): 102–9.

15. See Archives nationales, Paris, MS F12 677C, letter of the spy Le Turc (to Tolozan?), 4 Sept. 1786; see also the memoir of 27 Jan. 1786 for the quotation and a detailed description of the division of labor in mining, which included women. In a letter of 4 Feb. 1788 he says that even in three years of work an English worker does not see "un metier assemblé."

16. But no one should forget that A. E. Musson and Eric Robinson taught us to think in new ways in *Science and Technology in the Industrial Revolution*, 1969, second printing with foreword by this author, New York, Gordon and Breach, 1989. The present book, like *The Cultural Meaning*, builds on their work.

17. For an excellent comparison with Chinese culture, see Edgerton, *The Heritage of Giotto's Geometry*, cited earlier. Another valiant attempt at comparison can be found in Toby E. Huff, *The Rise of Early Modern Science. Islam, China, and the West*, Cambridge, Cambridge University Press, 1993; it is not, however, up to date on what historians are now saying about Western science.

18. Betty Jo Teeter Dobbs, "Newton as Final Cause and First Mover," *Isis*, 85 (1994): 633–43. On some questions Professor Dobbs and I differed slightly. How I wish she were still alive to disagree with me.

19. *Letters of Josiah Wedgwood, 1762–1772*, London, 1903, p. 165; see also p.24.

20. For the early use of the term (1799) "Industrial Revolution" see David Landes, "The Fable of the Dead Horse; or, The Industrial Revolution Revisited," in Joel Mokyr, ed., *The British Industrial Revolution*, Westview, Conn., Westview Press, 1993, pp. 133–34.

Chapter 1

1. Quoted in Jean Dietz Moss, *Novelties in the Heavens. Rhetoric and Science in the Copernican Controversy*, Chicago, University of Chicago Press, 1993, p. 33.

2. Michael Adas, *Machines as the Measure of Men. Science, Technology, and Ideologies of Western Dominance*, Ithaca, N.Y., Cornell University Press, 1989, chaps. 1 and 2.

3. Owen Gingerich, *The Eye of Heaven. Ptolemy, Copernicus, Kepler*, New York, American Institute of Physics, 1993, p. 200.

4. Galileo Galilei, *Two Chief World Systems*, trans. S. Drake, Berkeley, University of California Press, 1967, p. 207.

5. Ibid.

6. See the useful discussion in Fernand Hallyn, *The Poetic Structure of the World. Copernicus and Kepler*, New York, Zone Books, 1990, pp. 152–54.

7. Peter Kriedte, *Peasants, Landlords and Merchant Capitalists*, Leamington Spa, U.K., Berg Publishers, 1983, pp. 57–64; and for printing Elizabeth Eisenstein, *The Printing Press as an Agent of Change*, 2 vols., Cambridge, Cambridge University Press, 1978.

8. Letter of March 1615 from Galileo to Father Dini; reproduced in Richard J. Blackwell, *Galileo, Bellarmine, and the Bible*, South Bend, Ind., University of Notre Dame Press, 1991, p. 209.

9. For these developments see the extremely useful essay by Olaf Pedersen, "Galileo and the Council of Trent: The Galileo Affair Revisited," *Journal of the History of Astronomy*, 14, no. 39 (1983): 3–26. Some of the points made in this chapter were originally in a co-authored essay, "The Social Foundations of Modern Science: Historiographical Problems" by James R. Jacob and Margaret C. Jacob, presented to the American Historical Association, 1981.

10. Galileo Galilei, *Letter to the Grand Duchess Christina*, in Stillman Drake, ed., *Discoveries and Opinions of Galileo*, Garden City, N.Y., Doubleday, 1957, p. 177. Cf. " 'By an Orphean Charm': Science and the Two Cultures in Seventeenth Century England," in Phyllis Mack and Margaret C. Jacob, eds., *Politics and Culture in Early Modern Europe*, Cambridge, Cambridge University Press, 1986, pp. 231–32.

11. Drake, ed., *op. cit.*, p. 161.

12. Galileo, in Drake, ed., *Discoveries*, pp. 181–82 and 200 for all the quotations.

13. On the Jesuits see James M. Lattis, *Between Copernicus and Galileo. Christoph Clavius and the Collapse of Ptolemaic Cosmology*, Chicago, University of Chicago Press, 1994.

14. Quoted in Stillman Drake, ed., *Galileo Galilei's Dialogue Concerning the Two Chief World Systems*, Berkeley, University of California Press, 1967, p. xxv.

15. See E. A. Gosselin and L. S. Lerner, "Galileo and the Long Shadow of Bruno," *Archives internationales d'histoire des sciences*, 25 (1975): 222–46. The most famous interpreter of Bruno remains Frances Yates, *Giordorno Bruno and the Hermetic Tradition*,

Chicago, University of Chicago Press, 1964. On practical mathematicians advocating experimentation see J. A. Bennett, "The Mechanics' Philosophy and the Mechanical Philosophy," *History of Science*, 24 (1986): 1–28.

16. Mario Biagioli, *Galileo Courtier. The Practice of Science in the Culture of Absolutism*, Chicago, University of Chicago Press, 1993. His approach is nicely summarized in Mario Biagioli, "Scientific Revolution, Social Bricolage, and Etiquette," in Roy Porter, ed., *The Scientific Revolution in National Context*, Cambridge, Cambridge University Press, 1992. Cf. Olaf Pedersen, "Galileo and the Council of Trent: The Galileo Affair Revisited," *Journal for the History of Astronomy*, 14, no. 39 (1983): 6–24.

17. Here I am endorsing a modified version of the argument that is overstated but nonetheless important in Pietro Redondi, *Galileo Heretic*, Princeton, Princeton University Press, 1987.

18. Vincenzo Ferrone, *The Intellectual Roots of the Italian Enlightenment. Newtonian Science, Religion, and Politics in the Early Eighteenth Century*, Atlantic Highlands, N.J., Humanities Press, 1995, pp. 2–4.

19. Carlo Ginzburg, "High and Low: The Theme of Forbidden Knowledge in the Sixteenth and Seventeenth Centuries," *Past and Present*, no. 73 (November 1976): 28–41; cf. J. R. Jacob, " 'By an Orphean Charm,' " *op. cit.*, p. 240.

20. Now there is a book that develops the argument found in rudimentary form originally in *The Cultural Meaning*; see Julian Martin, *Francis Bacon, the State, and the Reform of Natural Philosophy*, Cambridge, Cambridge University Press, 1992.

21. Francis Bacon, *The Advancement of Learning*, in Arthur Johnston, ed., Oxford, Clarendon Press, 1974, pp. 70–71.

22. For gender identity in Bacon's thought, but with an argument that misses the reforming elements in his vision of a masculinity suitable for a new aristocracy and state, see Carolyn Merchant, *The Death of Nature: Women, Ecology and the Scientific Revolution*, San Francisco, Harper and Row, 1980. On the seventeenth century in general, and with an argument that misinterprets Descartes's understanding of mind and body, see the provocative essay of Susan Bordo, "The Cartesian Masculinization of Thought," *Signs*, 11, no. 3 (1986): 439–56.

23. Bacon, *The Advancement of Learning*, p. 42; see also p. 69.

24. On Bacon and the Apocalypse, see Katharine R. Firth, *The Apocalyptic Tradition in Reformation Britain, 1530–1645*, Oxford, Oxford University Press, 1979, pp. 204–7.

25. See J. R. Jacob, " 'By an Orphean Charm,' " in Mack and Jacob, eds., *op. cit.*, pp. 241–45. And see J. R. Jacob, "The Political Economy of Science in Seventeenth-Century England," in Margaret C. Jacob, ed., *The Politics of Western Science, 1640–1990*, Atlantic Highlands, N.J., Humanities Press, 1994, pp. 19–46.

26. On these points see Paolo Rossi, *Francis Bacon: From Magic to Science*, London, Routledge Kegan & Paul, 1968; and Charles Webster, *From Paracelsus to Newton: Magic and the Making of Modern Science*, Cambridge, Cambridge University Press, 1982.

27. P. M. Rattansi, "The Social Interpretation of Science in the Seventeenth Century," in Peter Mathias, ed., *Science and Society, 1600–1900*, Cambridge, Cambridge University Press, 1972, pp. 12–18.

28. See Rio Howard, "Guy de La Brosse: Botanique et chimie au début de la revolution scientifique," *Revue d'histoire des sciences*, 31 (1978): 325–26.

29. Alice Stroup, *A Company of Scientists. Botany, Patronage, and Community at the Seventeenth-Century Parisian Royal Academy of Sciences*, Berkeley, University of California Press, 1990, pp. 28–29.

30. Th. H. L. Scheurleer and G. H. P. Meyjes, eds., *Leiden University in the Seventeenth Century*, Leiden, Brill, 1975, p. 312; and see E. Kegel-Brinkgreve and A. M. Luygendijk-Elshout, eds., *Boerhaave's Orations*, Leiden, Brill and Leiden University Press, 1983, p. 177. See also *Nieuwen Atlas, Ofte Beschrijvinge van het noytmeer gevonden Eylandt van Bensalem*, trans. J. Williaemson, Dordrecht, 1656.

31. See also Franciscus Bacon, *De Proef-Stucken*, trans. Peter Boener, apothecary of Nijmegen—a translation of Bacon's moral and religious essays and his *Wisdom of the Ancients*. The copy at the University Library, Amsterdam, is from the library of Constantine Huygens. This is a very rare edition.

32. See *Neues Organon aus dem Lateinischen ubersetzt von George W. Bartoldy*, Berlin, 1793. See also Steven Turner, "The Prussian Professoriate and the Research Imperative 1790–1840," in H. N. Jahnke and M. Otte, eds., *Epistemological and Social Problems of the Sciences in the Early Nineteenth Century*, Dordrecht, Reidel, 1981, pp. 116–18.

33. Jack Morrell and Arnold Thackray, *Gentlemen of Science: Early Years of the British Association for the Advancement of Science*, Oxford, Clarendon Press, 1981, pp. 267–73. See also Richard Yeo, "An Idol of the Market-Place: Baconianism in Nineteenth Century Britain," *History of Science*, 23, no. 61 (1985): 251–98.

Chapter 2

1. See the account in Stephen Gaukroger, *Descartes. An Intellectual Biography*, Oxford, Clarendon Press, 1995, pp. 317–19; he argues that up until the condemnation of Galileo in 1633 Descartes was not that concerned about skepticism.

2. For a splendid discussion of the roots of seventeenth-century skepticism, see Richard Popkin, *The History of Skepticism from Erasmus to Descartes*, New York, Harper and Row, 1964, chaps. 1–3; p. 46 for the quotation.

3. Gaukroger, *op. cit.*, pp. 32–37.

4. Gaukroger, *op. cit.*, p. 33. The social argument, although not fully developed in this very helpful biography, complements the argument originally developed in *The Cultural Meaning*.

5. On Vanini, see Francesco P. Raimondi, ed., *Scuola e Cultura nella realtà del Salento. Annuario del Liceo Scientifico "G.C. Vanini" di Casarano*, n.p., Carra Editrice, 1994/95, pp. 9–62.

6. For a summary of the French reception (not particularly good on the Dutch side), see Nicholas Jolley, "The Reception of Descartes' Philosophy," in John Cottingham, ed., *The Cambridge Companion to Descartes*, Cambridge, Cambridge University Press, 1992, pp. 393–423. The volume is good for recent bibliography on Descartes.

7. T. de Renaudot, ed., *Recueil général des questions traitées et conférences de Bureau d'addresse*, 5 vols., Paris, 1658–66. For background see Geoffrey Vincent Sutton, "A Science for a Polite Society: Cartesian Natural Philosophy in Paris During the Reigns of Louis XIII and Louis XIV," Ph.D. dissertation, Princeton University, 1982.

8. Klaas van Berkel, *Isaac Beeckman (1588–1637) en de Mechanisering van het Wereldbeeld*, Amsterdam, Rodopi, 1983, p. 215. I am indebted to this work for my summary of Beeckman's career.

9. See Thomas A. McGahagan, "Cartesianism in the Netherlands, 1639–76," Ph.D. dissertation, University of Pennsylvania, 1976. On the English side of the story, see Alan

Gabbey, "Philosophia Cartesiana Triumphata: Henry More (1646–71)," in Thomas M. Lennon, et al., eds., *Problems of Cartesianism*, Kingston, Ontario, McGill-Queen's University Press, 1982, pp. 244–50. On the disputes among Dutch Calvinists, see J. van den Berg, "The Synod of Dort in the Balance," *Nederlands archief voor kerkgeschiedenis*, 69 (1989): 176–94.

10. Gaukroger, *op. cit.*, p. 322.

11. For an excellent study that is yet to be surpassed, see A. J. Krailsheimer, *Studies in Self-Interest: Descartes to La Bruyère*, Oxford, Clarendon Press, 1962, p. 32.

12. See Daniel Garber, *Descartes' Metaphysical Physics*, Chicago, University of Chicago Press, 1992, pp. 79–82; and on Descartes's critique of atomism see chap. 5.

13. See Bruce Stansfield Eastwood, "Descartes on Refraction: Scientific Versus Rhetorical Method," *Isis*, 75 (1984): 481–502.

14. As quoted in Eastwood, p. 486.

15. A. D. Lublinskaya, *French Absolutism: The Crucial Phase, 1620–29*, Cambridge, Cambridge University Press, 1968, p. 33. Cf. Orest Ranum, *Artisans of Glory: Writers and Historical Thought in Seventeenth Century France*, Chapel Hill, University of North Carolina Press, 1980, p. 119, on Descartes as a scathing critic of *ars historica*, the genre of historical writing that gloried in the heroic rather than in the rational vindication of royal authority. For an excellent discussion of recent scholarship on absolutism, see William Beik, *Absolutism and Society in Seventeenth-Century France*, Cambridge, Cambridge University Press, 1985, chap. 1. For the title page I have relied on the copy of the *Discourse* in the rare book room at Van Pelt Library, University of Pennsylvania.

16. Jonathan Dewald, *Aristocratic Experience and the Origins of Modern Culture. France, 1570–1715*, Berkeley, University of California Press, 1993, p. 140.

17. René Descartes, *Discourse on Method and the Meditations*, trans. F. E. Sutcliffe, Harmondsworth, U.K., Penguin, 1979, p. 27. For the reader's convenience I am using this edition for my explication. All page numbers in the text henceforth refer to it.

18. For a very nuanced discussion of Descartes's debt to stoicism, see Gaukroger, *op. cit.*, pp. 118–19.

19. Paul Zambelli, *La formazione filosofica di Antonio Genovesi*, Naples, Morano, 1972. Since this chapter was first written for *The Cultural Meaning* a new book contains some of the same insights; see Philippe-Jean Quillen, *Distionnaire politique de René Descartes*, Presses universitaires de Lille, Lille, 1994. For a highly philosophical and abstract approach to the political implications of Cartesianism, see Pierre Guenancia, *Descartes et l'ordre politique*, Paris, Presses Universitaires de France, 1983.

20. In the account that follows I am heavily indebted to an unpublished paper by David A. Smith, "Jacques Rohault and the Popularization of Cartesianism," 1992; see also Roger Hahn, *The Anatomy of a Scientific Institution: The Paris Academy of Sciences, 1666–1803*, Berkeley, University of California Press, 1971, pp. 10–15.

21. J. Rohault, *Traité de Physique*, 2 vols, Amsterdam, 1672; vol. 1, pp. 13–17; vol. 2, pp. 142–43. For his attack on the Aristotelians, see vol. 1, pp. 4–5.

22. [J. G. Padries and Rochon], *Lettre d'un philosophe à un Cartesien de ses amis*, Paris, 1672, p.5.

23. See the excellent discussion in Geoffrey Vincent Sutton, "A Science for a Polite Society: Cartesian Natural Philosophy in Paris During the Reigns of Louis XIII and Louis XIV," Ph.D. dissertation, Princeton University, 1982, chap. 1 and pp. 437–52. There is a paperback edition of the *Conversations* with an introduction by Nina Gelbart and published by the University of California Press.

24. See Erica Harth, *Ideology and Culture in Seventeenth Century France*, Ithaca, N.Y., Cornell University Press, 1983, p. 231 and *passim*. Cf. M. de Cordemoy, *Dissertation physique*, preface, Paris, 3rd ed., 1689.

25. See Ruth Perry, "Radical Doubt and the Liberation of Women," *Eighteenth Century Studies*, 18 (1985): 472–93; and Londa Schiebinger, *The Mind Has No Sex? Women in the Origins of Modern Science*, Cambridge, Mass., Harvard University Press, 1989, pp. 170–78.

26. See Schiebinger, *The Mind Has No Sex?* pp. 175–78; Siep Stuurman at the University of Rotterdam is writing a book on Poullain.

27. Pierre S. Regis, *Système de Philosophie*, Paris, 1690 (avec privilege du roy), with dedicatory preface to abbé de Louvois.

Chapter 3

1. Ilan Rachum, "The Meaning of 'Revolution' in the English Revolution (1648–1660)," *Journal of the History of Ideas*, 56 (1995): 195–215.

2. This entire chapter relies on Christopher Hill, *The Century of Revolution 1603–1714*, London, Nelson, 1961; the articles by S. F. Mason, H. F. Kearney, Christopher Hill, T. K. Rabb, Barbara Shapiro, and Margaret 'Espinasse that first appeared in *Past and Present*, collected since in Charles Webster, ed., *The Intellectual Revolution of the Seventeenth Century*, London and Boston, Routledge and Kegan Paul, 1974, pp. 197–316, 347–368; P. M. Rattansi, "The Social Interpretation of Science in the Seventeenth Century," in Peter Mathias, ed., *Science and Society 1600–1900*, Cambridge, Cambridge University Press, 1972, pp. 1–32; Margaret C. Jacob, *The Newtonians and the English Revolution 1689–1720*, Ithaca, N.Y., Cornell University Press, 1976; and J. R. Jacob, *Robert Boyle and the English Revolution*, New York, Burt Franklin, 1977. Portions of this chapter first appeared in James R. Jacob and Margaret C. Jacob, "The Anglican Origins of Modern Science: The Metaphysical Foundations of the Whig Constitution," *Isis*, vol. 71 (1980): 251–67. See also Robert K. Merton, *Science, Technology and Society in Seventeenth-Century England*, New York, Howard Fertig, 1970; and Charles Webster, *The Great Instauration: Science, Medicine and Reform, 1620–60*, London, Duckworth, 1975. Note the perceptive comments by Benjamin Nelson in Tom Bottomore et al., eds., *Varieties of Political Expression in Sociology*, Chicago, University of Chicago Press, 1972, pp. 202–210. For a recent defense of the Merton thesis, unreformed, see Gary A. Abraham, "Misunderstanding the Merton Thesis: A Boundary Dispute Between History and Sociology," *Isis*, 74 (1983): 368–87. See also Margaret C. Jacob, ed., *The Politics of Western Science*, Atlantic Highlands, N.J., Humanities Press, 1994. For sanity and light on recent historiography, see Nicholas Tyacke, "Anglican Attitudes: Some Recent Writings on English Religious History, from the Reformation to the Civil War," *Journal of British Studies*, 235 (1996): 139–67.

3. H. A. M. Snelders, "Science in the Low Countries During the 16th Century: A Survey," *Janus*, 70 (1983): 213–27; the great exodus of intellectuals out of the southern Netherlands after the Spanish conquest in 1585 had "a paralyzing effect on the culture" of the region. For the teaching of science in a typical Dutch academy of the seventeenth century, see Rijksarchief, Gelderland, MSS, Academie te Harderwij, no. 154. Catalogue of the library includes Gassendi and Bacon in 1671; Descartes and John Ray by 1698. Cf. Th. J. Meijer, "De historische achtergronden van wetenschappelijk onderzoek in Leids universitair verband," *Tijdschrift voor geschiedenis*, 85 (1972): 432–43.

Cf. Charles Webster, *The Great Instauration*, pp. 90–96, 259. For the role of religion in the revolution, without accepting its conclusions, see John Morrill, "The Religious Context of the English Civil War," *Transactions of the Royal Historical Society*, 5th ser., 34 (1984): 155–78. For a corrective to Morrill's views, see Christopher Hill, *The English Bible and the Seventeenth Century Revolution*, 1995.

4. P. M. Rattansi, "Paracelsus and the Puritan Revolution," *Ambix*, 11 (1963): 24–32.

5. For how this happened see Michael Mendle, "De Facto Freedom, De Facto Authority: Press and Parliament, 1640–43," *The Historical Journal*, 38, no. 2 (1995): 307–32.

6. For women in these movements see the splendid discussion in Phyllis Mack, *Visionary Women. Ecstatic Prophecy in Seventeenth-Century England*, Berkeley, University of California Press, 1992.

7. Thomas H. Jobe, "The Devil in Restoration Science: The Glanvill-Webster Witchcraft Debate," *Isis*, 72 (1981): 343–56. If the student wants to get at the original Hermetic texts turn to Brian P. Copenhaver ed., *Hermetica: The Greek 'Corpus, Hermeticum' and the Latin 'Asclepius' in a New English Translation*, Cambridge, Cambridge University Press, 1992.

8. Christopher Hill, *The World Turned Upside Down*, London, Temple Smith, 1972, chap. 14. On the appropriateness of the term "radical" see Gary S. de Krey, "Rethinking the Restoration: Dissenting Cases for Conscience, 1667–1672," *The Historical Journal*, 38 (1995): 53–83. See also Antonio Clericuzio, "From van Helmont to Boyle. A Study of the Transmission of Helmontian Chemical and Medical Theories in Seventeenth-Century England," *The British Journal for the History of Science*, 26 (1993): 303–34. For the view that the English Revolution was the pinnacle of radicalism see Christopher Hill, "Freethinking and Libertinism: The Legacy of the English Revolution, in R. Lund, ed., *The Margins of Orthodoxy*, Cambridge, Cambridge University Press, 1995: 54–70.

9. Robert Boyle, *Some Considerations Touching the Usefulness of Experimental Natural Philosophy*, London, part 1 (1663) and part 2 (1671). Both parts were written during the 1650s; see R. S. Westfall, "Unpublished Boyle Papers Relating to Scientific Method," *Annals of Science*, 12 (1956): 65; and Thomas Birch, ed., *The Works of the Honourable Robert Boyle*, 6 vols, London, 1972, vol. 3, p. 395. For a treatment of parts 1 and 2, see James R. Jacob, *Boyle*, pp. 104–18 and 141–43, respectively. See also Charles Webster, "The College of Physicians: 'Solomon's House' in Commonwealth England," *Bulletin of the History of Medicine*, 41 (1967): 393–412; J. J. O'Brien, "Commonwealth Schemes for the Advancement of Learning," *British Journal of Educational Studies*, 16 (1968): 30–42; and Christopher Wren, *Parentalia: Or Memoirs of the Family of Wrens*, London, 1950, p. 196.

10. J. R. Jacob, *Robert Boyle and the English Revolution*, New York, Burt Franklin, 1977, pp. 141–43; and Royal Society of London, Letter Book Supplement, A-B Copy, John Beale, pp. 348, 382, 389–90, 403–10. Cf. James R. Jacob, "The Political Economy of Science in Seventeenth Century England," in Margaret C. Jacob, ed., *The Politics of Western Science, 1640–1990*, Atlantic Highlands, N.J., Humanities Press, 1994, pp. 19–46.

11. Allen G. Debus, ed., *Science and Education in the Seventeenth Century: The Webster-Ward Debate*, London, Macdonald, 1970; Henry Stubbe, *A Light Shining out of Darkness*, London, 1659, which was "answered by H. F. [Henry Ferne?] but never printed," according to Anthony á Wood, *The History and Antiquities of the University of Oxford*, 3 vols., Oxford, Oxford University Press, 1792–1796, vol. 3, p. 695; for the

conservative reaction to Stubbe's attack on conventional religion and the universities: Anthony á Wood, *Athenae Oxoniensis*, ed., P. Bliss, 4 vols., London, 1813–1820, vol. 3, p. 1069. See also: *Sundry Things from Several Hands Concerning the University of Oxford*, London, 1659; and Charles Webster, "William Dell and the Idea of University," in Mikulas Teich and Robert Young, eds., *Changing Perspectives in the History of Science*, London, Heinemann, 1973, pp. 110–26.

12. Christopher Hill, *The Religion of Gerrard Winstanley*, supplement 5, Past and Present Society, Oxford, Oxford University Press, 1978, p. 18. For a continuation of radical activity see the work of Richard Greaves, for example, *Enemies under His Feet. Radicals and Nonconformists in Britain, 1664–1677*, Stanford, Calif., Stanford University Press, 1990.

13. For a good general introduction to Winstanley, see G. E. Aylmer, "The Religion of Gerrard Winstanley," in J. F. McGregor and B. Reay, eds., *Radical Religion in the English Revolution*, Oxford, Oxford University Press, 1984, pp. 91–120. For the beliefs of ordinary folk and their sects in this period, see also Margaret Spufford, *Small Books and Pleasant Histories: Popular Fiction and Its Readership in Seventeenth-Century England*, London, Methuen, 1981.

14. Marie Boas, *Robert Boyle and Seventeenth-Century Chemistry*, Cambridge, Cambridge University Press, 1958; Robert H. Kargon, *Atomism in England from Harriot to Newton*, Oxford, Clarendon Press, 1966, pp. 93–105. Cf. Steven Shapin and Simon Schaffer, *Leviathan and the Air-Pump: Hobbes, Boyle, and the Experimental Life*, Princeton, Princeton University Press, 1986. For my reservations see "Reflections on the Ideological Meanings of Western Science from Boyle and Newton to the Postmodernists," *History of Science*, **xxxiii** (December 1995): 333–57.

15. J. R. Jacob, *Boyle*, pp. 112–15.

16. Thomas Edwards, *Gangraena*, 3rd ed., London, 1646, div. 1, part 1, pp. 25–26; Hill, *The World Turned Upside Down*, chap. 6.

17. Edwards, *Gangraena*, pp. 15–19, 23–24, 28–29; J. R. Jacob, *Boyle*, chaps. 3 and 4; and M. C. Jacob, *The Newtonians*, chap. 1.

18. Thomas Sprat, *A History of the Royal Society*, London, 1667, pp. 343, 400, 408, 425–29.

19. [Thomas Tenison], "The Epistle Dedicatory," in *The Creed of Mr. Hobbes Examined*, London, 1671, pp. 7–8, 13–15; Joseph Glanvill, *A Blow at Modern Sadducism*, London, 1668, pp. 153–60; John Evelyn, *The History of Religion*, ed., R. M. Evanson, 2 vols., London, 1850, vol. 1, pp. xxvii–xxviii; and J. R. Jacob, "Civil Religion and Radical Politics: Stubbe to Blount," paper presented at the annual meeting of the American Historical Association, San Francisco, 1978.

20. For the attack on Hobbes see John Wallis to John Owen, 10 Oct. 1665, in Peter Toon, ed., *The Correspondence of John Owen (1616–1683)*, Cambridge, Cambridge University Press, 1970, pp. 87–88; John Wallis, *Hobbiani Puncti Dispunctio*, Oxford, 1657, pp. 42–43; and Robert Boyle, "The Preface," in *An Examen of Mr. T. Hobbes His Dialogus Physicus de Natura Aeris*, Oxford, 1662. Boyle, "The Preface," *An Examen*; and John Wallis, "The Epistle Dedicatory," in *Elenchus Geometriae Hobbianae*, Oxford, 1655; and for the argument developed in Wren, *Parentalia*, p. 196; and Royal Society, Letter Book Supplement, A-B Copy, John Beale, pp. 348, 382, 389–90, 403, 410. For a suggestive approach to politics and economic ideology in this period, see Joyce Appleby, *Economic Thought and Ideology in Seventeenth Century England*, Princeton, Princeton University Press, 1978, chap. 9.

21. See Robert Martin Krap, *Liberal Anglicanism: 1636–1647*, Ridgefield, Conn., Acorn Press, 1944; John F. H. New, *Anglican and Puritan: The Basis of Their Opposition, 1558–1640*, Stanford and London, Stanford University Press, 1964, pp. 16–21. For further evidence of preaching against predestination in Cambridge during the 1650s, see Spencer Research Library, University of Kansas, diary of Charles North, MS A.41, fol. 1, Dr. Cudworth of Clare Hall, "On 4 Esiah: 5"; also Dr. Arrowsmith and Dr. Love on the theme "faith without good works is dead." Cf. Gregory Memorandum, Gregory MSS, Edinburgh University Library, DC. 1.61, fol. 93; "When Dr Duport resigned the chair of Greek he recommended his pupil Mr. Barrow who . . . being suspected of Arminianism he could not obtain it and therefore in 1654 he . . . went first to France, in Paris he found his father attending the English Court."

22. Peter Pett, *A Discourse Concerning Liberty of Conscience*, London, 1661, p. 9. This was a tract commissioned by Boyle and representative of his views. Cf. G. R. Abernathy, "Richard Baxter and the Cromwellian Church," *Huntington Library Quarterly*, 24 (1961): pp. 227–31; and J. R. Jacob, *Boyle*, pp. 118–26. On the details of the church at the Restoration, see John Miller, *Charles II*, London, Weidenfeld and Nicolson, 1991, pp. 50–68.

23. For a discussion of the arguments put forward by the Catholic opponents of the new science, see Edward Grant, "In Defense of the Earth's Centrality and Immobility: Scholastic Reaction to Copernicanism in the Seventeenth Century," *Transactions of the American Philosophical Society*, 74, part 4 (1984): 11ff. For a valuable discussion of the response to Descartes and a review of the historiography, see C. Webster, "Henry More and Descartes: Some New Sources," *British Journal of the History of Science*, 4, no. 16 (1969): 359–77. Cf. Henry More, *Enchiridion Metaphysicum*, London, 1671.

24. Quoted in John Gascoigne, " 'The Holy Alliance': The Rise and Diffusion of Newtonian Natural Philosophy and Latitudinarian Theology Within Cambridge from the Restoration to . . . George III," Ph.D. dissertation, Cambridge University, 1981, p. 132; and Gascoigne, "The Universities and the Scientific Revolution: The Case of Newton and Restoration Cambridge," *History of Science*, 23 (1985): 391–434.

25. For a good exposition of Newton's notebook, see Gale E. Christianson, *In the Presence of the Creator: Isaac Newton and His Times*, New York, Free Press, 1984, pp. 55–56. For similar work see University Library, Cambridge, student notebook of John Smyth of Gonville and Caius in 1681, fol. 34ff.; on physics according to Descartes; University Library, Cambridge, MS 6160 notebook of William Bright, November 1645, e.g., 170–76ff. very similar to Newton's notes; these on God's power and prudence in the government of the world; on the style of these notes see the instructions found in Add. Mss. 6986 "Dr. Duport's Rules to Fellow Commoners," fol. 9: "When you are ye respondent evermore repeat ye syllogisme before you answer. . . . Write yr. logical and Philosophical rules, distinctions or questions in a little paper pocket book you may carry them about with you." And when the practice became formalized, and incidentally used to teach Newton's science, see *Quaestiones philosophicae in usum juventutis academicae*, Cambridge, 1732; and finally A. R. Hall, "Sir Isaac Newton's Note-Book, 1661–65," *Cambridge Historical Journal*, 9 (1948): 245–50.

26. See John Craig to John Conduitt, 7 April 1727, Cambridge University Library, MSS. Add. 4007, fol. 686. For context see Bodleian Library, Oxford, MS Rawlinson c. 146, fol. 132–37. Cf. John Gascoigne, "Politics, Patronage and Newtonianism: The Cambridge Example," *Historical Journal*, 27 (1984): 1–24. And see Newton's manuscript, which may date from either the 1660s or the 1680s: "De Gravitatione et ae-

quipondo fluidorum," in A. Rupert Hall and Marie Boas Hall, eds., *Unpublished Scientific Papers of Isaac Newton*, Cambridge, Cambridge University Press, 1962, pp. 142–44, 148. For the singularly important role of this repudiation of Descartes for the development of Newton's natural philosophy, see Richard Westfall, *Never at Rest: A Biography of Isaac Newton*, Cambridge, Cambridge University Press, 1980, p. 381.

27. Ronald Hutton, *Charles the Second. King of England, Scotland, and Ireland*, Oxford, Clarendon Press, 1991, pp. 183–84.

28. Newton manuscript, Burndy Library, Burndy MS 16, fol. 6, r-v. On Newton's alchemy, see Betty Jo Teeter Dobbs, *The Foundations of Newton's Alchemy*, Cambridge, Cambridge University Press, 1975; see p. 80 for Newton's link to Hartlibian circles.

29. Newton MS, University Library, Cambridge, Add. MS 3968.41, fol. 85r.

30. See Christopher Hill, *The Experience of Defeat: Milton and Some Contemporaries*, New York, Viking, 1984. Cf. J. R. Jacob, "Restoration Ideologies and the Royal Society," *History of Science*, 18 (Feb. 1980): p. 18.

31. David L. Wykes, "James II's Religious Indulgence of 1687 and the Early Organization of Dissent: The Building of the First Nonconformist Meeting-House in Birmingham," *Midland History*, xvi (1991): 86–102, p. 88 for the quotation from Ralph Thoresby.

32. Michael Hunter, *Science and Society in Restoration England*, Cambridge, Cambridge University Press, 1981, pp. 93, 117.

33. Royal Society MSS C.P. 18, item 8, fols. 66–80. On getting a patent, see Christine MacLeod, "Patents for Invention and Technical Change in England, 1660–1753," Ph.D. dissertation, Cambridge University, 1982, p. 247. Cf. Alan Smith, "Steam and the City: The Committee of Proprietors of the Invention for Raising Water by Fire, 1715–35," *Transactions of the Newcomen Society*, 49 (1977–1978): pp. 5–18.

34. For how land and industry interacted see Trevor Raybould, "Aristocratic Landowners and the Industrial Revolution: The Black Country Experience c. 1760–1840," *Midland History*, ix (1984): 59–86.

35. Frank E. Manuel, *The Religion of Isaac Newton: The Fremantle Lectures, 1973*, Oxford, Clarendon Press, 1974, pp. 99–100; for portions of Yahuda MS 1 by Newton, see Appendix to Manuel, *Isaac Newton, Historian*, Cambridge, Cambridge University Press, 1963, pp. 1–17.

36. Christopher Hill, "Sir Isaac Newton and his Society," in his *Change and Continuity in 17th Century England*, London, Weidenfeld and Nicholson, 1974, p. 274; cf. George Grinnell, "Newton's *Principia* as Whig Propaganda," in Paul Fritz and David Williams, eds., *City and Society in the 18th Century*, Toronto, Hakkert, 1973, pp. 181–92, which at least raises the issue of political motives, although I do not agree with Grinnell's conclusions.

37. On Halley and James II, see I. Bernard Cohen and Robert E. Schofield, eds., *Isaac Newton's Papers and Letters on Natural Philosophy*, Cambridge, Mass., Harvard University Press, 1958, pp. 397–424; on Halley and Tillotson, see British Library, MSS Add. 17017, fols. 143, 145–46; MSS Add. 4236, fols. 230, 233, 227.

38. W. R. Albury, "Halley's Ode on the *Principia* of Newton and the Epicurean Revival in England," *Journal of the History of Ideas*, 39 (1978): 27, 36–37.

39. For the letter see H. W. Turnbull, ed., *The Correspondence of Isaac Newton*, 7 vols., Cambridge, Cambridge University Press, 1961, vol. 3, 12–13, 279; cf. for Newton in Parliament see Millicent B. Rex, *University Representation in England, 1604–1690*, London, Allen and Unwin, 1954. For Newton's strong interest in his seat, see also A.

Rupert Hall and Laura Tilling, eds., *The Correspondence of Isaac Newton*, 7 vols., Cambridge, Cambridge University Press, 1977, vol. 7, pp. 436–37.

40. Citing Keynes MS 121, f.3 in Scott Mandelbrote, "Isaac Newton and the Writing of Biblical Criticism," *The British Journal of the History of Science*, 26 (1993): 288.

41. On Church thinking at the Revolution of 1688–89, see Mark Goldie, "The Political Thought of the Anglican Revolution," in Robert Beddard, ed. *The Revolutions of 1688*, Oxford, Clarendon Press, 1991, pp. 102–36.

42. Memorandum by David Gregory, 28 Dec. 1691, found in Turnbull, *Correspondence of Newton*, vol. 3, p. 191.

43. Scott Mandelbrote, *op. cit.*, p. 301.

44. For the Scottish context see Bruce P. Lenman, "The Scottish Nobility and the Revolution of 1688–90," in Beddard, *op. cit.*, pp. 137–62.

45. See Samuel Clarke, *A Demonstration of the Being and Attributes of God: More Particularly in Answer to Mr. Hobbes, Spinoza, and Their Followers*, London, 1705; cf. John Toland, *Socinianism Truly Stated: Being an Example of Fair Dealing in All Theological Controversys . . . by a Pantheist to an Orthodox Friend*, London, 1705; cf. Giancarlo Carabelli, *Tolandiana*, Florence, La Nuova Italia, 1975, pp. 119–20. For the framework of these ideas, see J. E. McGuire, "Existence, Actuality and Necessity: Newton on Space and Time," *Annals of Science*, 35 (1978): 470; on More and Newton as revealed in "De Gravitatione," pp. 471, 480–82; on Spinoza, p. 493. The quotation is derived from J. E. McGuire, "Newton on Place, Time and God: An Unpublished Source," *British Journal for the History of Science*, 11 (1978): 114–23, quoting from Cambridge University Library, MSS ADD. 3965, section 13, fols. 445r–446r. For the complexity of belief and unbelief see Silvia Berti, "At the Roots of Unbelief," *Journal of the History of Ideas*, 56 (1995): 555–75.

46. Quoted from Nicholas Robinson in Anita Guerrini, "Ether Madness: Newtonianism, Religion, and Insanity in Eighteenth-Century England," in Paul Theerman and Adele F. Seeff, eds., *Action and Reaction. Proceedings of a Symposium to Commemorate the Tercentenary of Newton's 'Principia'*, Newark, Del., University of Delaware Press, 1993, p. 240.

47. Norriss S. Hetherington, "Isaac Newton and Adam Smith: Intellectual Links between Natural Science and Economics," in P. Theerman and Adele F. Seeff, *op. cit.*, pp. 277–91.

Chapter 4

1. For a brilliant discussion of the crisis, see Paul Hazard, *The European Mind*, New Haven, Conn., Yale University Press, 1953. Some of these themes are examined in chap. 1 of Joyce Appleby, Lynn Hunt, and Margaret Jacob, *Telling the Truth about History*, New York, W. W. Norton, 1994.

2. For a more detailed discussion see John Hedley Brooke, *Science and Religion. Some Historical Perspectives*, Cambridge, Cambridge University Press, 1991, chaps. 5 and 6.

3. For a wider discussion than is possible here, see Christopher Fox, Roy Porter, and Robert Wokler, eds., *Inventing Human Science. Eighteenth-Century Domains*, Berkeley, University of California, 1995.

4. For an analysis of the working of censorship in France, see Joseph Klaits, *Printed Propaganda under Louis XIV: Absolute Monarchy and Public Opinion*, Princeton,

Princeton University Press, 1976. For how repression worked in Italy see Vincenzo Ferrone, *The Intellectual Roots of the Italian Enlightenment. Newtonian Science, Religion, and Politics in the Early Eighteenth Century*, Atlantic Highlands, N.J., Humanities Press, 1995, pp. 1–4. On religious persecution see Bernard Cottret, ed., *The Huguenots in England: Immigration and Settlement*, trans. P. and A. Stevenson, Cambridge, Cambridge University Press, 1991.

5. For another approach to the emergence of probability, see Barbara Shapiro, *Probability and Certainty in Seventeenth Century England*, Princeton, Princeton University Press, 1983.

6. See Jacques-Bénigne Bossuet, *Politique tirée des propres paroles de l'Ecriture sainte*, ed. by Jacques Le Brun, Geneva, 1967, originally published in 1709, p. 185. I owe the citation to Jeffrey Merrick.

7. Argument spelled out in greater detail in Margaret C. Jacob, "Reflections on the Ideological Meanings of Western Science from Boyle and Newton to the Postmodernists," *History of Science*, 33 (1995): 333–57.

8. A good example of the virulence of the campaign can be found in Aubrey Rosenberg, *Nicholas Gueudeville and His Work, (1652–172?)*, The Hague and Boston, Nijhoff, 1982, p. 61; Pierre J. W. van Malssen, *Louis XIV d'aprèes les pamphlets repandus en Hollande*, Amsterdam, H. Paris, 1936; Guy Howard Dodge, *The Political Theory of the Huguenots of the Dispersion*, New York, Columbia University Press, 1947; K. Malettke, *Opposition und Konspiration unter Louis XIV*, Göttingen, Vandenhoesch und Ruprecht, 1976.

9. See David Cressy, "Levels of Illiteracy in England, 1530–1730," in Harvey L. Graff, ed., *Literacy and Social Development in the West: A Reader*, Cambridge, Cambridge University Press, 1981, pp. 123–24. On Germany, see Gerald Strauss, *Luther's House of Learning: Indoctrination of the Young in the German Reformation*, Baltimore, Johns Hopkins University Press, 1978, p. 202.

10. On the Dutch side of this story see Rienk H. Vermij, *Secularisering en Natuurwetenschap in de zeventiende en achttiende eeuw: Bernard Nieuwentijt*, Amsterdam, Rodopi 1991.

11. For the manuscript version see Clark Library, Los Angeles, MS J43M3 A859, "Astrological Experiments Exemplified by Samuel Jeake"; cf. his diary, MS J43M3 D540, 1G94. The diary has now been edited by Michael Hunter.

12. Pierre Retat, *Le Dictionnaire de Bayle et la lutte philosophique au XVIIIe siècle*, Paris, Presse de Université de Lyon, 1971.

13. C. M. G. Berkevens-Stevelinck, *Prosper Marchand et l'histoire du livre*, Ph.D. dissertation, University of Amsterdam, 1978, pp. 2–16. To be supplemented by Margaret C. Jacob, *The Radical Enlightenment*, London, Unwin-Hyman, 1981.

14. Cf. G. Bonno, "Lettres inedites de Le Clerc à Locke," *University of California Publications in Modern Philosophy*, 52 (1959).

15. On Furly, see William Hull, *Benjamin Furly and Quakerism in Rotterdam*, Philadelphia: Swarthmore Monographs, 1941; for his library, see *Bibliotheca Furliana*, Rotterdam, 1714. On Locke as a refugee see John Marshall, *John Locke. Resistance, Religion and Responsibility*, Cambridge, Cambridge University Press, 1994, pp. 357–66.

16. Rex A. Barrell, ed., *Anthony Ashley Cooper. Earl of Shaftesbury (1671–1713)*, Lewiston, Edwin Mellon Foundation, 1989, pp. 92–93.

17. See British Library, MSS. ADD. 4283, fols. 265–66, and Furly's letters to William Penn at the Pennsylvania Historical Society, Locust St., Philadelphia.

18. Balthasar Bekker, *De Philosophia Cartesiana admonitis candida et sincera*, Vesaliae, 1668, pp. 14–18.

19. Balthasar Bekker, *Uitlegginge van den Prophet Daniel*, Amsterdam, 1688. The preface is dated 14 May 1688, and is clearly written under the impact of the outfitting of the Dutch fleet for what many assumed would be a war against France. Cf. K. H. D. Haley, "Sir Johannes Rothe: English Knight and Dutch Fifth Monarchist," in Donald Pennington and Keith Thomas, eds., *Puritans and Revolutionaries: Essays in Seventeenth-Century History Presented to Christopher Hill*, Oxford, Clarendon Press, 1978, pp. 310–32.

20. Balthasar Bekker, *De Betoverde Weereld*, 1691, preface and p. 656.

21. Balthasar Bekker, *Le monde enchanté*, Amsterdam, 1694, vol. 4, pp. 296, 719. On journalistic propaganda in support of Bekker, see J. J. V. M. de Vet, *Pieter Rabus (1660–1702)*, Amsterdam, Holland University Press, 1980. Cf. Jacques Revel, "Forms of Expertise: Intellectuals and 'Popular' Culture in France (1650–1800)," in Steven L. Kaplan, ed., *Understanding Popular Culture: Europe from the Middle Ages to the Nineteenth Century*, Berlin, Mouton, 1984, pp. 255–73.

22. Erica Harth, *Ideology and Culture in Seventeenth Century France*, Ithaca, N.Y., Cornell University Press, 1983, pp. 290–92, 297, on Denis Vairasse.

23. R. H. Campbell and A. S. Skinner, eds., *The Origins and Nature of the Scottish Enlightenment*, Edinburgh, Donald, 1982, p. 70, found in Christine M. Shepherd, "Newtonianism in Scottish Universities in the Seventeenth Century."

24. John Colerus, *The Life of Benedict de Spinosa, Done out of French*, London, 1706, pp. 3, 7. To be used with some caution, as Colerus is an essentially hostile source.

25. On the career of pantheism as derived from Spinoza and others, see Paul Verniere, *Spinoza et la pensée française avant la revolution*, 2 vols., Paris, Presses Universitaires de France, 1954.

26. See Margaret C. Jacob, *The Radical Enlightenment*, London, Unwin-Hyman, 1981, p. 244; and see also Jonathan Israel, *The Dutch Republic. Its Rise, Greatness, and Fall 1477–1806*, Oxford, Clarendon Press, 1995, pp. 916–33, where this thesis is expanded upon.

27. See Margaret C. Jacob, "The Knights of Jubilation: Masonic *and* Libertine," *Quaerendo*, 14 (1984): 63–75.

28. University Library, Amsterdam, MS. coll.hss. V 84.

29. Aubrey Rosenberg, *Tyssot de Patot and His Work, 1655–1738*, The Hague, Nijhoff, 1972; and Rosenberg, "An Unpublished Letter of Tyssot de Patot," *Vereeniging tot Beoefening van Overijsselsch Regt en geschiedenis*, 96 (1981): 71–76. Cf. Alan Gabbey, "Philosophia Cartesiana Triumphata: Henry More (1646–71)," in Thomas M. Lennon et al., eds., *Problems of Cartesianism*, Kingston, Ontario, McGill-Queen's University Press, 1982, p. 246.

30. Koninklijk Huisarchief, The Hague, MS G 16-A29, fol. 14, Allamand to M.M. Rey, 1762.

31. Agatha Kobuch, "Aspekte des aufgeklarten burgerlichen Denkens in Kursachsen in der ersten Halfte des 18. Jh. im Lichte der Bucherzensur," *Jahrbuch für Geschichte*, Berlin, 1979, pp. 251–94.

32. Anon., *War with Priestcraft or, the Freethinkers' Iliad: A Burlesque Poem*, London, 1732, pp. 36–37.

33. On this literature and its debt to science see Margaret C. Jacob, "The Materialist World of Pornography," in Lynn Hunt, ed., *The Invention of Pornography*, New York, Zone Books, 1994.

34. Ruth Perry, *Women, Letters and the Novel*, New York, AMS Press, 1980.

35. For a splendid description of this new culture, see Roy Porter, "Science, Provincial Culture and Public Opinion in Enlightenment England," *British Journal for Eighteenth Century Studies*, 3, no. 1 (1980): 20–46. For a fascinating account of the earliest applications of Newtonian science, see Larry Stewart, "The Selling of Newton: Science and Technology in Early Eighteenth-Century England," *Journal of British Studies*, 25 (1986): 178–92.

36. *The Freethinker*, (London), no. 16 (16 May 1718), pp. 69–72. Cf. Harry Payne, *The Philosophes and the People*, New Haven, Yale University Press, 1976.

37. See Michael Adas, *Machines as the Measure of Men. Science, Technology, and Ideologies of Western Dominance*, Ithaca, N.Y., Cornell University Press, 1989.

38. *Oeuvres diverses de Pierre Bayle*, 3 vols. in 4, Hildesheim, 1968, vol. 4, pp. 794–95.

39. J. van der Berg, "Eighteenth century Dutch translations of the works of some British latitudinarian and enlightened theologians," *Nederlands archief voor kerkgeschiedenis*, n. s. vol. 59, no. 2 (1979): 198–206.

40. For a gossipy account of in-fighting among journalists, see Anne Goldgar, *Impolite Learning. Conduct and Community in the Republic of Letters, 1680–1750*, New Haven, Conn., Yale University Press, 1995.

41. A. C. de Hoog, "Some Currents of Thought in Dutch Natural Philosophy," Ph.D. dissertation, Oxford University, 1974, pp. 300–301. Jean T. Desaguliers sponsored this edition, and its translator told Toland that it was aimed against him. For Desagulier's being courted by publishers, see Bibliothèque Cantonale et Universitaire, Lausanne, Fonds de Crousaz, IS 2024II/137.

42. *The Englishman*, no. 42 (26 Jan. 1714), cited in James E. Force, *William Whiston: Honest Newtonian*, Cambridge, Cambridge University Press, 1985, p. 162–63n.

43. James Force, *Whiston, Honest Newtonian*, pp. 135–36.

44. Judith Colton, "Kent's Hermitage for Queen Caroline at Richmond," *Architecture*, 2 (1974): 181–91. Occasionally Newtonians could be Jacobites; see Andrew Cunningham, "Sydenham vs. Newton: The Edinburgh Fever Dispute of the 1690's . . ." *Medical History*, suppl. 11 (1981): 71–79.

45. René Pomeau, *La Religion de Voltaire*, Paris, Nizet, 1956.

46. Voltaire, *Traité de Metaphysique (1734)*, ed. H. Temple Patterson, Manchester, Manchester University Press, 1957, pp. 17–19.

47. Voltaire, *The Elements of Sir Isaac Newton's Philosophy*, trans. John Hanna, London, 1738, pp. 182–83.

48. Ibid., p. 236n.

49. For s'Gravesande's statement, see J. N. S. Allamand, ed., *Oeuvres philosophiques et mathematiques de M. W. J. s'Gravesande*, Amsterdam, Marc Michel Rey, 1774, vol. 2, pp. 316–17. The sphere was seen by an English woman tourist in 1726, Clark Library, MS J86Z, n.f. Wednesday, 16 June. According to one account, this was a "fine Copernican sphere with 1500 wheels, made by Tracy an English Man Living at Rotterdam which not only shews the different motions of the heavenly bodies but the year, month, day. . . ."; Los Angeles, Clark Library, MS Phillips 9356.

50. W. A. Speck, "Politicians, Peers and Publication by Subscription, 1700–50," in Isabel Rivers, ed., *Books and Their Readers in Eighteenth Century England*, Leicester, Leicester University Press, 1982, p. 64.

51. J. R. Clarke, "The Royal Society and the Early Grand Lodge Freemasonry," *Ars Quatuor Coronatorum*, 80 (1967): 110–19.

52. See J. A. van Reijn, "John Theophilus Desaguliers, 1683–1983," *Thoth*, no. 5 (1983): 165–203.

53. *The Constitutions of the Freemasons*, London, 1723, p. 50.

54. On women's freemasonry see Janet Burke and Margaret C. Jacob, "French Freemasonry, Women and Feminist Scholarship," *Journal of Modern History*, forthcoming in v. 68, 1996.

55. Quoted in M. C. Jacob, *The Radical Enlightenment*, pp. 243–44. The quotation is by Rousset de Missy.

56. V. Mandey, *Mechanick Powers; or the Majesty of Nature and Art Unvail'd*, London, 1702.

57. E. Truesdell, "Reactions of Late Baroque Mechanics to Success, Conjecture, Error, and Failure in Newton's *Principia,*" in Robert Palter, ed., *The "Annus Mirabelis" of Sir Isaac Newton, 1666–1966*, Cambridge, Mass., MIT Press, 1970, p. 209.

58. Francis Hauksbee, *Physico-Mechanical Experiments in Various Subjects . . .*, London, 1719.

59. J. U. Nef, *The Rise of the British Coal Industry*, 2 vols., London, 1966, Cass reprint of 1932 edition, vol. 2, p. 126–28.

60. Fitzwilliam Museum, Cambridge, MS 37-1947, William Strutt to Maria Edgworth, 1823. Similar sentiments are to be found in the Strutt MSS, Derby Local Library, Derbyshire.

61. Fitzwilliam, MS 48-1947, manuscript by Joseph Strutt, "On the relative advantages and disadvantages of the English and Scottish Universities," 1808. The next quotation is also from the Strutt correspondence.

Chapter 5

1. For the business cards that are stuck in a manuscript volume see JWP, BPL, MS C4/B28; for the letters of James Watt to his brother in the same collection, C4/A4, letter book for 1740–41. His account books also comprise many volumes.

2. Article by Simon Schaffer in John Brewer and Roy Porter, eds., *Consumption and the World of Goods*, New York, Routledge, 1993, p. 492.

3. Daniel Garber, *Descartes' Metaphysical Physics*, Chicago, University of Chicago Press, 1992, p. 182, citing the preface to part III of his *Principles.*

4. JWP, BPL, MS C4/B29, n.f.

5. Muirhead MSS, BPL, MIV/box 14/1. "Essai d'une Nouvelle Theorie du Choc de Corps par Gravesande 1722," appears in a margin.

6. JWP, BPL, C4/B32, dated 1682 on cover. For background see Ann Geneva, *Astrology and the Seventeenth Century Mind. William Lilly and the Language of the Stars*, New York, Manchester University Press, 1995; and on Pordage see Christopher Hill *The World Turned Upside Down*, London, Penguin, 1972, pp. 224–26.

7. For a concise summary of mechanistic concepts at work see Carlo Cipolla, ed., *The Emergence of Industrial Societies*, Fontana Economic History of Europe, Hassocks, Sussex, Harvester Press, 1976, in particular the essay by Phyllis Deane.

8. For a good critique of rational choice economics that pervade the older model see in particular, David S. Landes, "Introduction: On Technology and Growth" in Patrice Higonnet, David S. Landes, and Henry Rosovsky, eds., *Favorites of Fortune.*

Technology, Growth and Economic Development since the Industrial Revolution, Cambridge, Mass., Harvard University Press, 1991, pp. 9–17; in the same volume see the example of failure in the case of Ulster in the essay by Joel Mokyr, "Dear Labor, Cheap Labor, and the Industrial Revolution."

9. David S. Landes, "Introduction: On Technology and Growth" in Patrice Higgonet, David S. Landes, and Henry Rosovsky, eds., *Favorites of Fortune. Technology, Growth and Economic Development since the Industrial Revolution*, Cambridge, Mass., Harvard University Press, 1991, p. 9.

10. Larry Stewart, *The Rise of Public Science. Rhetoric, Technology, and Natural Philosophy in Newtonian Britain, 1660–1750*, Cambridge, Cambridge University Press, 1992. For the teaching of applied mathematics, i.e., hydrostatics, geometry, astronomy, surveying, and gunnery in Edinburgh as early as the Restoration period and its growing popularity, see R. H. Houston, "Literacy, Education and the Culture of Print in Enlightenment Edinburgh," *History*, (October 1993): 373–92. See also Richard S. Tompson, "The English Grammar School Curriculum in the Eighteenth Century," *British Journal of Educational Studies*, 29 (1971): 32–39. By the end of the century the French perceived even the average English soldier as being possessed of "de plusiers procédés de fabrique, nécessaires et inconnus en France" and sought to have English prisoners interrogated for the information. See AN F 12 2195, François Bardel to Ministry of the Interior [year V?]. Kindly supplied by Jeff Horn.

11. For an example of the kind of trial and error to which I refer see Basil Harley, "The Society of Arts' Model Ship Trials, 1758–1763," *The Newcomen Society for the Study of the History of Engineering and Technology. Transactions*, 63 (1991–92): 53–71. For a similar, but eighteenth century discussion of how innovation works see Thomas Barnes cited in note number 50. For a good survey of the role of technology in eighteenth century science texts see Donald Beaver, "Textbooks of Natural Philosophy: The Beatification of Technology," in J. L. Berggren and B. R. Goldstein, eds. *From Ancient Omens to Statistical Mechanics*, Copenhagen, University Library, 1987, pp. 203–13.

12. The phrase comes from the otherwise excellent introduction by Patrick O'Brien and Roland Quinault, eds., *The Industrial Revolution and British Society*, Cambridge, Cambridge University Press, 1993, p. 4.

13. Quoted from Denys Papin, *Nouvelle manière pour lever l'eau par la force du feu*, Cassel/Frankfurt, 1707, pp. 3–6, by Alan Smith, " 'Engines Moved by Fire and Water'. The Contribution of Fellows of the Royal Society to the Development of Steam Power, 1675–1733," unpublished paper dated March 10, 1995, kindly communicated by J. R. Harris.

14. For a good summary of this argument as it stood in the 1970s see D. S. L. Cardwell, "Science, Technology and Industry," in G.S.Rousseau and Roy Porter, eds., *The Ferment of Knowledge*, Cambridge, Cambridge University Press, 1980, pp. 449–83, with good insight into Smeaton. Further research has enabled historians to expand on and nuance Cardwell's arguments.

15. A visiting French engineer in 1784 [L'Ecole des Ponts et Chaussées, Paris, 1784, MS 48, Le Sage, f.51] noted how in the decision to construct a road, the locals bring in an engineer; they then go to Parliament, not for permission to construct it . . ."car les particuliers pourraient l'arreter entre eux; mais pour obtenir le droit d'etablir un Peage. . . ." For a description of the Bristol harbor by a visiting French engineer see L'Ecole des Ponts et Chaussées, Paris, MS 85, Ports d'Angleterre par Mr Cachin, 1785, f.15. Note that this French observer makes mention of "un nombre considérable de français fugitifes, qui y ont établi des manufactures superbes . . .", i.e., Huguenots.

16. For a general approach to the themes presented here see Joel Mokyr, *The Lever of Riches. Technological Creativity and Economic Progress*, New York, Oxford University Press, 1990; the phrase belongs to Ian Inkster, *Science and Technology in History. An Approach to Industrial Development*, London, Macmillan, 1991, chap. 2; Jan Golinski, *Science as Public Culture. Chemistry and Enlightenment in Britain, 1760–1820*, Cambridge, Cambridge University Press, 1992; a similar approach also found in Eric Dorn Brose, *The Politics of Technological Change in Prussia. Out of the Shadow of Antiquity, 1809–1848*, Princeton, Princeton University Press, 1993; and in Svante Lindqvist, *Technology on Trial. The Introduction of Steam Power Technology into Sweden, 1715–1736*, Uppsala, Almqvist & Wiksell, 1984. I do not mean to endorse the kinds of arguments found in Lawrence E. Harrison, *Who Prospers? How Cultural Values Shape Economic and Political Success*, New York, Basic Books, 1992.

17. For a recent discussion of aspects of the French scene, see C. Comte and A. Dahan-Dalmedico, "Mécanique et physique: Euler, Lagrange, Cauchy," in R. Rashed, ed., *Sciences a l'époque de la révolution française. Recherches historiques*, Paris, Blanchard, 1988, pp. 329–444. Cf. Antoine Picon, *L'Invention de l'ingenieur moderne. L'Ecole des Ponts et Chaussées 1747–1851*, vol. 1. Paris, Presses d l'École nationale des Ponts et Chaussées, 1992.

18. For his argument see the important essay that summarizes the work of Terry Shinn, "Science, Tocqueville, and the State: The Organization of Knowledge in Modern France," *Social Research*, 59 (1992): 533–66; reprinted in Margaret C. Jacob, ed., *The Politics of Western Science, 1640–1990*, Atlantic Highlands, N.J., Humanities Press, 1994. Reinforcing Shinn's approach is Eda Kranakis, "Social Determinants of Engineering: A Comparative View of France and America," *Social Studies of Science*, 19 (1989): 5–70. For a summary of current research on the French Academy in the eighteenth century see the opening chapter in Maurice Crosland, *Science under Control. The First Academy of Sciences 1795–1914*, Cambridge, Cambridge University Press, 1992. For a further example of how French science drew its character from the requirements of the state, see James McClellan III, *Colonialism and Science. Saint Domingue in the Old Regime*, Baltimore, Johns Hopkins University Press, 1992; and for a comparative overview of the European academies of science in the eighteenth century, see James McClellan III, *Science Reorganized. Scientific Societies in the Eighteenth Century*, New York, Columbia University Press, 1985. For a contemporary observer who compared the French and English academies and came to a similar conclusion see John Nicholls, *Remarques sur les avantages de la France et de la Grand Bretagne*, Leiden, 1754, [trans. from the English], pp. 50–54: "If you examine the different objects that occupy the academies the preference is for those things which are unuseful." Louis Bergeron sees this social dimension that worked against application being to a certain degree reasserted by Napoleon and continuing, but to a lesser degree than before the Revolution, into the nineteenth century: "Ce qui est certain, c'est que la formation, les ambitions ou les exigences du polytechnicien furent pendant longtemps en discordance avec l'attente, les besoins ou les possibilités de la plupart des entreprises. Intelligence trop théorique, tendances autoritaires héritées de l'administration, esprit de caste. . . ." See Louis Bergeron, *Les capitalistes en France (1780–1914)*, Paris, Gallimard, 1978, p. 70. Cf. B. Belhoste, A. Picon, J. Sakarovitch, "Les exercices dans les écoles d'ingénieurs sous l'ancien régime et la révolution," *Histoire de l'éducation*, 46 (1990): 53–109, esp. 62.

19. JWP, BPL, Smeaton to Boulton and Watt, 5 Feb. 1778. Underlining in the original.

20. See Musson and Robinson, *Science and Industry in the First Industrial Revolution*, [1989], chap. 5.

21. For example, see the letters in Birmingham City Library, M.II/4/2/1-34; JW to AW, 7 Jan. 1787, Paris, on his privilege being confirmed; and in the letter of JW to AW, 8 Mar. 1787, "unfortunately Mr Calverts rotative gadgeon twisted broke off just within the coupling brasses of the link. . . ." For a refreshing approach to the issue of the private and the public spheres among the middle class, see Dror Wahrman, " 'Middle-Class' Domesticity Goes Public: Gender, Class, and Politics from Queen Caroline to Queen Victoria," *Journal of British Studies*, 32, no. 4 (1993): 396–432.

22. Discussed briefly in "Memoir of Gregory Watt. Son of the Great Engineer," by James Patrick Muirhead, ms in the James Watt Papers, Birmingham Public Library.

23. Thomas Mortimer, *Everyman His Own Broker: or, A Guide to Exchange-Alley*, London, 1775.

24. David Cressy, "Literacy in Context: Meaning and Measurement in Early Modern England," in John Brewer and Roy Porter, eds., *Consumption and the World of Goods*, New York, Routledge, 1993, pp. 314–15, diagram 17.3. But Cressy doubts that there was an "industrial revolution." For the periodicals see Eliza Haywood's *The Female Spectator* of the 1740s and *The Ladies' Diary*; and see F. Algarotti, *Sir Isaac Newton's Philosophy Explained for the Use of Ladies*, London, 1739.

25. Cited in *The Cultural Meaning of the Scientific Revolution*, pp. 232–33; from Bristol Record Office, Bright MSS, 11168(3), 15 Nov. 1790.

26. Boulton and Watt MSS, BPL, Boulton to Count Wassilieff, 19 March 1806. And see AN, Paris F17 1344/1, Prof Vivalieu [?] from the Allier department: "Ici il sera impossible de suppleir par des figures au defaut de Machines, d'Appareils, de produits de la nature et de l'art, de drogues de toute . . . les descriptions verbales sont bien insuffisantes dans les sciences où l'on ne l'instruit pour ainsi dire que par une manipulation continuelle." Cf. AN, F17 1344/1 Prof. Derrien from Dept. du Finistère on being reduced to teaching theory; in same book see the report from Verdun for Desaguliers.

27. See James Watt Papers, BPL, C4/C6 for a printed copy of its Rules and Regulations dated April 1793 with a list of members.

28. Manchester College Library, Oxford, Truro MSS, MB to Wilson, 10 Feb. 1788.

29. This source remains basic: Nicolas Hans, *New Trends in Education in the Eighteenth Century*, London, Heinemann, 1951. See also AN, Paris F17 1344/1 for complaints in the 1790s on the lack of mathematical knowledge on the part of students as young as 15 and as old as 40.

30. See John Money, "Teaching in the market-place or 'Caesar adsum jam forte: Pompey aderat': the retailing of knowledge in provincial England during the eighteenth century," in John Brewer and Roy Porter, eds., *Consumption and the World of Goods*, New York, Routledge, 1993, p. 338; and Diana Harding, "Mathematics and Science Education in Eighteenth-Century Northamptonshire," *History of Education*, I (1972): 139–59, showing that by 1729 mechanics was being taught to second-year students who for the most part would have been 17; by the 1730s mechanical apparatus was used in some schools.

31. James Watt Papers, Birmingham City Library, LB/1, to James Watt, Jr., 1785.

32. Ibid., LB/1, letters to James Watt, Jr., 3 March 1785 and 3 March 1785.

33. Alan Smith, " 'Engines Moved by Fire and Water.' The Contributions of Fellows of the Royal Society to the Development of Steam Power," summary of paper in *The Newcomen Society for the Study of the History of Engineering and Technology. Transactions*, 63 (1991–92): pp. 229–30; see also Barbara Smith, ed., *Truth, Liberty, and Religion. Essays Celebrating Two Hundred Years of Manchester College*, Manchester College,

Oxford, 1986; in particular, see Jean Raymond and John Pickstone, "The Natural Sciences and the Learning of English Unitarians: an Exploration of the Role of Manchester College," pp. 127–64. One such academy at Spitalfields is currently being studied by Larry Stewart.

34. Birmingham Public Library, U.K., Watt MSS, MIV/14/1, a notebook entitled "Mechanic Principles" in the hand of John Watt.

35. Bristol Record Office, White MSS, no.08158, 73–81ff. It is worth noting that visiting French engineers in 1789–90 who observed carpenters and rope makers believed them to work better by virtue of their education and "national character." They also observed (8f.): "Nous avons adopté en France, l'usage des Entreprises qui quoi qu'il ait de grands inconveniens, offre néanmoins de grands avantages, capables de faire pencher la balance en faveur de ce sistême; mais nous ne tirons pas dans tous nos ports un égal parti de cette forme de service. Ordinairement les Entreprises sont faites par du Contre-maître ou du constructeurs du Commerce. Delors les ouvriers travaillent à la journée, et n'ont point ce stimulant qui les porterait à developper plus de zèle et d'intelligence." This comment appears in MS 1899, L'Ecole les Ponts et Chaussés, Paris, Mémoire de M.M. Forfait et Lescallier . . . sur La Marine pendant leur Séjour en Angleterre." Note they also comment at length on new inventions for pumps and pulleys seen in their English travels, 26–7f. In this same manuscript they dwell on the superiority of English and Dutch rope-making (37–39ff.).

36. For the most radical of these and their curriculum, which in science differed not at all from the others, see Ruth Watts, "Revolution and Reaction: 'Unitarian' academies, 1780–1800," *History of Education*, 20 (1991): 307–23.

37. *Lectures,* "Address to my Pupils," n.p.

38. Preface.

39. G. Gregory, *The Economy of Nature Explained and Illustrated on the Principles of Modern Philosophy*, London, 1804, 3 vols; vol. I, p. viii. Gregory was largely self-educated.

40. See note 10.

41. See their letters to James Watt and James Watt, Jr., in James Watt Papers, Birmingham City Library, C6/1/9; January 11, 1811, R. E. to J. W.; C6/1/37 M. E. to J. W. Oct. 1, 1811; C6/2/96, R. E. to J. W., 7 August 1813; C6/10 J.W., Jr., to M. E. 21 May 1820 (she is in Paris). And see hers of Jan. 1820 to J. W., Jr., C6/10. For a somewhat heavyhanded account of Maria and Richard Edgeworth see Elizabeth Kowaleski-Wallace, *Their Fathers' Daughters. Hannah More, Maria Edgeworth and Patriarchal Complicity*, New York, Oxford University Press, 1991, pp. 95–101, 144–45.

42. JWP, BPL, C6/2/96, 7 August 1813 to James Watt.

43. Fitzwilliam Musuem, Cambridge, Strutt MS 48–1947; letter of 1808.

44. Royal Society, London, MSS C.P. 18, item 8, 66–80ff. Cf. Christine MacLeod, *Inventing the Industrial Revolution. The English Patent System, 1660–1800*, Cambridge, Cambridge University Press, 1988, pp. 159–60.

45. Richard Biernacki, *The Fabrication of Labor in Germany and Britain, 1640–1914*, Berkeley, University of California Press, 1995, pp. 222–23; cf. Richard Olson, *The Emergence of the Social Sciences, 1642–1792*, New York, Twayne, 1993, chap. 5.

46. P. Langford, *Public Life. . .*, p. 71. And for how science played into the seventeenth-century interests of the propertied, see James R. Jacob, "The Political Economy of Science in Seventeenth-Century England," in Margaret C. Jacob, ed., *The Politics of Western Science, 1640–1990*, Atlantic Highlands, N.J., Humanities Press, 1994, pp. 19–46.

47. *A Course of Experimental Philosophy*, London, 1744, vol. II, pp. 530–31. A French student engineer in 1791 when writing a treatise on the steam engine began his discussion: "on sait que la Vapeur et de l'eau bouillante, suivant les experiences du docteur Desaguliers est 14000 fois plus rare que l'eau." L'Ecole des Ponts et Chaussées, Paris [EPNC], Ms. 100 by M. Fay, student, 1791.

48. Stanley Chapman, *Merchant Enterprise in Britain. From the Industrial Revolution to World War I*, Cambridge, Cambridge University Press, 1992, pp. 58–68.

49. For Parliament and improvement see P. Langford, *Public Life . . .*, pp. 139–43. See Manchester College, Oxford, exam papers, 1823, for political philosophy among Dissenters. Dissenters could not, however, sit in Parliament.

50. The quotation is from a report and address given by Thomas Barnes, D.D. "On the Affinity subsisting and extending Manufactures, by encouraging those Arts on which Manufactures principally depend," *Memoirs of the Literary and Philosophical Society of Manchester*, vol. I, Warrington, 1785, pp. 72 et. seq.

51. Boulton and Watt MSS, BPL, Russian Mint/2 L. Copy MB to Count Woronzow, 11 August 1799. Soho was their Birmingham factory.

52. Quoted in William Chapman, *Address to the Subscribers to the Canal from Carlisle to Fisher's Cross*, Newcastle, 1823, pp. 2–3,7.

53. See L. Mulligan, "Self-Scruting and the Study of Nature . . . ," *Journal of British Studies*, 35 (1996): 311–42.

Chapter 6

1. JWP, BPL, LB/1 Watt to Robison, 10/30 1783: "I am almost unknown except among a *very few* men of science. . . ."

2. Eric Hopkins, "Boulton before Watt: The Earlier Career Re-considered," *Midland History*, ix (1984): 43–58. For background see Leonore Davidoff and Catherine Hall, *Family Fortunes. Men and Women of the English Middle Class, 1780–1850*, London, Hutchinson, 1987, pp. 247–52. It is not the case that Watt Jr. served no apprenticeship. He worked with Manchester manufacturers but did not last long with them.

3. JWP, BPL, 6/46; list of his tools in a letter to his father, from London, 19 June 1756. In 1784 when he advised a friend what her son needed to know to become an engineer, he put drawing first, then geometry, algebra, arithmetic, the elements of mechanics; see same collection, Letter Book, 30 May 1784 (last name not given).

4. These details are drawn from that report; JWP, BPL, 4/53, 11 April 1775, Committee on . . . Mr. Watt's Engine Bill. On why he chose to go before Parliament, see Christine MacLeod, *Inventing the Industrial Revolution. The English Patent System, 1660–1800*, New York, Cambridge University Press, 1988, p. 73.

5. JWP, BPL, 4/76, Edinburgh, 13 March 1775, Cochrane to Watt. See also James Hutton to Watt in 1774 on approaching Parliament: "your friends are trying to do something for you what success will attend their endeavours time only will show—every application for publick employment is considered as a job and to be carried into execution requires nothing but a passage thro the proper channels; it is then a well digested plan; the honestest endeavour must to succeed put on the face of roguery but what signifies the dress of a rogue unless you have the address of a wise man; come and lick some great mans arse and be damned to you." And see John Gascoigne, *Joseph Banks and the English Enlightenment. Useful Knowledge and Polite Culture*, New York, Cambridge University Press, 1994, pp. 211–12.

6. JWP, BPL, W/6, see for example letter of 13 March 1791, Manchester, from James Watt, Jr., to his father on orders of his engine and competitors at work in the town. See MS C2/10 item 3 list of all Watt engines at work in Manchester in 1797.

7. AN, Paris, Marine G 110, dossier 1 and 2; ff.146–201; including a list of 1778 from Boulton and Watt on all the engines installed in Britain to date (27 on this list).

8. Boulton and Watt Papers, BPL, James to Annie Watt, from House of Commons, 3 April 1792.

9. Boulton and Watt MSS, BPL, MII/4/4/28, James to Annie, 28 Feb. 1792.

10. JWP, BPL, LB/1, May 1782 Watt to Wedgwood.

11. JWP, BPL, W/6, James Watt, Jr., to his father, 19 April 1791, Manchester: "I am extremely concerned to see by your letter . . . the low state of spirits that your late misfortunes in business have thrown you into. I wish you could treat them with more indifference and rather look forward to future prospects, than suffer your mind to be depressed by reflecting on the past." As early as 1762 Watt suffers from depression as a letter from his fiancee, Margaret Miller, shows (MS 4/4, 1762, signed "Miss Millar").

12. JWP, LB/1 11 July 1782 Watt to Wedgwood.

13. JWP, MS L/B1, Watt to de Luc, 8 Oct. 1786.

14. Ibid., James Jr. to James Watt, 19 April 1791.

15. JWP, BPL, James to his father in Scotland, 12 June 1755 arrives in York ("thank God") and visits the Cathedral; the one in Durham "Magnificent"; "ridiculous manner of worship of Prebends and canons" who were laughing at the time they "were addressing the most high." He is quite shocked. He likes England but thinks the people are "very sharp."

16. Papers of Matthew Boulton, BPL, Box 357, 1 Sept. 1777, Annie Watt to Mrs. Boulton; Annie to Matthew Boulton on Watt's depression, 15 April 1781.

17. JWP, MS 4/4, 1767. His wife, Margaret Miller, is pregnant with their first son.

18. JWP MS, BPL, James Watt to his father, 21 July 1755, "my hand is shaking after working." On the life of the London apprentice see Peter Earle, *The Making of the English Middle Class. Business, Society and Family Life in London, 1660–1730*, London, Methuen, 1989, pp. 100–105.

19. The preceding and following paragraphs draw details from JWP, BPL, MS 4/11 letters to father, 1754–74, October 1756, James now back in Glasgow; has got some instruments from Jamacia. He is getting mail at the College. Young Watt is working on the foundations of the observatory. The uncle, John, is in straights for money and had to draw from a bank. Sorry to hear that his brother Jockey has not got employment, 9 Jan. 1758: "you should not give any fee with him as one of his age that understands bookkeeping ought rather to be getting." See letter of 31 May 1758; Jockey wants to go abroad after he has served his time, "a foolish notion" James tells their father. See bill of 1762 detailing Watt's debts to his father. See MS C4/A7 for his father's account books for 1748–49. On the slow development of banking among the middling sorts see Leonore Davidoff and Catherine Hall, *Family Fortunes*, pp. 245–46.

20. JWP, BPL, 3/69, report dated 1774 to the Lords of the Police for Scotland.

21. JWP, MS 4/11, letter of 8 October 1765 to his father; MS C1/15 correspondence with Lind on his electrial machine.

22. JWP, BPL, MS 3/18, letter of 16 Feb. 1782 to Boulton: "I am certain that with proper loads such an engine can easily make 30 strokes per minute when not impeded by vis inertia or gravity."

23. JWP, MS 3/18 to Boulton 9 Feb. 1782, on a competitor: "as his theories are all abstract and run only on the commonly known properties of steam as an elastic fluid I cannot conceive anything wherein he can surpass us particularly as he seems to be greatly divested of geometrical principles." Then follows a long mechanical discussion. See MS 3/69, his report dated 1774 where he has used trigonometry to try to estimate the volume of Lough Ness.

24. JWP, BPL, Letter Book, 30 Oct. 1783 to Mr. Robison.

25. JWP, W/5, Watt letter to Black, no date but probably 1780, "he [the French spy Magellan] made many enquirys about your latent heat, which I answered in so far as was expedient—he wants to know when you invented it I answered I could not tell but that you taught it before the year 1763."

26. JWP, MS 6/14, Annie Watt to Gregory Watt, 27 April 1793. See the university notes kept by Gregory MS 6/3; translations from the Greek; speeches against superstition and barbarism.

27. Quoted in A. E. Musson and Eric Robinson, *Science and Technology in the Industrial Revolution*, pp. 210–11, Boulton to his son, 1787.

28. Boulton and Watt MSS, BPL, London, 1 Feb. 1792.

29. JWP, BPL, Letter Book Nov. 30 [1783] to Mr De Luc.

30. JWP, Gregory's exercise book, C4/C18A.

31. For the survival of revolutionary sentiments see the superb essay by Kathleen Wilson, "A Dissident Legacy: Eighteenth Century Popular Politics and the Glorious Revolution," in J. R. Jones, ed., *Liberty Secured? Britain Before and After 1688*, Stanford, Calif., Stanford University Press, 1992, pp. 299–334.

32. JWP, BPL, LB/1, Watt to James Jr., 16 Jan. 1784.

33. JWP, C1/20 letter of 8 July 1791, a draft letter written just six days before the Birmingham riots. For the hint of a class element in the riots see P. Langford, *Public Life. . .*, p. 245.

34. Boulton and Watt MSS, BPL, MII/4/4/10, March 1792, James to Annie. On the slave traders same folder, letter of 30 March 1792.

35. Ibid., MII/4/4/27, James to Annie, 26 April 1792.

36. Boulton and Watt MSS, MII/4/4/1–51; letter from Watt to Annie, 10 Nov. 1792; see letter of Nov. 5 on the retribution of divine justice.

37. JWP, MS 6/14 20 Nov. 1794 Annie Watt to Gregory; same to same, late 1794 on burning in effigy of Thomas Paine.

38. JWP, C2/12, Gregory to James Watt, Jr., 3 August 1802.

39. JWP, BPL, W/6, 7 July 1791, Manchester, James Watt, Jr., to his father: "Upon a revision of the motives which gave rise to my journey to Scotland [to see his sister], I cannot find any thing deserving of the severe reprehension you bestow upon it, and although deeply hurt by the severity of your remarks. . . ."

40. JWP, MS LB/2, 25 April 1791, to Peggy; LB, 30 May 1784 on Peggy as dull.

41. JWP, MS W/6, Nantes, 17 Oct. 1792, James Jr. to his father.

42. JWP, BPL, James Watt, Jr., private letter book; letter to Cooper no date on the machine set in motion; 16 Sept. 1794 to Stephen Delesart [?], on the revolution.

43. JWP, MS W/6, James Jr. to his father, from Naples, 8 May 1793. Cf. John Money, *Experience and Identity. Birmingham and the West Midlands, 1760–1800*, Montreal, Mc-Gill-Queen's University Press, 1977, chap. 9.

44. For background see Ian R. Christie, *Riots and Revolutions. Britain, 1760–1815*, Cambridge, Mass., Harvard University Press, 1982, pp. 215–29.

45. JWP, BPL, MS 4/11. Letter in 1766; at the time they are selling flutes. On the rediscovery of women's role in enterprise see Davidoff and Hall, *op. cit.*, p. 279.

46. See the moving letters in JWP, MS 6/14; 24 Feb. 1795 on living to improve.

47. On James Jr.'s education see chap. 5 in A. E. Musson and Eric Robinson, *Science and Technology in the First Industrial Revolution* (1969), reprinted, New York, Gordon and Breach, 1989.

48. JWP, W/6 James Jr. to father, 5 Nov. 1793 writing from France.

49. On the Oberkampfs see Serge Chassagne, *Le Coton et ses patrons. France, 1760–1840*, Paris, Éditions de l'école des hautes études en sciences sociales, 1991, and particularly p. 369 for the delay in implementing steam in cotton manufacturing throughout France. His manuscript instructions to his son are at the Archives nationales, Paris, 44 AQ 1 (93 M 1); "Regles generales pour la conduite du commerçant" wherein the date 1780 appears.

50. Boulton and Watt Papers, BPL, James to Annie Watt, 6 April 1792.

51. See Max Weber, *The Protestant Ethic and the Spirit of Capitalism*, New York, Scribner, 1953 [originally published in German in 1904].

52. JWP, MS 4/76, James Hutton to Watt, dated only 1774.

53. See Margaret C. Jacob, "The Materialist World of Pornography," in Lynn Hunt, ed., *The Invention of Pornography. Obscenity and the Origins of Modernity*, New York, Zone Books, 1994, pp. 157–202.

54. JWP, MS C2/2, list of books and prints bought in France.

55. Boulton and Watt Papers, BPL, MI/6/9, for a list.

56. For Gregory see JWP, MS C2/15, which also provides a good account of Watt's total assets in 1804; for Watt himself see MI/6/12, dated 7 July 1819.

57. Jan Golinski, *Science as Public Culture. Chemistry and Enlightenment in Britain, 1760–1820*, New York, Cambridge University Press, 1992, pp. 176–94.

58. See "William Strutt—A Memoir," a typescript, Derby Local Library, no. 3542, p. 60; and Fitzwilliam Library, Cambridge, Strutt MS 48-1947.

59. J. Gascoigne, *op. cit.*, p. 245. For Watt Jr. and the Manchester club see JWP, MS W/6, his letter to his father, Paris, 22 April 1792. For political troubles in the Lunar Society see JWP, L/B 1, Watt to Dr. Black, 23 Nov. 1791.

60. JWP, LB/1, Watt to James Jr., 13 March 1785, advising him to never lose sight of the "Christian precept do unto others as you would have them do unto you. I am your true friend."

Chapter 7

1. *A Collection of Dissertations Issued by Dutch Universities . . . Leiden, Utrecht, Groningen, Hardewijk*, 42 theses in all housed in the Rare Book Room, Van Pelt Library, University of Pennsylvania, Philadelphia.

2. Ester Boserup, *Population and Technology*, Oxford, Blackwell, 1981, p. 4. For a popular text that does now at least acknowledge the need for education, see Simón Teitel, *Industrial and Technological Development*, published by the Inter-American Development Bank, distributed by Johns Hopkins University Press, Washington, D.C., 1993, pp. 241–43.

3. Shelby T. McCloy, *French Inventions of the Eighteenth Century*, Lexington, University of Kentucky Press, 1952, p. 13.

4. British Library, London, MSS ADD. 33, 564, diary of Samuel Bentham while in Russia, fol. 21. The machine shown was for driving piles. For a much more comprehensive treatment of Russian science than is possible here, see Valentin Boss, *Newton and Russia: The Early Influence, 1698–1796*, Cambridge, Mass., Harvard University Press, 1972.

5. D. S. L. Cardwell, *The Organization of Science in England*, London, Heinemann, 1972, pp. 17–18.

6. For a sophisticated statement of the lead, see G. Timmons, "Education and Technology in the Industrial Revolution," *History of Technology*, 8 (1983): 135–49. For a clear statement of how the "new" economic history discounts the entrepreneur, see Clive Trebilcock, *The Industrialization of the Continental Powers, 1780–1914*, London, Longman, 1981, p. 141; cf. pp. 63–65 on the critically important role of science and technology to late nineteenth-century German industrial development.

7. The John Rylands Library, Manchester, ENG MSS 1110, to Josiah Wedgwood from Prof. Pictet, Geneva, November 28, 1787.

8. Siegfried Giedion, *Mechanization Takes Command*, New York, Norton, 1969, p. 35. For a balanced account of Vaucanson, see Charles C. Gillespie, *Science and Polity in France at the End of the Old Regime*, Princeton: Princeton University Press, 1980, pp. 414–17.

9. D. Todericiu, "Jean Hellot (1685–1766), savant chimiste, fondateur de la technologie chimique en France au XVIIIe siècle," *Comptes rendus du Congres National des Societés Savants*, Caen, 1980, pp. 201–11.

10. Abbé Nollet, *Leçons de Physique experimentale*, Amsterdam and Leipzig, 1754, vol. 1, preface, pp. xxii–xxv.

11. Ibid., vol. 1, p. 44.

12. Ibid., vol. 3, pp. 1–5.

13. L. W. B. Brookliss, "Aristotle, Descartes and the New Science: Natural Philosophy at the University of Paris, 1600–1740," *Annals of Science*, 38 (1981): 57–58, 67–68; cf. for a good general discussion, Henry Guerlac, *Newton on the Continent*, Ithaca, N.Y., Cornell University Press, 1981.

14. L. W. B. Brockliss, *French Higher Education in the Seventeenth and Eighteenth Centuries*, Oxford, Clarendon Press, 1987, pp. 353–58, 376–80, 366 for the quotation. There was still, however, a strong emphasis on mathematical skills in university courses. The French colleges are the nearest equivalent to the Dissenting academies. In the year XI, the first *Bulletin de la société pour l'industrie nationale*, Paris, p. 179, complained that "on s'est peu occupé en France de technologie, et jamais cette étude n'a fait partie de l'instruction publique." Supplied by Jeff Horn.

15. R. R. Palmer, "The Central Schools of the First French Republic: A Statistical Survey," in *The Making of Frenchmen: Current Directions in the History of Education in France, 1679–1979*, Donald N. Baker and Patrick J. Harrigan, eds.; a special issue of *Historical Reflections*, vol. 7, Waterloo, Can., Historical Reflections Press, 1980, pp. 230–31. For the *ancien regime* he is relying on the figures of Taton; in the 1790s adults, both men and women, began to seek education in physics, and these schools had pupils ranging from age 15 to 30. By this time the courses in physics and mechanics are remarkably uniform and employ the textbooks of Brisson, Nollet, and Chaptal or Fourcroy in chemistry. Where there were no machines professors drew descriptions of them and

they sometimes indicated their application in manufacturing; see AN, Paris, 17 1344/1, the entire box. In year 7 Brisson was teaching 200 students; most of the other 800 (approx.) respondents are teaching about 25 to 40; we will take 30 as the average.

16. The John Rylands Library, Manchester, ENG MS 724, John Walsh's Diary, "Journey to France, 1772," entry for June 17; entry for June 18 on instruments.

17. Shelby J. McCloy, *French Inventions of the Eighteenth Century*, Lexington, University of Kentucky Press, 1952, pp. 30–31, 112–13.

18. R. Rappaport, "Government Patronage of Science in Eighteenth Century France," *History of Science*, 8 (1969): 119–36.

19. James E. McClellan, "Un Manuscrit inedit de Condorcet: Sur l'utilité des académies," *Revue d'histoire des sciences*, 30 (1977): 247–48; cf. Keith Baker, *Condorcet*, Chicago, University of Chicago Press, 1975, pp. 2–28, 401. For science in eighteenth-century Spain, see David Goodman, "Science and the Clergy in the Spanish Enlightenment," *History of Science*, 21 (1983): 111–40.

20. James McClellan III, *Science Reorganized*, pp. 9–10.

21. Heilbron, *Electricity*, pp. 115–17.

22. Daniel Roche, *Le Siècle des lumières en Province*, Paris, Mouton, 1978, vol. 1, p. 329.

23. Dorinda Outram, "The Ordeal of Vocation: The Paris Academy of Sciences and the Terror, 1793–95," *History of Science*, 21 (1983): 254–55.

24. Library of the University, Strasbourg, MS 1432, 1785; cf. Margaret C. Jacob, *Living the Enlightenment. Freemasonry and Politics in Eighteenth Century Europe*, New York, Oxford University Press, 1991, pp. 199–202.

25. John Hubbel Weiss, *The Making of Technological Man: The Social Origins of French Engineering Education*, Cambridge, Mass., MIT Press, 1982, pp. 13–24.

26. Jean Dhombres, "L'enseignement des mathématiques par la 'methode révolutionnaire.' Les leçons de Laplace à l'Ecole normale de l'an III," *Revue d'histoire des sciences*, 33 (1980): 315–48.

27. Janis Langins, "Sur la première organisation de l'Ecole polytechníque. Texte de arreté du 6 frimaire an III," *Revue d'histoire dessciences*, 33 (1980): 289–313.

28. Denis Diderot, *Oeuvres completes*, Paris, 1875, vol. 3: "Plan d'une université pour le gouvernement de Russie," p. 429, for "leur mère commune et leur infatigable ennemie"; and p. 457.

29. Charles C. Gillespie, *Science and Polity in France at the End of the Old Regime*, Princeton, Princeton University Press, 1980, p. 90.

30. R. Rappaport, "Government Patronage of Science in 18th Century France," *History of Science*, 8 (1969): 119–36.

31. C. Stewart Gillmore, *Coulomb and the Evolution of Physics and Engineering in Eighteenth Century France*, Princeton, Princeton University Press, 1971, pp. 12–14. In The Netherlands, too, military engineering was much more highly developed than was civil; see Harry Lintsen, *Ingenieurs in Nederland in der negentiende eeuw*, The Hague, Nijhoff, 1980, pp. 23–28. For a good illustration of the French "style" of scientific inquiry versus the British, see Richard Gillespie, "Ballooning in France and Britain, 1783–1786," *Isis*, 75 (1984): 249–68.

32. See the student notebooks of Eleuthère Irénée du Pont (b. 1771), Hagley Museum and Library, Delaware, Longwood MSS, Series B Box 10, course notes taken at the Collège Royal in the period 1784–89, on natural history, physics, pneumatics,

botany, and notes from books by Desaguliers, Nollet, and Franklin; lesson of 5 Feb. 1789 on simple and complex pumps; copy book for 1787 on specific gravity of water and gravity in general. Compare M. Sigorgne, de la Maison & Société de Sorbonne, Professeur de Philosophie en l'Université de Paris, *Institutions Newtoniennes, ou introduction a la philosophie de M. Newton*, Paris, 1747, with this later text, which illustrates the change that occurred in the next half century: Mathurin-Jacques Brisson, *Traité élémentaire, ou principes de physique*, Paris, An VIII, p. v: "Cet ouvrage, qui est destiné à la jeunesse de l'un et l'autre sexe, comprend toutes les questions relatives à la Physique. . . ." It is complete with illustrations that could have been out of Desaguliers, and it made physics and mechanics accessible to any highly literate reader.

33. Jacques Payen, *Capital et machine à vapeur au xviiie siècle. Les frères Périer et l'introduction en France de la machine à vapeur de Watt*, Paris, Mouton & Co., 1969, p. 129.

34. On the French engineering corps see Anne Blanchard, *Les ingénieurs du "roy" de Louis XIV à Louis XVI*, Montpellier, l'Université Paul-Valéry, 1979, pp. 182–94; note the absence of any machinery or mechanical instrumentation in the description of the curriculum in mathematics, mechanics, and hydraulics. Note also (p. 236) the increasingly noble character of engineering corps after 1748. She builds on and confirms the work of Roger Chartier, "Un recrutement scolaire au xviiie siècle. L'ecole royale du génie de Mézières," *Revue d'Histoire Moderne et Contemporaine*, 20 (1973): 353–75.

35. Margaret Bradley, "Engineers as Military Spies? French Engineers Come to Britain, 1780–1790," *Annals of Science* 49, no.2 (March 1992): 137–61.

36. Blanchard, *op. cit.*, pp. 289–311.

37. Blanchard, *op. cit.*, pp. 453–61, p. 465.

38. Richard L. Gawthrop, *Pietism and the Making of Eighteenth-Century Prussia*, Cambridge, Cambridge University Press, 1993, pp. 55–57.

39. James Watt Papers, BPL, MS W/6, James Watt, Jr., to his father, Naples, 15 Jan. 1793.

40. G. Vanpaemel, "Rohault's *Traité de Physique* and the Teaching of Cartesian Physics," *Janus*, 71–74 (1984): 31–40. See also by the same author, *Echo's van een wetenschappelijke revolutie. De mechanistische natuurwetenschap aan de Leuvense Artesfaculteit (1650–1797)*, Brussels, Verhandelingen van de Koninklijke Academie voor Wetenschappen, Letteren en Schone Kunsten van België, 1986.

41. A. Rupert Hall, "Further Newton Correspondence," *Notes and Records of the Royal Society of London*, 37, no. 1 (1982): p.32. I owe the point about Pitcairne to Anita Guerrini.

42. J. L. Heilbron, *Electricity in the Seventeenth and Eighteenth Centuries: A Study of Early Modern Physics*, Berkeley, University of California Press, 1979, p. 142. On decline in the Dutch universities, see also J. Israel, *The Dutch Republic*, pp. 1050–51.

43. Heibron, *Electricity*, p. 26.

44. J. T. Desaguliers, *De Natuurkunde uit Ondervindingen*, Amsterdam, Isaak Tirion, 1751; first edition, 1736. Cf. Edward G. Ruestow, *Physics at Seventeenth and Eighteenth Century Leiden: Philosophy and the New Science in the University*, The Hague, Nijhoff, 1973, pp. 143–44; cf. C. de Pater, *Petrus van Musschenbroek (1692–1761) een Newtonians natuuronderzoeken*, Utrecht, Elinkwijk, 1979.

45. See D. van der Pole, "De introductie van de Stoommachine in Nederland," in J. de Vries, ed., *Ondernemende Geschiedenis*, The Hague, 1977.

46. Royal Library, The Hague, MS 128 B. 3., s'Gravesande MSS. Cf. J. N. S. Allamand, *Catalogus van eene aanzienlijke Verzameling van allerleije . . . Instrumenten*, Amsterdam, 1788, which includes a list of s'Gravesande's instruments, among them copies of windmills and water mills, electrical devices, etc.

47. Royal Society, MS 702, e.g., s'Gravesande, Justus van Effen, Sallengre, St. Hyacinthe, William Bentinck. On Sallengre and Newton, see A. Rupert Hall, "Further Newton Correspondence," p. 26.

48. University Library, Leiden, Marchand MS 2, 15, 7, *bre*, 1723, from Surinam; Jac. de Roubain to P. Marchand: "Vous pourrez en etre plus particulièrement informée le plan que j'ai ici joint, et si vous vouliez abjurer le Newtonnisme je suis aussi puis d'abjurer le Carthesianisme."

49. A. C. de Hoog, "Some Currents of Thought in Dutch Natural Philosophy," Ph.D. dissertation, Oxford University, 1974, p. 295. On Fahrenheit, see University Library, Leiden, MS BPL 772; and Pieter van der Star, ed. and trans., *Fahrenheit's Letters to Leibniz and Boerhaave*, Amsterdam, Rodopi, 1983, p. 13.

50. Harry Lintsen and Rik Steenaard, "Steam and Polders. Belgium and The Netherlands, 1790–1850," *Tractrix. Yearbook for History of Science, Medicine, Technology and Mathematics*, 3 (1991): 122–26. These authors favor purely economic explanations. For a count of French engines see AN, F12 2200, memoir dated 8 April 1817.

51. For a detailed description of factories in Gouda, Amsterdam, Haarlem (poor houses particularly), Schiedam, Utrecht, Zaandam, making paper, biscuits, refining salt (seasonal, run by women), camphor, purifying borax, grinding corn, making pipes (one factory making over 5 million white clay pipes a year), bricks, etc., complete with drawings see L'Ecole des Ponts et Chaussées, Paris, MS 3013 (2), Sganzin, a French engineer, whose reports from 1795, approx. 100 ff. include conversations with Dutch engineers. Men, women, and children can be found working in most of these factories. On the polders and windmills see the folder labeled "extrait du voyage . . . machines à epuiser." Note that invariably the French engineers regarded the Dutch as "industrious" and the Belgians as "careless." See AN, Paris, F12 508 for a list of every *fabrique* and windmill for water in the country in 1810.

52. For more detail see Margaret C. Jacob, *The Cultural Meaning*, pp. 189–92. For a typical philosophical society in the Republic at this time see M. J. van Lieburg, *Het Bataafsch Genootschap der Proefondervindelijke Wijsbegeerte te Rotterdam 1769–1984; een bibliografisch en documenterend overzicht* in *Nieuwe Nederlandse Bijdragen tot de Geschiedenis der Geneeskunde en Natuurwetenschappen*, vol. xviii, Amsterdam, 1985. On the division of opinion on the merits of mechanization, see the essays submitted to the Hollandsche Maatschappij der Wetenschappen, Haarlem, for 1827 and 1837, folder #370, found in the archives of the society in Haarlem.

53. Archives générales du Royaume, Brussels, Conseil privé, MS 1097 B, Vincent Mousset described as an engineer and mechanician.

54. Birmingham City Library, Birmingham, U.K., Boulton and Watt MSS, Box 36/17 J.D.H. van Liender to Watt, 21 Oct. 1790. Dutch scientific education discussed in greater detail in Margaret C. Jacob, *The Cultural Meaning. . .*, McGraw-Hill, 1988, chap. 6. See also I. Inkster, "The Public Lecture as an Instrument of Science Education for Adults—The Case of Great Britain, c. 1750–1850," *Paedogogica historica*, 20 (1981): 80–112, and see note 4. For an engine bought by the province of Utrecht for drainage see description in ENPC, Paris, MS 3013 (1), loose page with drawing. In a letter to Watt in May 1786 Van Liender describes how Dutch patenting works.

55. Provincial Archives, Middleburg, on the various efforts made in fits and starts, see archives of Zeeuwsch Genootschap der Wetenschappen, 1769–1969, for 1782, prize essay of October 1806 by T. Speleveld, 1809 on the commission, 1815 another commission, new harbor of 1817, etc. On one of the key engineers of the period, Jan Blanken, see R. M. Haubourdin, *Inventaris van Kaarten, tekeningen en modellen van de waterbouwkundige ingeniers*, The Hague, 1984.

56. C. A. Davids, *Zeewezen en wetenschap: De wetenschap en de ontwikkeling van de navigatie techniek in Nederland tussen 1585 en 1815*, Amsterdam, 1986. For the commercial ideology of one of its spokesmen, see Wyger R. E. Velema, *Enlightenment and Conservatism in the Dutch Republic. The Political Thought of Elie Luzac (1721–96)*, Maastricht, Van Gorcum, 1993, pp. 124–32.

57. Middleburg, Gemeente Archief, Register ten Rade, deel 2, f.365.

58. Rijksarchief, Arnhem, MSS of the Academy of Harderwijk, nos. 154, 153, 155, 156, 157, 141.

59. Rijksarchief, Arnhem, MSS of J. van Leeuwen, nos. 5 and 6; note praise of Freemasons (no. 6, fol. 10 ff.).

60. Willem Frijhoff, "Deventer en zijn gemiste universiteit, Het Athenaeum, in de sociaal-culturele geschiedenis van Overijssel," *Vereeniging tot Beoefening van Overijsselsch regt en geschiednis, Verslagen en Medeelingen*, 97 (1982): 71.

61. Thomas Schwenke, *Noodig bericht over de Inventinge der Kinderpokjes*, The Hague, 1756, p. 15; he was able to inoculate only 41 prominent citizens in a city of approximately 35,000.

62. Rijksarchiv Friesland, Leeuwarden, FA Van Sminia 1944a, diary of Hessel Vegelin van Claerbergen, see 41 f; et seq. for a rich portrait of Allamand.

63. Rijsarchiv Friesland, Leeuwarden, FA Van Sminia MS 1944a, 40–81 ff.

64. For example, by Phyllis Deane, "Industrial Revolution in Great Britain," in Carlo Cipolla, ed., *The Emergence of Industrial Societies*, Hassocks, Sussex, Harvester Press, 1976, p. 177, where technological know-how in the Dutch republic (p. 174) is vastly exaggerated. For a good summary of the various and older Dutch contributions to this question, see J. G. van Dillen, "Omstandigheden en psychische factoren in de economische geschiedenis van Nederland," in *Mensen en achtergronden*, Groningen, Wolters, 1964, pp. 53–79.

65. For a list of these societies, see J. H. Buursma, *Nederlandse Geleerde Genootschappen opgericht in de 18deeuw*, The Hague, Discom, 1978; cf. James E. McClellan III, *Science Reorganized: Scientific Societies in the Eighteenth Century*, New York: Columbia University Press, 1985, pp. 9–10.

66. For the transactions of this society, see *Verhandelingen uitgegeeven door de Hollandse Maatschappij der Wetenschappen te Haarlem*, vol. 1 (1754) to vol. 11. Cf. MSS of the society, at its offices in Haarlem, "Notulen 1752–67"; see also R. J. Forbes, ed., *Martinius van Marum, Life and Work*, Haarlem, Teyler's Museum, 1969; and J. A. Bierens de Haan, *De Hollandsche Maatschappij den Wetenschappen, 1752–1952*, Groningen, Willink, 1977.

67. Anon., *Aanspraak gedann aan de Goede Burgeren, die tot Welzyn van stad en land, op den 9 Augustus 1748, op den Cloveniers Doelen vergadert zyn geweest*, Amsterdam, 1748, p. 1: "de Konsten en Wetenschappen zyn onbeloond van ons gevlooden; de Koophandel is haare Stief-Vaders ontvlugt; de Fabriquen, die onuitputbaare Goudmynen der Volkeren, en waarop deeze STAAT met regt zig voormaals dorft beroemen, en waarop dezelve is gevest, zyn naar andere Natien overgegaan."

68. Marten G. Bruist, *At Spes non Fracta. Hope & Co. 1770–1815: Merchant Bankers and Diplomats at Work*, The Hague, Nijhoff, 1974, p. 9.

69. University Library, Amsterdam, MS. X.B.1, "Leçons de Physique de Mr le Prof. Koenig qu'il a donne à la Haye, 1751–52," 348 ff. These lectures were almost certainly for the circle around the Bentinicks and the court, given the opening remarks and the use of French. On Koenig and du Châtelet see Keiko Kawashima, "Les idées scientifiques de Madame du Châtelet dans ses *Institutions de physique*," *Historia scientiarum*, 3 (1993): 63–69.

70. Royal Library, The Hague, MS 75. J. 63, "Leçons d'Arithmétique et d'Algebre a l'usage . . . Le Prince d'Orange," May 1759, fol. 34 ff.

71. See Giles Barber, "Aspects of the Booktrade Between England and the Low Countries in the 18th Century," *Documentatieblad werkgroep achttiende eeuw*, no. 34–35 (1977): 47–63; and Robert Schofield, *Mechanism and Materialism: British Natural Philosophy in an Age of Reason*, Princeton, Princeton University Press, 1970, pp. 137–40, on B. Nieuwentyt's *The Religious Philosopher* (1718–1719) and its many English editions; the translator was Desaguliers, who compared the author to John Ray and William Derham. Cf. de Hoog, "Dutch Natural Philosophy," p. 295 ff. on Nieuwentyt.

72. Rijksarchief in Gelderland, Familiearchief Van Eck 82; brought to my attention by Arianne Baggerman who along with Rudolf Dekker is doing an edition of the diary. Cf. Rudolf Dekker, *Uit de Schaduw in 't grote licht. Kinderen in egodocumenten van de Gouden Eeuw tot de Romantiek*, Amsterdam, WereldBibliotheek, 1995.

73. Simon Schama, *Patriots and Liberators: Revolution in the Netherlands, 1780–1813*, New York, Knop, 1977, p. 50.

74. Dr. William's Library, London, Wodrow-Kenrick correspondence, MS. 24. 157, fol. 41; dated 1760.

75. See MS of Concordia et Libertate, Gemeente Archief, Amsterdam, P.A.9.1–10.

76. For an Orangist society see A. J. J. Ph. Haas, "De Saturdagse Krans 1718–93. Een gezellige vereeniging van Amsterdamsche Regenten in de 18de eeuw," *Koninklijk Oudheidkundig genootschap Amsterdam*, 77 (1934–1935): 66–79.

77. I. K. van der Pols, "Early Steam Pumping Engines in the Netherlands," *Transactions of the Newcomen Society*, 46–47 (1973–1976): 13–16. See also Peter Mathias, "Skills and the Diffusion of Innovations from Britain in the Eighteenth Century," *Transactions of the Royal Historical Society*, 25 (1975): 99, where we also learn that Dutch artisans were prominent in technology transfer, but to Spain and Russia (p. 94). On use of the steam engine by the Austrian government, see M. Teich, "Diffusion of Steam-, Water-, and Air-Power to and from Slovakia During the 18th Century and the Problem of the Industrial Revolution," *Colloques Internationaux, Centre National de la Recherche Scientifique*, no. 538. On steam in the Republic, see also H. W. Lintsen, ed., *Techniek in Nederland. De wording van een moderne samenleving 1800–1890*, vol. 4, Zutphen, Walburg Pers, 1993, pp. 131–148.

78. *De Koopman*, 1 (1768): 40, 332–333.

79. Ibid., 4 (1773): 172.

80. See MSS of Felix Meritis, Gemeente Archief, Amsterdam, P.A. 59. 19.

81. Anon., *Redenvoering over het algemeen nut der Wetenschappen, fraaije letteren en konsten . . . Felix Meritis*, 1788; bound with J. H. van Swinden, *Redenvoering en aanspraak ter . . . inwijling van het gebouw der maatschappij Felix Meritis*, Amsterdam, 1789, pp. 29–30.

82. H. A. M. Snelders, "Het Department van natuurkunde van de Maatschappij van verdiensten Felix Meritis in het eerste kwart van zijn bestann," *Documentatieblad werkgroep achttiende eeuw*, 15 (1983): 200.

83. Benjamin Bosma, *Gronden der Natuurkunde*, Amsterdam, 1764. The edition of 1793 states the author's pride at having continued this tradition of lecturing for so many decades. Concordia et Libertate gave money to the radical reformers in 1748.

84. Benjamin Bosma, *Redenvoering over de Wijsbegeerte*, Amsterdam, 1767, and *Redenvoering over de Natuurkunde*, Amsterdam, 1762, pp. 5–8.

85. *Beknopte aanspraak, van den Heere Martinus Martens, uitgesprooken volpens jaarlykse gewoonte op den 6 Februari 1741*, Amsterdam, 1741, pp. 6, 12, 15, 17.

86. *Korte Beschrijving van de samenstelling en werking der Vuur of Stoommachine volg. Watt en Boulton. Met het rapport van J. H. van Swinden en C. H. Damen daarover*, 1789; University of Amsterdam, Library, sign 473.A 13. Cf. H. A. M. Snelders, "Lambertus Bicker (1732–1801), An Early Adherent of Lavoisier in the Netherlands," *Janus*, 67 (1980): 104–22n. For another example of the link between industrial interests and the *patriotten* movement, see C. Elderink, *Een Twentsch Fabriqueur van de achttiende eeuw*, Hengelo: Broekhuis, 1977, pp. 73–74.

87. On the Athenaeum, see *Gedenkboek van het Athenaeum en de Universiteit van Amsterdam, 1632–1932*, Amsterdam, 1932. I am very grateful to Mrs. Feiwel for her assistance with these archives.

88. For example, *Van Vaderlandsche Mannen en Vrouwen uit de zuidelijke provincien: Een Schoolboek. Uitgegeven door de Maatschappij tot Nut van 't Algemeen*, Leiden, Deventer, and Groningen, 1828; many subsequent editions. On educational reforms after the revolution and the importance attached to science, see Aart de Groot, *Leven en Arbeid van J. H. van der Palm*, Utrecht, University of Utrecht, 1960.

89. "Journal der reize van den agent van Nationale economie der Bataafsche Republick," *Tijdschrift voor Staathuishoudkunde en statistiek*, 18, 19 (1859–1860).

90. Quoted and discussed in Margaret C. Jacob, "Radicalism in the Dutch Enlightenment," in Margaret C. Jacob and Wijnand Mijnhardt, eds., *The Dutch Republic in the Eighteenth Century. Decline, Enlightenment and Revolution*, Cornell University Press, Ithaca, N.Y., 1992, pp. 229–40.

91. Quoted in C. R. Boxer, *The Dutch Seaborne Empire 1600–1800*, London, Hutchinson, 1965, p. 271. On navigational technology see C. A. Davids, *Zeewezen en Wetenschap. De wetenschap en de ontwikkeling van de navigatietechniek in Nederland tussen 1585 en 1815*, Amsterdam: De Bataafsche Leeuw, 1986. I wish to thank Dr. Davids for his helpful comments.

92. Ijsbrand van Hamelsveld, *De zedelijktoestand der Nederlandsche natie, op het einde der achttiende eeuw*, Amsterdam, 1791, p. 285; see also p. 244, where he calls for taking uncorrupted youths (from north Holland) and educating them "in art or science."

93. For background see Harry Lintsen, *Ingenieurs in Nederland in de negentiende eeuw*, The Hague, Nijhoff, 1980; C. Elderink, *Een Twentsch Fabriqueur van de achttiende eeuw*, Hengelo, 1977; Jonathan Irvine Israel, *Dutch Primacy in World Trade, 1585–1740*, New York, Oxford University Press, 1989; Margaret C. Jacob and W. W. Mijnhardt, eds., *The Dutch Republic in the Eighteenth Century. Decline, Enlightenment, and Revolution*, Ithaca, N.Y., Cornell University Press, 1992.

94. René Leboutte, "From Traditional Know-How to Technical Skill. The Process of Training and of Professionalization in the Belgian Coal-Mining Industry, 1700–1850," *History and Technology*, 12 (1995): 95–108.

95. Mons, Archives d'etat, MS A.E.M.Charbonnages B•ois du Luc, 51–87 ff., from the 1730s to 1780; the decision to install and the actual installation. In 1750 when horses are still being used they made representations "au Sieur Biseau en sa qualité de Seigneur . . . Houdeng. La justice et necessite de faire par lui certain moderation sur droit. . . ." For this company see also J. Plumet, "Une Société. . .," *Annales du Cercle Archeologique de Mons*, 57 (1940): 89–95. On fire and steam engines in the Belgian mines see Hervé Hasquin, *Une Mutation le "Pays de Charleroi aux XVIIe et XVIIIe siècles. Aux origines de la Révolution industriele en Belgique*, Université Libre de Bruxelles, 1971.

96. Rijksarchief Limburg, Maastricht, archieven en de handschriften der abdij Kloosterrade, MS 1091, film no. 12.

97. J. Breuer, "Matériaux pour l'histoire du Corps du Génie dans les Pays-Bas autrichiens de 1717 à 1756," *Revue Internationale d'Histoire Militaire*, 6 (1960–66): 337–54.

98. See Max Barkhausen, "Government Control and Free Enterprise in Western Germany and the Low Countries in the Eighteenth Century," in Peter Earle, ed., *Essays in European Economic History, 1500–1800*, Oxford: Clarendon Press, 1974, pp. 248–50. And Hervé Hasquin, *Le "Pays de Charleroi" aux XVIIe et XVIIIe siècles. Aux origines de la révolution industrielle en Belgique*, Brussels, Université libre de Bruxelles, 1971, p. 80; for interest in the Newcomen engine with a mastery of its operation, see pp. 138–39n.

99. D. Droixhe, "Noblesse éclairée, bourgeoisie tendre dans la principauté de Liège au XVIIIe siècle," *Études sur le XVIIIe siècle*, 9 (1982): 9–47, especially, 24–31.

100. Hervé Hasquin, ed., *La vie culturelle dans nos provinces au XVIIIe siècle*, Brussels, Credit Communal, 1983, pp. 132–33.

101. Annette Andre-Felix, *Les débuts de l'industrie chimique dans les Pay-Bas autrichiens*, Brussels, Université libre de Bruxelles, 1971.

102. A copy of this plan can be found in Rijksarchief Limburg, archief Kloosterrade, in the papers of S. P. Ernst; no. 2061 on film 51; *Plan provisionnel d'études ou instructions pour les professeurs des classses respectives dans les pensionnaits, colleges ou ecoles publiques aux pays-bas*, 1777 and intended for education in both Flemish and French. It was possible in 1740 for a French professor of hydrography to write to the Academy of Science in Paris attacking the Coperican system; see AN, Paris, G 94 (Marine), 74–84 ff.

103. See for example Friedrich Gren, *Grundriss der Naturlehre zum Gebrauch akademischer Vorlesungen*, Halle, 1788; see also industrial school projects discussed in *Göttingisches Magazin zur Industrie und Armenpflege*, 1 (1789), and annually thereafter.

104. *Programm . . . Joachimsthalsches Gymnasium*, Berlin, 1735, Staatsbibliothek, AH 15768; for girls' education see Johann J. Hecker, *Teutsches Programma von den Verdiensten Kaysers Karl des Grossen*, 1749.

105. *Ankündigung der Vorlesungen und Uebungen* . . ., 1771; found in Staatsbibliothek, Berlin (records formerly housed in the DDR). I thank Axel Utz for his work on this section. See also J. A. G. Einem, *Feierliche Ankündigung der Schulprüfung*. . ., Berlin, 1764, p. 15.

106. I rely here on the account in Richard L. Gawthrop, *Pietism and the Making of Eighteenth-Century Prussia*, Cambridge, Cambridge University Press, 1993, *passim*.

107. For policies later in the eighteenth century in one of the smaller absolutist states, see Robert Uhland, "Karl Friherr von Kerner: Offizier, Techniker, Erneuerer des württembergischen Berg- und Hüttenwesens," in *Ludwigsburger Geschichtsblätter*, 29

(1977): 5–68. On Halle, see Richard L. Gawthrop, *Pietism*, p. 61, and the academy, p. 65.

108. Collège Royal Francois, *Relation de l'école de charité*, 1781, Staatsbibliothek, AH 15753, no. 38.

109. Johan Julius Hecker, *Mit der Jugend welche in den Schulanstalfen der Dreyfaltigkeits-Kirche*, Berlin, 1748.

110. Johann Julies Hecker, *Nachricht von einer Oeconomisch-Mathematischen Real-Schule welche bey den Schul-Anstalten der Dreyfaltigkeits-Kirche*, Berlin, 1747.

111. Andreas J. Hecker (possibly the son of J. Hecker), *Geschichte der Königliches Realschule*, January 1797, Berlin, found in Staatsbibliothek, AY 15288.

112. See [Anon.] *Vorläuffige Nachricht*, 1745, and *Anzeige der Vorlesungen und Uebungen*, 1745, both found in Staatsbibliothek, Berlin.

113. R. Gawthrop, *op. cit.*, p. 221.

114. See, for example, R. Rey, "La circulation des idées scientifiques entre la France et l'Allemagne: Le cas Cuvier," in J. Mondot, J-M. Valentine, V. Jürgen, eds., *Deutsche in Frankreich, Franzosen in Deutschland, 1715–1789*, Sigmaringen, Jan Thorbecke Verlag, 1992.

115. Marita Hein, "Wissenschaftstransfer zwischen Deutschland und dem belgischen Raum im 18. und frühen 19. Jahrhundert: Kontakte der Brüsseler Akademie und einzelner Gelehrter im Grenzgebiet Maas und Rhein," *Rheinische Vierteljahresblätter*, no. 56, 1992, 206–228.

116. See Pamela H. Smith, *The Business of Alchemy. Science and Culture in the Holy Roman Empire*, Princeton, Princeton University Press, 1994, pp. 247–62.

117. Martina Lorenz, "Der Einfluss Christian Wolffs (1679–1754) auf das Physikverständnis der Naturforscher und den protestantischen deutschen Universitäten der Aufklärungszeit," in Friedrich-Schiller-Universität, Jena, *Wissenschaft und Schulenbildung*, Jena, Universitätsverlag, 1991, pp. 114–19.

118. For these developments see Hans-Peter Müller and Ulrich Troitzsch, eds. *Technologie zwischen Fortschritt und Tradition: Beiträge zum internationalen Johann Beckmann-Symposium, Göttingen 1989*, Frankfurt-am-Main, Peter Lang, 1992.

119. Peter Lundgreen, "Education for the science-based industrial state? The case for nineteenth-century Germany," *History of Education*, 13 (1984): 59–67. For the forces of reaction, see Robert M. Berdahl, *The Politics of the Prussian Nobility. The Development of a Conservative Ideology, 1770–1848*, Princeton, N.J., Princeton University Press, 1988.

120. Manuscript of these lectures, possibly by a student named Pruninger, to be found at the Bakken Library, Minneapolis, dated 1795.

121. W. Weber, "Friedrich Anton von Heynitz," in Wilhelm Treue and Wolfgang König, eds., *Berlinische Lebensbilder*, vol. 6, *Techniker*, Berlin, Colloquium Verlag, 1990, pp. 15–28.

122. Eric Dorn Brose, *The Politics of Technological Change in Prussia. Out of the Shadow of Antiquity, 1809–1848*, Princeton, Princeton University Press, 1993.

123. Friedrich Klemm, *A History of Western Technology*, Ames, Iowa State University Press, 1991 [1954], p. 244, quoting from Johann Beckmann, *Anleitung zur Technologie*, Göttingen, 1777. Cf. Karl Hufbauer, *The Formation of the German Chemical Community*, Berkeley, University of California Press, 1982, and Lars U. Scholl, *Ingenieure in der Frühindustrialisierung: Staatliche und privat Techniker im Königreich Hannover und an der Ruhr (1815–1873)*, Göttingen, Vanderhoeck & Ruprecht, 1978.

124. See records in AN, Paris, F12 2204, 17 Brumaire to Ministre, Conseiller de Regence à Berlin, signed Plümieke who had been a paper manufacturer: "Il est bien triste, mais tres fondé, qu'en général les souverains de l'Allemagne, soutiennent rarement avec vigueur les fabriques & les manufactures, qui sans contredit sont la base la plus resurre des bien-être des états." He wants to encourage cotton manufacturing.

125. See Herbert Kisch, *From Domestic Manufacture to Industrial Revolution. The Case of the Rhineland Textile Districts*, Oxford, Oxford University Press, 1989.

126. Winfried Speitkamp, "Educational Reforms in Germany between Revolution and Restoration," *German History*, 10 (1992): 1–23.

127. Vincenzo Ferrone, *The Intellectual Roots of the Italian Enlightenment. Newtonian Science, Religion, and Politics in the Early Eighteenth Century*, Atlantic Highlands, N.J., Humanities Press, 1995, pp. 1–16.

128. Ferrone, *op. cit.*, p. 4.

129. Paola Zambelli, "Antonio Genovesi and Eighteenth Century Empiricism in Italy," *Journal of the History of Philosophy*, 16 (1978): 198–99.

130. Zambelli, "Antonio Genovesi," p. 208.

131. See Vincenzo Ferrone, "Tecnocrati militari e scienziati nel piemonte dell'antico regime. Alle origini della reale accademia della scienze di torino," *Rivista storica italiana*, 96, no. 2 (1984): 414–509. Note the presence here of freemasonry.

Chapter 8

1. Conservatoire des Arts et Metiers, Paris, MS U 216 Le Turc to Citoyen, 14 Nivoise An 3 [December, 1794]. Le Turc was born in 1748 and in the 1780s as an engineer and spy he traveled extensively in England describing techniques and recruiting workers. I owe this splendid quotation to the kindness of J. R. Harris.

2. The phrase belongs to Philippe Minard, *L'inspection des manufactures en France, de Colbert à la Révolution*, doctorat nouveau régime, Université Paris-1 Panthéon-Sorbonne, December 1994, vol. II, p. 467, referring to correspondence from Trudaine to Tolozan. Between 1740 and 1789 the government spent 5 million and a half livres on subventions for inventions (p. 475). Made available through the kindness of Daniel Roche. On the early development of the division of labor in Britain, see Peter Earle, *The Making of the English Middle Class. Business, Society, Family Life in London, 1660–1730*, London, Methuen, 1989, pp. 18–34.

3. In passing the following essay makes the same point: Ian Inkster, "Technology as the Cause of the Industrial Revolution: Some Comments," *The Journal of European Economic History*, 12 (1983): 651–55; also writing from a cultural perspective is Thomas C. Cochran, "Philadelphia: The American Industrial Center, 1750–1850," *The Pennsylvania Magazine of History and Biography*, (July 1982): 323–40. According to Philippe Minard, *op. cit*, vol. II, p. 470, the French had sent industrial spies to England as early as the 1730s.

4. Archives nationales (AN), Paris, F12 502, a survey of French industry dated 1807. When the same administration tried to set up a school for public works to train engineers, its library began with the works of Newton. See the archives of the École des Ponts et Chaussées, (hereafter ENPC), MS 3013, list of books coming from the Library of the Stadholder, beginning with mathematics and astronomy.

5. See Alice Stroup, "Louix XIV as Patron of the Parisian Academy of Sciences," in David Lee Rubin, ed., *Sun King. The Ascendancy of French Culture during the Reign of Louis XIV*, Cranbury, N.J., Associated University Presses, 1991, pp. 221–337.

6. For the day-to-day working of one such bureau see Harold T. Parker, *An Administrative Bureau during the Old Regime. The Bureau of Commerce and Its Relations to French Industry from May 1781 to November 1783*, Newark, University of Delaware Press, 1993.

7. Paul Langford and Christopher Harvie, *The Eighteenth Century and the Age of Industry*, vol. IV in *The Oxford History of Britain*, New York, Oxford University Press, 1992, p. 78.

8. For archives see AN, Paris, Marine G 106, on pumps, 38–190 ff.; one of the earliest descriptions concerns a pump in a mine at Guadalcanal (Spain) done by an English company in 1731, 38 f. In the same archive a description of pumps installed in gardens in London by Newsham, 1743 (42 f.); 69 f. a pump of 1736 described as being able to elevate water in the English manner. By the 1770s (215–16 ff., 253) it is overwhelmingly clear that English pumps are superior. See also Marine G 108 Mémoires et Projets, Machines, 1768–81, 87 f. on water supply for Paris compared to superior London system and discussion of cost of coal in Paris, which is higher. On the silk industry in Lyon and John Badger, see AN, F12 1442 and letter of 23 Oct. 1753 on trying to stay on the "good side" of Mr. Montigny from the Académie des Sciences; F12 993 on bringing English technology in cotton to Rouen; note report of 1747 from Mons on English techniques complete with a sample of cloth. As early as 1758, if not earlier, French ministers were in contact with English steam engineers and making inquiries about getting coal for the new engines; see AN, Marine G 110, 133f., London 1758 letter of T. Stephens to Mr. Kavanagh. Note also that according to one French report, the King of Prussia had an agent in London "to instruct the state on different manufactures"; see AN F12 657/9, dated 1776. For a general survey of changes after 1789, see *Scientifiques et sociétés pendant la Révolution et l'Empire. Actes du 114e Congrès national des sociétés savantes*, Paris, 3–9 avril 1989, Paris CTHS, 1990. Cf. Jacques Payen, *Capital et machine à vapeur au xviiie siècle. Les frères Périer et l'introduction en France de la machine à vapeur de Watt*, Paris, Mouton & Co., 1969, p. 102n.

9. On this complex system of subsidies and grants see Liliane Hilaire-Pérez, "Invention and the State in 18th-Century France," *Technology and Culture*, 32, no.4 (1991): 911–31. This article cites other secondary sources where it is claimed that French administrators "did not feel that English industry was much more advanced than their own, and other historians have said much the same thing." None of this research, however, has been actually comparative, and in addition there is a wealth of primary source material that contradicts the assessment. See David S. Landes's useful introduction to *Favorites of Fortune* (1993), for a good corrective (p. 13): "foreign contemporaries of the Industrial Revolution were anxiously aware that something momentous was going on in Britain that threatened to upset not only commercial relationships but the international order."

10. ENPC MS 48 (fol.), "Journal. Notes et Observation sur l'Angleterre . . . 1784." The building (presumably New Jonathan's) was 200 ft by 170; Le Sage's notes do not even mention the sociology of spatial arrangement he drew, only the items he saw: the statues (of Charles I and II), the cafe, the registers of ships arriving, etc. See Thomas Mortimer, *Everyman His Own Broker: or, A Guide to Exchange-Alley*, London, 1775, pp. 43–50; 58–61 suggests that certainly national rivalries were quite real on the floor; p. 81 for mention of Jews. After this chapter was written I discovered the work of social scientists who seemed to be thinking about culture and economic life in ways somewhat similar to my own: Walter W. Powell and Paul J. DiMaggio, eds., *The New Institutionalism in Organizational Analysis*, Chicago, University of Chicago Press, 1991, pp. 1–37.

11. ENPC, Paris, MS 2465, dated 13 Mars 1782 and written in his hand. He worked with M. Macquer. There is a discussion of efforts to acquire English skilled workers, particularly English Catholics. His job was to oversee "toutes les Découvertes, Inventions, Machines, procedés utiles aux arts qui peuvent interesser le Commerce du Royaume." He also tried to "éclairer les pratiques des artistes en leur donnant des Connoissances de Théorie qui souvent sont audessus de leur portée." He further states: "les Magistrats faute être instruit dans les détails des arts et du Commerce, dans la Connoissances des Métiers, des Machines des Procédés sur les métaux et minéraux etoient souvent dupés par les Charlatans et laissoient engager le Public dans des Entreprises ruineuses en leur accordant des Privilèges dont ils abusoient et avec lesquels ils vivoient aux dépendes d'autorité." For John Badger's fear of de Montigny see AN, Paris F12 1442.

12. Ian Roy, "The Profession of Arms," in Wilfrid Prest, *The Professions in Early Modern England*, London, Croom Helm, 1987, pp. 209–15.

13. Robin Briggs, "The *Académie royale des sciences* and the pursuit of utility," *Past and Present*, no. 131, (May 1991): 38–87.

14. Here I refer to a letter of one Thomas Stephens to Mr Kavenagh, 29 August 1758 in AN, Paris Marine G 110, 138 f.; and see 33 f. for objections raised by the Académie to another proposal. For an example of a dubious proposal sent to the government see AN Marine G 105, no. 1, about S. Darles de Linière who has an invention by which men may use their arms to better augment the force of gravity with a pump. Some of his work was, however, quite useful. See also AN F 12 2201: the letter of 25 Dec. 1775 from Fleury D'Ardois to Turgot on how the high and the mighty "rien neglige pour diminuer le merité de sa découverte." The attack is on the Jurés Gardes du Bureau de la fabrique. See AN Microfilm 13, 5–7, 10 July 1783 on being intimidated by the Académie in Paris. See also the report by the spy, Leturc, dated 30 June 1797 in AN F 12 2204, against his treatment by the Académie before 1789, but given the date to be used with caution. Desaguliers's text is being used as late as the 1790s; see AN, Paris, F17 1344/1 Cours de Physique experimentale, Ecole Centrale, Dept. de la Meuse.

15. AN, F12 661, April 1778, "Reponse du Sieur Clicquot Blervanche aux questions proposés par M. de directeur Général des finances relativement aux reglemens concernant les manufactures." On Desaguliers's role in spreading the Newcomen engine, see G. J. Hollister-Short, "The Introduction of the Newcomen Engine into Europe," *Transactions of the Newcomen Society for the Study of the History of Engineering and Technology*, 48 (1976–77): 11–22.

16. AN, Paris, Marine G 105, 16 f.; this inventor managed to get his privilege, but it was a touch and go negotiation.

17. AN, Paris, F12 2195; item 460, 1788: "Cependant cet academicien ne prétend point exclure la nouvelle méthode proposée par le S. Ainavet; mais il observe qu'il n'y a qu'un long usage qui puisse décider la question."

18. AN, Paris, Marine G 117, 102 f., Gilbert de Marette, 21 June 1775 to secretary of the Marine: "ainsi il ne reste plus qu'à constater par l'experience la realité de ma décourverte. . . ." See also 46 f., 31 Octobre, 1736: Ciceri to the Ministry of the Marine: "Il ne s'agit point icy de Science mais bien de Genie & de mecanique et d'un fait qui concerne La Marine les gens de mer peuvent en juger sainement." See also AN, Microfilm 13 5–7, 10 July 1783.

19. AN, Paris F12 1442 note the letter of John Kay to Badger, 23 Oct. 1753, on how all these people have to be carefully handled.

20. AN, F12 1442, report of October 1758 to M. Trudaine. I am grateful to J. R. Harris for his comments here. Badger did have trouble getting his factory constructed, and had to rely on Vaucauson for assistance. Badger's letters indicate a minimal literacy. For the art of the calender see Eric Kerridge, *Textile Manufactures in Early Modern England*, Manchester, Manchester University Press, 1985, pp. 173–74.

21. Here I rely on Terry Shinn, "Science, Tocqueville, and the State: The Organization of Knowledge in Modern France," *Social Research*, 59, no.3 (1992): 533–66; reprinted in Margaret C. Jacob, ed., *The Politics of Western Science, 1640–1990*, Atlantic Highlands, N.J., Humanities Press, 1994. For further evidence of the mentality of the officials, see Edward A. Allen, "Business Mentality and Technology Transfer in Eighteenth-Century France: The *Calandre Anglais* at Nîmes, 1752–92," *History and Technology*, 8 (1990): 9–23.

22. ENPC, Paris, archives et manuscrits non catalogués, Carton "Concours de Style, 1789–1803." I am immensely grateful to Mme. M. Deschamps for leading me to this rich collection at the ENPC; there are 25 answers for 1789 and about the same for 1802. In 1789 there would have been slightly less than 400 engineers employed by the crown. In ENPC, Carton: "Concours de Style, 1778–1812" there are about 25 answers for 1778 on a question about the value of the school for commerce and agriculture. Roads, canals, and bridges are seen to facilitate both; manufacturing, mentioned only occasionally, will also be assisted. "The active circulation of commerce is the first and principal cause of the splendor of the state," is a sentiment commonly voiced. For 1784 there are 22 answers to the question: "the advantages and disadvantages of the equality of conditions in a great society." These show that enlightened thought had permeated deeply into the school; women's equality is occasionally discussed, but generally the engineers, including the one who got first prize, do not think that the arts and sciences would flourish in conditions of equality.

23. Picon, p. 51. Cf. Nicole et Jean Dhombres, *Naissance d'un pouvoir: sciences et savants en France (1793–1824)*, Paris, Payot, 1989, p. 560.

24. "Un entrepreneur qui trop peu instruit se chargeroit à vil prix d'un ouvrage dont le devis auroit été trop legérèment fait; ou bien forcer un entrepreneur ambitieux et peu délicat à se contenter d'un gain légitime et autorisé par le gouvernement." Quoted in "Cours de stile, 1789–1803," École nationale ponts et chausées, Paris.

25. Quoted in M. Bradley, *op. cit.*, p. 145.

26. AN, Paris F12 2204 to Citoyen [?] 1791[?], [dossier le Turc]. See also F12 677C and in same file letters of 13 June 1786 from London where he makes clear that he is not actually going to do the manufacturing. Supplied by the kindness of J. R. Harris.

27. JWP, BPL, W/5, Birmingham, 13 Jan. 1779 to Dear Doctor [Black]. See also AN, Paris, F12 2205.

28. Ibid., Watt to Black, no date but placed among other letters from 1780.

29. Ibid., 13 Jan. 1779 Watt to Black.

30. AN, Paris, Marine G 110, dossier 1 and 2; 146–201 ff.; including a list of 1778 from Boulton and Watt on all the engines installed in Britain to date (27 on this list). Jary signs himself as "concessionaire des Mines de Nantes." See 183 f. for evidence that he "seems to be possessed, in an eminent degree, of the necessary previous knowledge." This new evidence from the archives of the Marine supplements the account found in Jacques Payen, *Capital. . .*, pp. 102–7.

31. Note that in the Dutch Republic the main importer of Watt's engine, Van Liender, set up just the kind of trial that Watt wanted in a polder near Haarlem where

there was a Newcomen engine already installed; Boulton and Watt MSS, BPL, Box 36, letter from Van Liender to Watt, 9 Oct. 1787, written in French. Watt's letters to Jary and the French authorities from the period 1778 to 1782 can be found in the same Birmingham archives of Boulton and Watt; 16 Oct. 1778, Watt to Magellan on Jary: "In the first place he can certify that he hath seen many of our machines actually at work and that they are very much superior to the common Engine . . . that Bretagne is a more proper place to make the trial of comparison in, than Paris, because there are two Engines actually at work & the great expence of erecting an Engine at Paris will thereby be avoided [Jary wanted to buy an engine at this time]." And "End of December 1778 or 1 Jan. 1779 . . . Mr Magellan . . . 1. The superiority of our Engine over those of the old construction can be ascertained only by a comparison. But as there is no old Engine at Paris with which the comparison can be made . . . 3. Mr Jary Concessionaire etc. has applied to us to erect one of our Engines at or near Nantes in Bretagne in place of an old one already erected there. . . . Watt." Magellan's despairing letter is 197 f. Magellan is well aware of the concern about Perier trying to steal the privilege. Watt recounted the outlines of this story and Perier's failure to pay him for designs submitted in a letter to M. Genet, Chef du Bureau des Affaires Etrangères, 31 August 1783; found in the Albany Institute of History and Art and kindly supplied through Eric Robinson. See AN, Paris, Minutier Central, XXX, 459, for Perier brothers and list of members in the society; Lettres patentes dated 7 Feb. 1777 and registered with Parlement 16 July 1778. Neither brother used an acute when signing his name, and so I continue their practice. As late as 1817 French commentators were complaining about their backwardness in steam engines; see AN, Paris, F 12 2200, 8 April 1817 Albert to M. Becquey.

32. JWP, Watt to Wedgwood, 16 Feb. 1784: "We have had these two days past a visit of no less than six French engineers and iron masters who have come over in hopes we would teach them to make fire engines and that some other benevolent people would teach them how to improve their cast iron. We treated them with all manner of civility but took care to show them nothing but what they knew before. . . . I believe they do not intend to visit the pottery but if they do you are warned that they are clever scientific people and one of them Mr. Perier an excellent mechanic." In the same collection a letter from Joseph Banks dated 12 August 1784 to Watt: "I cannot resist the desire I have of recommending Mr. Bertier Intendant de la Généralité de Paris, etc, etc. a man of no small consequence in France considerable proficiency in usefull knowledge & unwearied patronage of the usefull arts to your good offices . . . some companions of his journey among whom is Dr. Broussonet. . ." Originally supplied by the kindness of Eric Robinson; these manuscripts are now at the Birmingham City Library.

33. James E. McClellan III, *Colonialism and Science. Saint Domingue in the Old Regime*, Johns Hopkins University Press, Baltimore, 1992, p. 74.

34. The archives of the Marine supplement the account found in Charles Ballot, *L'Introduction du machinisme dans l'industrie française*, Slatkine Reprints, Geneva, 1978, pp. 390–403 [original in 1923]. Cf. AN, *Innovations techniques dans la Marine, 1641–1817. Mémoires et projets reçus par le département de la Marine (Marine G 86 à 119)*, Paris, 1990. This account does not contradict but it does supplement what appears in J. R. Harris, "Michael Alcock and the Transfer of Birmingham Technology to France before the Revolution," *Journal of European Economic History*, 15, no. 1 (1986): 7–59. See also J. Payen, *Capital et machine à vapeur au xviiie siècle*, Paris, 1969, pp. 102–4.

35. For testimony to this effect see the diary of Prof. Salomon de Monchy's trip to Paris, 1790; Rotterdam, Gemeente Archief, familie de Monchy, no. 51, 39–41 ff.

He also saw Van Liender on this trip who was now in Paris with his sister. For a condensed version of this whole story see Jacques Payen, *Les Frères Périer et l'introduction en France de la machine à vapeur de Watt*, Conférence . . . Palais de la Découverte, 1968. See ENPC, MS 100, for a student discussion of the engine dated 1791.

36. Archives departmentales, Loire-Atlantique, Nantes, C 129, privilege dated 1746 to Simon Jarry [sic]; in 1765 it was extended another thirty years with the right to pass it on to his children. On these concessions and the confusions around them see Gwynne Lewis, *The Advent of Modern Capitalism in France, 1770–1840. The Contribution of Pierre-François Tubeuf*, Oxford, Clarendon Press, 1993.

37. Catherine Blanloeil, "La Société académique de Nantes et de la Loire-Inférieure de 1798 à 1825," in Jean Dhombres, ed., *La Bretagne des savants et des ingénieurs, 1750–1825*, Rennes, Editions Ouest-France, 1991, p. 69.

38. Anne Brulé, "L'exemple des mines," in Jean Dhombres, ed., *op. cit.*, p. 147 citing AD35 C 1473 for 1783.

39. Jacques Payen, *Capital. . .*, p. 31. For the capitalist side of his operation, see Louis Bergeron, *Banquiers, négociants et manufacturiers parisiens du Directoire à l'Empire*, Paris, Mouton, 1978, pp. 301–4. For the memoir see Bib. Historique de la Ville de Paris, ms. nouv. acq. 147, 446–69 f., et. seq.

40. AN, Microfilm 13 5–7, 14 May 1772; exclusive privilege given to D. de Auxiron; Perier was involved here. I am grateful to J. R. Harris for his comments on this section.

41. Note the discussion of Tubeuf's rival, de Castries, in G. Lewis, *op. cit.*, pp. 133–37.

42. JWP, BPL, W/11, Letter from Brunelle, de Salins en Franche Comté, 1 Sept. 1788. "je n'ai pu obtenir de l'intendant des finances qui devoit mécouter, un quart d'heure de rendés vous pour lui faire mon rapport. c'est pour le roi que j'ai travaillé et je n'ai pu obtenir audience!" This is almost certainly from the father of the famous French émigré engineer, Brunel.

43. For a picture of 1789 see J-F. de Tolozan, *Memoire sur le commerce de la France et de ses colonies*, Paris, Moutard, 1789 [BN microfiche V.17731], pp. 24–25: "Nous avions autrefois une supériorité bien marquée sur toutes les Fabriques étrangeres dans plusieurs especes de toiles." For the earlier career of Tolozan see Harold T. Parker, *op. cit.*, pp. 17 *et. seq.*

44. One such institution, La Société d'Encouragement pour l'Industrie Nationale, had an entirely industrial and mechanical focus; see AN F12 502, for the founding documentation and the debt to Chaptal. English observers were also convinced of this gap; see James Watt, Jr., to his father, dated Rouen 16 June 1792 on cotton manufacturing in the town: "New improvements I have seen none, not any processes which we have not in England, on the contrary, they are considerably behindhand here in the Manufactory, but yet not as much as I expected." Boulton and Watt collection, BPL. Watt Jr. was impressed by the size of Oberkampf & Co. with 150 tables employing upwards of 1,200 persons. Cf. Dominique Julia, *Les trois couleurs du tableau noir. La Révolution*, Paris, Belin, 1981, chap. 8: "L'avènement de l'ingénieur." See also for the similar policies of Chaptal's revolutionary predecessor, François de Neufchâteau, *Circulaire aux Administrations centrals de département*, 9 Fructidor, Year V, in *Recueil de lettres, circulaires, instructions, programmes . . . du Ministre de l'Intérieur*, 2 vols., Paris, Imprimerie de la République, years VII–VIII, vol. I, pp., 102–3, 155, and p. xxx. I owe the point about Neufchâteau to Jeff Horn, who is now working on the industrial expositions.

45. AD, Hérault, D 186, 215–28 ff. On his Newtonianism see J. A. Chaptal, *Élémens de Chymie*, 3rd ed., Paris, 1796, introduction; and his *Mes souvenirs de Napoleon*, Paris, 1893, p. 19 for application of mathematics to the study of the human body and the attack on hypotheses. For recent historiography on Chaptal see Michel Pérronnet, ed., *Chaptal*, Bibliothèque historique Privat, Paris, 1988; and M. Peronnet, "Un chimiste en politique: J. A. Chaptal à Montpellier (1788–1794)," in *Actes du 114e Congres National des sociétés savantes, scientifiques et sociétés pendant la révolution et l'empire*, Paris, Editions du CTHS, 1990, pp. 145–60 .

46. John Graham Smith, *The Origins and Early Development of the Heavy Chemical Industry in France*, Oxford, Clarendon Press, 1979, pp. 20–24.

47. For his club see Archives départementales, Hérault L 5498, entry for 9 May 1790; request made by "Soze, associé etranger du Club," permission given on 21 May. The club only began in February. For Chaptal's thinking see J. Chaptal, *Essai sur le perfectionnement des arts chimiques en France*, Paris, 1800, p. 50. See also his immensely knowledgeable, *Programme des Prix proposés par le Ministre de l'Intérieur pour le perfectionnement des Machines à ouvrir, peigner, carder et filer la laine*, 22 Messidor, year IX, with extensive information about English practices and found in AN F12 2208. Note also the minute detail with which Chaptal described the divided practices of his workers, step by step, in a factory that was not mechanized but did employ his chemical techniques: *L'Art de la teinture du coton en rouge*, Paris, 1807, especially chap. 4.

48. J. A. Chaptal, *Catéchisme a l'usage des bons patriotes*, Montpellier, 1790, pp.12–13. Cf. a journal founded by Chaptal and his colleagues, *Annales des arts et manufactures, ou mémoires technologiques sur Les Découvertes modernes concernant les Arts, les Manufactures, l'Agriculture et le Commerce*, year viii [1800], Paris, p. 35.

49. AN, Paris, F12 1556, dated 29 Xbre 1791; the report was logged in his office on 6 Jan. 1792 as item #121 from M. Cahier, Minister of Interior. For a useful overview see Jean-Pierre Hirsch, "Revolutionary France, Cradle of Free Enterprise," *American Historical Review*, 94 (1989): 1281–89. For Chaptal's anticlericalism see AD, Hérault, L 5498, his manuscript speech to the club, 16 July 1790, and his printed revolutionary catechism; J. A. Chaptal, *Catéchisme à l'usage des bons patriotes*, 1790; copy available in Bibliothèque de la ville de Montpellier. The ideas in this document owe much to Chaptal's freemasonry.

50. For Chaptal and this point see the old but still useful, Jean Pigeire, *La Vie et l'oeuvre de Chaptal (1756–1832)*, Thèse pour le Doctorat, Paris, Editions Domat-Montchrestien, 1931, p. 133.

51. AN, Paris, F12 2195 6 Ventose Year 5; F. Bardel, Manufacturier et membre du conseil des arts et manufactures, to Minister of the Interior: "Je vais etablir une manufacture de mousselines, d'organdis et de toiles de coton proprès a L'impression des Indiennes. J'ai deja pris en Angleterre des notions exactes sur la main d'oeuvre et les apprêti de ces articles, qui son peu Connus en France." The accompanying report indicated that these exact notions included the better construction of machines; the economy of combustion to be used in bleaching; the renewing of surfaces of different substances under the action of steam; the ability to calculate the different times needed for different materials; the degree of pressure to give to the steam; the use of alkali in the right quantity and quality; other substances that can be employed in bleaching and their various costs.

52. Denis Woronoff, *L'industrie sidérurgique en France pendant la Révolution et l'Empire*, Paris, Éditions de l'École des hautes Études, 1984, pp. 32–33.

53. *Essai sur le perfectionnement des arts chimiques en France*, Paris, 1800, pp. 3, 16–20.

54. J. A. Chaptal, *Rapport et project de loi sur l'instruction publique*, Paris, AN9 [1801], pp. 92–93. Machines to demonstrate physical and mechanical principles are being demanded by professors in the new central schools by the year 7; see AN, Paris, F17 1344/1.

55. F. de Neufchâteau, *Discours prononcé par le Ministre de l'Interieur, le 5 prairial, an 7*, in *Recueil de lettres, circulaires, instructions, programmes. . .*, vol. 3, 1799, p. 243.

56. M. le Comte Chaptal, *De l'Industrie française*, Paris, 1819 [2 vols in one], vol. 2, p. 32. For a new edition see Louis Bergeron, ed., with intro. published by Imprimerie Nationale, Paris, 1993. A similar perspective can be found in Ternaux, see L. M. Lomüller, *Guillaime Ternaux, 1763–1833. Createur de la première intégration industrielle française*, Académie nationale de Reims, Les Editions de la Cabro d'Or, Paris, 1977, p. 124.

57. On the German side of this story with mixed results see Herbert Kisch, *From Domestic Manufacture to Industrial Revolution. The Case of the Rhineland Textile Districts*, New York, Oxford University Press, 1989, pp. 190–91, 202–3.

58. AN, Paris, F 17 1098, for Brussels university faculty; on the struggle in Liège (dossier 4, 50 f.) between "la partie des mathematiques pures" and those who want applied sciences; report dated 9 November 1810 to "le Grand-Maître de l'Université" on the need for certain sciences, i.e., physics, chemistry, and natural history "because of their application to the arts and manufacturing." Note also evidence of trying to enforce the teaching of Catholic doctrines in the Dutch Republic. These policies were first noted a long time ago by L. Brummel, "De Zorg voor kunsten en wetenschappen onder Lodewijk Napoleon," *Genootschap voor Napoleontische Studien*, The Hague, 1951, pp. 11–26.

59. For the Maastricht story see J. P. L. Spekkens, *L'École Centrale du département de la Meuse-Inférieure. Maëstricht 1798–1804*, Maastricht, Ernest van Aelst, 1951, pp. 62–64. For the archives see AN F17 1088; 17 1276; 17 1344, 3; 17 1428.

60. Alois Schumacher, *Idéologie révolutionnaire et pratique politique de la France en Rhénanie de 1794 à 1801*, Paris, Annales Littéraires de l'Université de Besançon, 1989, pp. 138–43.

61. AN, Paris, F17 1098, report on Brussels dated 1808.

62. AN, ibid., an invaluable set of reports on the universities and academies in the Netherlands both north and south.

63. AN, Paris, MS AP/147, papers of Jacques François Piou, an engineer employed in Belgium to build a canal between Mons and Brussels; letter to his wife of 15 Prairial, year 13. On education in the Austrian Netherlands (i.e., Belgium) see F. Macours, "L'enseignement technique à Liège au xviiie siècle," *Bulletin de l'Institut archeologique liègeois*, 69 (1952): 131–85; and Claude Sorgeloos, "Les Savants à l'école. Le case du Hainaut," in G. Van de Vyver et J. Reisse, eds., *Les Savants et la Politique à la fin du xviiie siècle*, in *Études sur le xviiie siècle*, 7 (1991): 85–88.

64. Louis Bergeron, *France under Napoleon*, trans. R. R. Palmer, Princeton University Press, Princeton, N.J., 1981, pp. 173–74, 182, 188–90.

65. See *Almanach des muses de l'ecole centrale du département des deux-sèvres*, Niort, year VI, p. 21, listing professors of mathematics and experimental physics for pupils over 14. For the nonexisting level of mathematical education for girls; see Martine Sonnet, *L'education des filles au temps des Lumières*, Paris, Les Éditions du Cerf, 1987. On both

sides of the channel the public culture of science as it developed offered more opportunities to women than did traditional institutions. See also T. P. Bertin, translator from English, *Le Neuton de la Jeunesse, ou Dialogues instructifs et amusans entre un père et sa petite famille*, Paris, 1808, dialogues that begin with mechanics for a girl and boy. See Nicole et Jean Dhombres, *Naissance d'un pouvoir: sciences et savants en France (1793–1824)*, Paris, Payot, 1989, pp. 218–22. Women attended meetings of the National Institute as spectators; see Maurice Crosland, ed., *Science in France in the Revolutionary Era. Described by Thomas Bugge*. . . , Cambridge, Mass., MIT Press and Society for the History of Technology, 1969, p. 89.

66. See for example, Lucas Oling, *Rekenkundige voorstellen*, Amsterdam and Leeuwarden, 1809; Gottfried Grosse, *Technologische Wandelingen of Gesprekken van een Vader met zyne Kinderen over eenige der belangrykste Uitvindingen*, (trans. from German), Zutphen, 1801. For the earlier and dominant physico-theology see, for example, L. C. Schmahling, *De Natuurkunde, ten gebruike in de Schoolen*, Amsterdam, 1798.

67. See Eda Kranakis, "Social Determinants of Engineering Practice; A Comparative View of France and America in the Nineteenth Century," *Social Studies of Science*, 19 (1989): 5–70; Charles P. Kindleberger, "Technical Education and the French Entrepreneur," in Edward C. Carter II, Robert Forster and Joseph Moody, *Enterprise and Entrepreneurs in 19th and 20th Century France*, Baltimore, Johns Hopkins University Press, 1976, pp. 1–39; For earlier advances in Britain see Richard S. Tompson, "The English Grammer School Curriculum in the 18th Century: A Reappraisal," *British Journal of Educational Studies*, 19 (1971): 32–39; and Diana Harding, "Mathematics and Science Education in Eighteenth-Century Northamptonshire," *History of Education*, I (1972): 139–59. For the relevant French documents see Bronislaw Baczko, ed., *Education pour la democratie*, Paris, 1982.

68. AN, Paris, F17 1344/1; writing from Moulins, 22 fructidor, year 6. Cf. Janis Langins, *La République avait besoin de savants. Les débuts de l'École polytechnique: l'École centrale des travaux publics*. . . , Paris, Belin, 1987.

69. AN, Paris, F17 1344/1 Lenormand at l'école centrale du Tarn, year 7. On him as an inventor see AN, Paris, F 12 2200, dated year 8. He also wrote extensively on the Paris expositions.

70. Archives départementales de l'Hérault, L 5787, documents pertaining to his arrest.

71. L.S. le Normand & J. G. V. de Moléon, *Description des expositions des produits de l'industrie française faites a Paris depuis leur origine jusqu'a celle de 1819 inclusivement*. . . , 4 vols., Paris, 1824, p. 19. Cf. for educational policy formation see Charles R. Day, *Education for the Industrial World: The Ecoles d'Arts et Métiers and the Rise of French Industrial Engineering*, Cambridge, Mass., MIT Press, 1987. For an overview of these exhibitions and engravings see *Comite Français des Expositions et Comité National des Expositions coloniales . . . 1925*, Cinquantenaire 1885–1935, Paris 1935.

72. For a description of all the arcades see *Première exposition des produits de l'industrie française*. [Paris 1798, located in Bibliothèque historique de ville de Paris].

73. L. M. Lomüller, *Guillaume Ternaux 1763–1833*, Paris, 1977, p. 109.

74. For a succinct statement of the ideological relationship see Prof. Le Normand to Neufchâteau, AN, Paris, F17 1344/1. Cf. Bruno Belhoste, "Les caractères généraux de l'enseignement secondaire scientifique de la fin de l'Ancien Régime à la Première Guerre mondiale," *Histoire de l'education*, no. 41 (1989): 1–45.

75. See their petition "Au Roy . . . 1777," Archives départmentales, Loire-Atlantique, Carton 1 C.630, côtes 1-4. On the Nantes harbor in the AN, Paris, see H543; F14 172 a and b; F14 735 Port de Nantes; F14 102 1757–58 canal from Nantes to ocean. Cf. Pierre Lelièvre, *Nantes au XVIIIe siècle. Urbanisme et architecture*, Picard, Paris, 1988, pp. 81–110, esp. 83, and 110. Perronet is the engineer in one of the controversies.

76. AD, Gard, C310–353 for the years 1697–1757; AD, Hérault C7530, C7572, C7556, C7590 for 1762, 1768, 1773, 1777. Two commissions, one for public works and the other for manufactures, concern us. Note that in the 1780s the *États* did consult Chaptal whom we may reasonably describe as an engineer of sorts; John Graham Smith, *op. cit.*, pp. 22–23.

77. Gwynne Lewis, *The Advent of Modern Capitalism in France, 1770–1840*, p. 39 for the engineer Renaux.

78. D. Woronoff, *L'industrie siderurgique en France. . .*, pp.49–60.

79. See Edward A. Allen, "Business Mentality and Technology Transfer in Eighteenth-Century France: The *Calandre Anglais* at Nîmes, 1752–92," *History and Technology*, 8 (1990): 9–23.

80. L. Bergeron, *Banquiers. . .*, p. 305. See the beginnings of this process in MS 100, EPNC, 24 ff.; where the technical treatise on a Watt-type engine goes on to calculate the cost of constructing two such engines in 1791: 26,491 francs exclusive of installation, housing for the machine, etc. This is still a machine for public works.

81. For the new world of bankers and capitalists after 1800, see ibid., pp. 46–48. On Chaptal's influence on Napoleon, see p. 213.

82. Joel Mokyr, *The Lever of Riches. Technological Creativity and Economic Progress*, New York, Oxford University Press, 1990, pp. 111–12.

83. Eric Dorn Brose, *The Politics of Technological Change in Prussia. Out of the Shadow of Antiquity, 1809–1848*, Princeton, N.J., Princeton University Press, 1993, p. 261. Cf. Kees Gilpen, *New Profession, Old Order. Engineers and German Society, 1815–1914*, Cambridge, Cambridge University Press, 1989.

Chapter 9

1. For these sorts of arguments see Peter Mathias, *The First Industrial Revolution: An Economic History of Britain, 1700–1914*, London, Methuen, 1983, pp. 128–29; or see E. A. Wrigley, "The Supply of Raw Materials in the Industrial Revolution," *Economic History Review*, 15 (1962): p. 4: For a useful corrective see D. S. L. Cardwell, *The Organisation of Science in England*, London, Heinemann, 1972, pp. 13–18; Alan Smith, "Steam and the City: The Committee of Proprietors of the Invention for Raising Water by Fire," *Transactions of the Newcomen Society*, 49 (1977–1978): 5–18, on the Royal Society and the steam engine. For one of the first cogently argued attacks on the view represented by Wrigley, see A. E. Musson and E. Robinson, "Science and Industry in the Late Eighteenth Century," *Economic History Review*, 2nd ser., 13 (1960–1961): pp. 222–44, especially pp. 241–42 for further evidence of scientific lecturing in Bristol and Sheffield.

2. William Chapman, *Address to the Subscribers to the Canal from Carlisle to Fisher's Cross*, Newcastle, 1823, pp. 2–3, 7. This essay was written as a result of a series of breakdowns in relations between an engineer of the next generation and the canal company. Emphasis in the quotation from Smeaton was added by Chapman.

3. Watt carefully preserved the testimony to be found in JWP, BPL, MS 4/53. This document gives a fuller account than had existed in the past about exactly how far Watt had got in developing his engine.

4. Chapman, *Address to the Subscribers*, p. 2.

5. Anthony Burton, *The Canal Builders*, London, David and Charles, 1981, pp. 157–58; and R. W. Malcolmson, *Life and Labour in England 1700–1780*, London, Hutchinson, 1981, pp. 83–93.

6. For a general discussion of Bristol in this period, see B. D. G. Little, *The City and County of Bristol: A Study in Atlantic Civilization*, London, Werner Laurie, 1954.

7. See Thomas A. Ashton, *Iron and Steel in the Industrial Revolution*, Manchester, Manchester University Press, 1963, pp. 21–30, 41–42; Brian Bracegirdle, *The Darbys and the Ironbridge Gorge*, London, David and Charles, 1974; and Isabel Grubb, *Quakerism and Industry Before 1800*, London, Williams and Norgate, 1930, pp. 50–51, 151–55.

8. Bristol Central Library, MS 20095, "Diary of William Dyer," vol. 1, 1760, f. 116. For an outline of the lectures Ferguson gave up and down the country, see James Ferguson, F.R.S., *Lectures on Select Subjects in Mechanics, Hydrostatics, Hydraulics*, 6th ed., London, 1784, an overtly Newtonian course, very similar to those discussed in the previous chapter. I am grateful to Jonathan Barry for information on Dyer.

9. "Diary of William Dyer," vol. 1, 1760, fol. 111, for this description of her; 1763, fol. 116, for the evening in question.

10. Ibid., fol. 126.

11. Bristol Record Office, White MS, no. 08158, fols. 73–81.

12. See Roy Porter, "Alexander Catcott: Glory and Geology," *British Journal for the History of Science*, 1977.

13. Bristol Central Library, MSB 26063, correspondence of Rev. A. S. Catcott and A. Catcott, letter of 23 June 1774, to A. Catcott.

14. Bristol Central Library, Bristol Library MSS, "Books proposed 1774," written in a variety of hands. For later developments, see Michael Neve, "Science in a Commercial City: Bristol 1820–60," in Ian Inkster and Jack Morrell, eds., *Metropolis and Province: Science in British Culture 1780–1850*, London, Hutchinson, 1983, pp. 179–204. For Ferguson see Fitzwilliam Museum, Cambridge, Perceval Bequest A.72; letter dated 21-5-1774.

15. Of the 155 pupils at Bristol Grammar School from 1710 to 1717, 53 became merchants and mariners. For the considerable education given to the sons of wealthier merchants, see W. Minchinton, "The Merchants of Bristol in the Eighteenth Century," *Societés et groupes sociaux en Aquitaine et en Angleterre*, Bordeaux, Federation historique du Sud-Ouest, 1979, pp. 190–91.

16. Alan F. Williams, "Bristol Port Plans and Improvement Schemes of the 18th Century," *Transactions of the Bristol and Gloucestershire Archaeological Society*, 81 (1962): 144.

17. Alexander Pope, *Letters to Martha Blount*, 1732, quoted in Williams, "Bristol Port Plans," p. 142.

18. For a general history of this body, with an excellent chapter pertaining to the river and harbor problems, see Patrick McGrath, *The Merchant Venturers of Bristol*, Bristol, Society of Merchant Venturers of the City of Bristol, 1975, especially pp. 150–53; and for the meeting records, see the Society of Merchant Venturers, Clifton, Bristol, Merchants' Hall Book of Proceedings, records for May 1776.

19. Williams, "Bristol Port Plans," p. 178.

20. See Nicholas Rogers, "The Urban Opposition to Whig Oligarchy, 1720–60," in Margaret C. Jacob and James R. Jacob, eds., *The Origins of Anglo-American Radicalism*, Atlantic Highlands, N.J., Humanities Press, 1991, pp. 138, 142–45.

21. Williams, "Bristol Port Plans," p. 148.

22. See, for example, *Observations on the Dangers and Inconveniences Likely to Attend the Execution of the Proposed Scheme of Building a Dam Across the River Avon*, Bristol, 1791.

23. Bristol Record Office, Proposal of 1765, MSS of Richard Bright.

24. Williams, "Bristol Port Plans," p. 147.

25. Bristol Record Office, MS 111689(3), proposal from A. Walker, 1791.

26. Pamela Bright, *Dr Richard Bright 1789–1858*, London, The Bodley Head, 1983, pp. 13–18, on this Bright, the father of her subject. See also Royal Society, B.L.A. b. ff. 325–29.

27. Bristol Record Office, Bright MSS, 11168(3), a long list in Bright's possession that estimates the number of ships using Bristol harbor, with direct comparisons to Liverpool.

28. Bristol Record Office, Bright MSS, 11168(3), letter of 16 Nov. 1791, Thomas Percival to Richard Bright. See Arnold Thackray, "Natural Knowledge in Cultural Context: The Manchester Model," *American Historical Review*, 79, no. 3 (June 1974): pp. 672–709.

29. Bristol Record Office, Bright Mss, MS 11168(3) "opinion tendered by Dr Falconer." Bright did profess his deep concern that no "injury should arise to health," see R.S. B.L.A. b. fol. 327.

30. Bristol Record Office, Bright MSS, 11168, Henry Cavendish to Richard Bright.

31. Bristol Record Office, Bright MSS, 11168(3), 15 Nov. 1790.

32. Bristol Record Office, Bright MSS, 11168(1)e. The plan was first submitted on 25 Feb. 1790.

33. Society of Merchant Venturers, Clifton, Bristol, MS Letter Book 1781–1816, for example, entry for 20 May 1792, the society to Mr. James Allen, on his architectural plans not to be preferred to what has been submitted; H.B. microfilm 4, 6 Dec. 1786, a meeting where a variety of engineers appeared and presented their ideas; MS Letter Book, 15 August 1815, to William Jessop: "Your plan of the proposed Crane has been submitted to the Society. . . . Upon examining it with that of Messrs Stewart and Ramsden the Radius described by your Crane does not appear to be equal to theirs. The Arm of the Crane does not reach so far out by two feet and taking a perpendicular or plomb line from any given point of the Brace C to the level of the Wharf there is a considerable difference in the height." See also Bristol Record Office, Bright MSS, 11168(66–68), Bright's notebooks.

34. Society of Merchant Venturers, MS Letter Book, entry for 17 July 1792, to Mr Faden, engraver, St. Martin's Lane; see also letter dated 18 August 1815, to Jessop, from which it is clear that the society's committee has once again altered his plans.

35. Ibid., f. 206, 1792.

36. Ibid., Jessop to Osborne, 11 Jan. 1793; for a comparison of the complexity of such plans versus those available a hundred years earlier, see Bristol Central Library, Southwell MS, undated handbill at end of the volume from the 1690s.

37. *Felix Farley's Bristol Journal*, 21 March 1807, quoted in R. A. Buchanan, "The Construction of the Floating Harbour in Bristol: 1804–1809," *Trans. BGAS*, 83 (1969): p. 199.

38. Little, *Bristol*, p. 167.

39. Bristol Central Library, MSS of the Bristol Library and Philosophical Institution, 1825. Cf. Charles H. Cave, *A History of Banking in Bristol from 1750 to 1899*, Bristol, 1899.

40. For a good description of the earliest partnership in canal building, which involved James Brindley, a mechanic of little or no scientific training, and a landed aristocrat, the duke of Bridgwater, see Francis Henry Egerton, *The First Part of a Letter to the Parisians, and, the French Nation, upon inland Navigation*, Paris, 1818; for James Brindley's orderly mind, see his diaries, 1759 to 1763, Central Library, Birmingham.

41. *The History of Inland Navigations. Particularly those of the Duke of Bridgewater in Lancashire and Cheshire*, London, 1766, p. 34.

42. Anthony Burton, *The Canal Builders*, London, David and Charles, 1981, p. 50; see also Derbyshire Record Office, D258/50/13/p, 19 March 1789, on canvassing Bishop Llandarff to support a canal bill, "He is a Liberal, though a Bishop." For a discussion of some of the complexities of this Whig commercialism, see J. G. A. Pocock, "Radical Criticisms of the Whig Order in the Age Between Revolutions," in Margaret C. Jacob and James R. Jacob, eds., *The Origins of Anglo-American Radicalism*, London and Boston, Allen and Unwin, 1984, pp. 42–43. On the social composition of the early Industrial Revolution, see Harold Perkin, *The Origins of Modern English Society 1780–1880*, London, Routledge and Kegan Paul, 1969, pp. 67–68. See also Peter Buck, "People Who Counted: Political Arithmetic in the Eighteenth Century," *Isis*, 73, no. 266 (1982): 32, on court Whigs favoring a national census in 1753.

43. See R. B. Schofield, "The Construction of the Huddersfield Narrow Canal 1794–1811: With Particular Reference to Standedge Tunnel," *Transactions of the Newcomen Society*, 53 (1981–1982): 17–38.

44. See Philip Riden, *The Butterley Company, 1790–1830: A Derbyshire Ironworks in the Industrial Revolution*, Chesterfield, 1973, p. 3 ff., for Benjamin Outram.

45. See, for example, Derbyshire Record Office, D258/50/14 w, E. Darwin to P. Gell, 22 April 1789.

46. R. B. Schofield, "The Promotion of the Cromford Canal Act of 1789: A Study in Canal Engineering," *Bulletin of the John Rylands University Library of Manchester*, 64 (1982): 246–47. Cf. R. S. Fitton and A. D. Wadsworth, *The Strutts and the Arkwrights 1758–1830*, Manchester, Manchester University Press, 1958, pp. 62, 80.

47. Derbyshire Record Office, D258/50/14 y, to Philip Gell from his brother in London, 7 July, n.a.

48. Derbyshire Record Office, D258/50/14 ta.

49. Schofield, "Promotion of the Cromford Canal Act," p. 268.

50. Derbyshire Record Office, D258/50/14 v, B. Outram to P. Gell. Cf. Schofield, "Promotion of the Cromford Canal Act," p. 274.

51. Schofield, "Promotion of the Cromford Canal Act," p. 270, quoting a letter from John Gell to Philip Gell. There is no evidence that committee members were chosen for their particular expertise; see O. Cyprian Williams, *The Historical Development of Private Bill Procedure and Standing Orders in the House of Commons*, London, HMSO, 1948, vol. 1, pp. 41–46.

52. House of Lords Record Office, Main Papers, H.L., 26 May 1789, et. seq.

53. House of Lords Record Office, Main Papers, 24 May 1791, evidence on Birmingham Canal Bill.

54. House of Lords Record Office, Main Papers, 26 May 1789, Cromford Canal.

55. House of Lords Record Office, Main Papers, 19, 20 May 1809, Kennet and Avon Canal Bill, examination of John Rennie, Esq.

56. House of Lords Record Office, Main Papers, 19 May 1809, Kennet and Avon Canal Bill. This is a bill to permit the raising of more money for a canal that is partially completed.

57. T. S. Ashton, *An Eighteenth Century Industrialist: Peter Stubs of Warrington 1756–1806*, Manchester, Manchester University Press, 1939, p. 41.

58. James H. Rieuwerts, "A Technological History of Drainage of the Derbyshire Lead Mines," Ph.D. dissertation, University of Leicester, 1981, pp. 145–49. Cf. Roy Porter, *The Making of Geology*, Cambridge, Cambridge University Press, 1976.

59. Sheffield City Library, Bagshawe Collection, MS 494, John Barker's Letter Book, 1765–1811, entry for 30 Sept. 1794, on a mine subject to a great deal of flooding.

60. Derbyshire Record Office, 503/D103, William Jessop to Mr. Godwin, Butterley Ironworks, 9 Sept. 1815, and 14 Dec. 1815.

61. Sheffield City Library, Bagshawe Collection, C. 654(1–116), letter of William Milner to George Barker on steam engine with the approval of Sir Joseph Banks, 21 Sept. 1807. Cf. Lynn Willies, "The Barker Family and the Eighteenth Century Lead Business," *Derbyshire Archaeological Journal*, 93 (1973): 68, on Wyatt taking over the failing business of the Barkers and revitalizing it.

62. Sheffield City Library, Bagshawe Collection, C. 587/(30), fol. 1, estimate with technical description of engine, from R. Smith to W. Wyatt, 9 Dec. 1836; fol 3, W. Sneyd to W. Wyatt for a 60-horsepower engine; fol. 8, another estimate with details. The cost involved is between £2,000 and 3,000; see 9 Feb. 1837 for sums.

63. Sheffield City Library, Bagshawe Collection, MS 587(30), fol. 4, William Wyatt to Mr. Cope, Bakewell, 31 Jan. 1837. Cf. N. Kirkham, "Steam Engines in Derbyshire Lead Mines," *Transactions of the Newcomen Society*, 38 (1965–1966): 72–73, 76–77, on Wyatt as innovator.

Bibliography

Comparative studies in the history of science and culture are few and far between. Inspiration can be found in Richard Biernacki, *The Fabrication of Labor in Germany and Britain, 1640–1914*, Berkeley, University of California Press, 1995. For another example of comparative work, but with a very different set of problems from those found in this book, see Lewis Pyenson, *Cultural Imperialism and Exact Sciences. German Expansion Overseas 1900–1930*, New York, Peter Lang, 1985. For a sense of what people knew about nature in general before 1600 see William Eamon, *Science and the Secrets of Nature: Books of Secrets in Medieval and Early Modern Culture*, Princeton, Princeton University Press, 1994. For a splendid discussion of how alchemists worked, see Pamela H. Smith, *The Business of Alchemy. Science and Culture in the Holy Roman Empire*, Princeton, Princeton University Press, 1994. If students wish to know about individual scientists discussed in this text, they should consult Charles C. Gillispie, ed., *Dictionary of Scientific Biography*, 16 vols., New York, Scribner, 1970. For complex ideas in philosophy, there is the helpful guide by Philip P. Wiener, ed., *Dictionary of the History of Ideas*, New York: Scribner, 1973. Western technology is usefully surveyed in Donald Cardwell, *The Norton History of Technology*, New York, W. W. Norton, 1995. The culture and science of the less educated, which increasingly came to be dismissed as magic, have been illuminated in Keith Thomas, *Religion and the Decline of Magic*, New York, Scribner 1971; Alan Macfarlane, *Witchcraft in Tudor and Stuart England*, London, Harper & Row, 1970; and Carlo Ginzburg, *The Cheese and the Worms,* Harmondsworth, U.K., Penguin, 1982, about the fascinating cosmology of a miller who ran afoul of the Roman Inquisition. See also C. Ginsburg, "High and Low: The Theme of Forbidden Knowledge in the Sixteenth and Seventeenth Centuries," *Past and Present*, no. 73 (1976): 28–41. And not least, to find out what ordinary folk read, see Margaret Spufford, *Small Books and Pleasant Histories: Popular Fiction and Its Readership in Seventeenth-Century England*, Athens, University of Georgia Press, 1981. A general introduction to the field of science and gender can be found in the popularizing book by Margaret Wertheim, *Pythagoras' Trousers. God, Physics, and the Gender Wars*, New York, Times Books, 1995.

Chapter 1

Galileo's miseries with the church are gone over in minute detail in Rivka Feldhay, *Galileo and the Church. Political Inquisition or Critical Dialogue?*, Cambridge, Cambridge University Press, 1995. Students need not resort to complex notions like "cultural field" or "discourse" to follow the main outlines of the story. Primary sources can be found in Maurice A. Finocchiaro, ed., *The Galileo Affiar. A Documentary History*, Berkeley, University of California Press, 1989. Everyone should read Pietro Redondi, *Galileo Heretic*, Princeton, Princeton University Press, 1989. Italy does not get as much attention as it deserves in this book; try the fascinating account in Paula Findlen, *Possessing Nature. Museums, Collecting, and Scientific Culture in Early Modern Italy*, Berkeley, University of California Press, 1994. For all the background and science left out of this chapter, especially for Kepler who is sadly missing, see Owen Gingerich, *The Eye of Heaven. Ptolemy, Copernicus, Kepler*, New York, American Institute of Physics, 1993. It is also a good place to go for Copernicus. One of the most important and fascinating topics to emerge in the study of science and culture since the 1960s is the role of magic in the new science. The *locus classicus* of those studies is Frances Yates, *Giordano Bruno and the Hermetic Tradition*, Chicago, University of Chicago Press, 1964. Perhaps the most interesting link between magic and scientific practice occurs in early modern medicine. There the leading figure is Paracelsus. See A. G. Debus, *The English Paracelsians*, London, Oldbourne, 1965. Francis Bacon is so very important in the story that links the new science to the reform of learning as well as to technology. The best places to begin with Bacon are Paolo Rossi, *Francis Bacon: From Magic to Science*, Chicago, University of Chicago Press, 1968; and B. Farrington, *The Philosophy of Francis Bacon*, Liverpool, Liverpool University Press, 1964. Bacon's influence is everywhere present in Charles Webster, *The Great Instauratian: Science, Medicine and Reform, 1626–1660*, London, Duckworth, 1975. And he was an inspiration to the founding of the Royal Society; see J. R. Jacob, "Restoration, Reformation and the Origins of the Royal Society," *History of Science*, 13 (1975): 155–76, which is a basic essay on the social and ideological origins of the society. For an essay that places Bacon into the context of economic ideology, see James R. Jacob, "The Political Economy of Science in Seventeenth-Century England," in Margaret C. Jacob, ed., *The Politics of Western Science, 1640–1990*, Atlantic Highlands, N.J., Humanities Press, 1994, pp. 19–46.

Chapter 2

Finally Descartes has a good biography in English. See the account in Stephen Gaukroger, *Descartes. An Intellectual Biography*, Oxford, Clarendon Press, 1995. A provocative study of Descartes's psychology appears in John R. Cole, *The Olympian Dreams and Youthful Rebellion of René Descartes*, Urbana/Chicago, University of Illinois Press, 1992. The argument that up until the condemnation of Galileo in 1633 Descartes was not that concerned about skepticism seems a bit strained. To get through the complexities of Descartes's metaphysics, turn to

Daniel Garber, *Descartes' Metaphysical Physics*, Chicago, University of Chicago Press, 1992. For insight on a way to read texts that anchors them within their social milieux, see Bruce S. Eastwood, "Descartes on Refraction: Scientific Versus Rhetorical Method," *Isis*, 75 (1984): 481–502. There is also much wisdom in A. J. Krailsheimer, *Studies in Self-Interest: Descartes to La Bruyère*, Oxford: Clarendon Press, 1962. Do not forget the now old, but always valuable Martha Ornstein, *The Role of Scientific Societies in the Seventeenth Century*, Chicago, University of Chicago Press, 1928. One of the best studies on French science is Roger Hahn, *The Anatomy of a Scientific Institution: The Paris Academy of Sciences, 1666–1803*, Berkeley, University of California Press, 1971.

Chapter 3

The Royal Society commands a large literature summarized without much interpretative framework in Michael Hunter, *Establishing the New Science. The Experience of the Early Royal Society*, Woodbridge, U.K., Boydell Press, 1989; to be used with caution. To begin a survey of the Merton thesis and its enormous influence try I. Bernard Cohen, ed., *Puritanism and the Rise of Modern Science: The Merton Thesis*, edited with . . . K. E. Duffin and Stuart Strickland, New Brunswick, N.J., Rutgers University Press, 1990. Henry More has a biography in A. Rupert Hall, *Henry More: Magic, Religion and Experiment*, Oxford, Blackwell, 1990. For a short account of Newton and his influence see Betty Jo Teeter Dobbs and Margaret C. Jacob, *Newton and the Culture of Newtonianism*, Atlantic Highlands, N.J., Humanities Press, 1995. Many scholarly works on Newton's science abstractly conceived exist. One place to start is a collection of essays from various decades by A. Rupert Hall, *Newton, His Friends and His Foes*, Aldershot, U.K., Ashgate Publishing, 1993. For background and handy identifications try Derek Gjertsen, *The Newton Handbook*, New York, Routledge & Kegan Paul, 1986. There is also a more technical work than any of the others, but it is helpful: Paul Theerman and Adele F. Seeff, eds., *Action and Reaction. Proceedings of a Symposium to Commemorate the Tercentenary of Newton's "Principia"*, Newark, University of Delaware Press, 1993. On Boyle there is the brilliant study by James R. Jacob, *Robert Boyle and the English Revolution*, New York, Burt Franklin, 1977. Locke now has an all encompassing study in John Marshall, *John Locke. Resistance, Religion and Responsibility*, Cambridge, Cambridge University Press, 1994. For an intelligent discussion of Hobbes's absolutism see Johann P. Sommerville, *Thomas Hobbes. Political Ideas in Historical Context*, New York, St. Martin's Press, 1992.

Chapter 4

The period when science becomes a major intellectual force within Western culture can be dated as roughly 1680–1730, the so-called crisis of the European mind. The student can begin with the old but classic, English translation of Paul Hazard, *The European Mind: 1680–1715*, New Haven, Yale University Press, 1953. There are many minor yet wonderfully fascinating historical characters that make up the story of the crisis. There was also the redoubtable Henry Stubbe in England; see James R. Jacob, *Henry Stubbe: Radical Protestantism and the Early Enlightenment*, Cambridge, Cambridge University Press, 1983. One other essay takes an approach to the crisis that rightly emphasizes its relationship to the English Revolution: J. G. A. Pocock, "Post-Puritan England and the Problem of the Enlightenment" in Perez Zagorin, ed., *Culture and Politics: From Puritanism to the Enlightenment*, Los Angeles, University of California Press, 1980. For a figure in both worlds, presecular and scientific, see James E. Force, *William Whiston: Honest Newtonian*, Cambridge, Cambridge University Press, 1985. To show how complex this period can be take a look at Andrew C. Fix, *Prophecy and Reason. The Dutch Collegiants in the Early Enlightenment*, Princeton, Princeton University Press, 1991.

Chapter 5

There is a book that is basic to this chapter, Larry Stewart, *The Rise of Public Science. Rhetoric, Technology, and Natural Philosophy in Newtonian Britain, 1660–1750*, Cambridge, Cambridge University Press, 1992. For Scotland and the depth of scientific learning there see "Literacy, Education and the Culture of Print in Enlightenment Edinburgh," *History* (October 1993): 373–92; and S. Shapin, "The Audience for Science in Eighteenth Century Edinburgh," *History of Science*, 12 (1974): 95–121; and S. Shapin, "Property, Patronage and the Politics of Science: The Founding of the Royal Society of Edinburgh," *British Journal for the History of Science*, 7 (1974): 1–41. For a good survey of eighteenth-century science in the British Isles but also in Europe, consult M. Crosland, ed., *The Emergence of Science in Western Europe*, London, Macmillan, 1975. For the complexity of economic life see Roy Porter and John Brewer, eds., *Consumption and the World of Goods*, New York, Routledge, 1993. There is also the helpful general study that puts the Royal Society in perspective: James E. McClellan III, *Science Reorganized: Scientific Societies in the Eighteenth Century*, New York, Columbia University Press, 1985. The larger question of science and industrial growth is tackled and somewhat downplayed in Peter Mathias, "Who Unbound Prometheus? Science and Technical Change, 1600–1800," in Peter Mathias, ed., *Science and Society*, Cambridge, Cambridge University Press, 1972. There is much more work to be done on the British literary and philosophical societies, and there are various model studies that can be imitated—for example, R. B. Schofield, *The Lunar Society of Birmingham*, Oxford, Clarendon Press, 1963; E. Robinson, "The Derby Philosophical Society," *Annals of Science*, 9 (1953): 359–67. Someone needs to write about the eighteenth- and early nineteenth-century engineers as the real but peculiar type of enlightened philosophes they were. Scientific culture in Continental Europe during the eighteenth century needs work, and that of course requires a knowledge of various European languages. For further reading, as opposed to research, see J. L. Heilbron, *Electricity in the Seventeenth and Eighteenth Centuries: A Study of Early Modern Physics*, Berkeley, University of California Press, 1979. For a discussion of the lives of some of those who prospered in eighteenth-century Britain, and how much they could prosper through overseas trade, see David Hancock, *Citizens of the World. London Merchants and the Integration of the British Atlantic Community, 1735–1785*, Cambridge, Cambridge University Press, 1995.

Chapter 6

There are now many good biographies of leading eighteenth-century British scientists, although the Watt family needs to be done again. The family papers have now all made their way to the Birmingham City Library, and the latest collections have been used to paint the portrait found in this chapter. Any work on this period should begin with A. E. Musson and Eric Robinson, *Science and Industry in the First Industrial Revolution*, New York, Gordon and Breach, 1989 [reprint of 1969 edition]. *Sir Joseph Banks 1743–1820*, London, British Museum, 1988, by Harold B. Carter, is found in any good research library. Banks has another very good biography in John Gascoigne, *Joseph Banks and the English Enlightenment. Useful Knowledge and Polite Culture*, New York, Cambridge University Press, 1994. There is an older book that must be used with caution: J. G. Crowther, *Scientists of the Industrial Revolution*, London, Cresset Press, 1962. For social history see Leonore Davidoff and Catherine Hall, *Family Fortunes. Men and Women of the English Middle Class, 1780–1850*, London, Hutchinson, 1987. I have benefitted from the context provided in Jan Golinski, *Science as Public Culture: Chemistry and Enlightenment in Britain, 1760–1820*, Cambridge, Cambridge University Press, 1992. For more detail than I have been able to give here about original science in the social circle of Watt, see David Knight, *Humphry Davy. Science and Power*, Cambridge, Cambridge University Press, 1992. For the Scotland of the Watt's see R. A. Houston, *Social Change in the Age of Enlightenment* Oxford, Clarendon Press, 1994.

Chapter 7

There have been no good comparative studies of educational systems at the beginning of modernity. So bits and pieces have to be borrowed to fill in the picture. The best overview of physics and mechanics for the period remains J. L. Heilbron, *Electricity in the Seventeenth and Eighteenth Centuries: A Study of Early Modern Physics*, Berkeley, University of California Press, 1979. For Spain see David Goodman, "Science and the Clergy in the Spanish Enlightenment," *History of Science*, 21 (1983): 111–40. Germany now can be approached through the excellent work of Richard L. Gawthrop, *Pietism and the Making of Eighteenth-Century Prusssia*, Cambridge, Cambridge University Press, 1993. The Dutch Republic can be approached initially in Margaret C. Jacob and Wijnand Mijnhardt, eds., *The Dutch Republic in the Eighteenth Century. Enlightenment, Decline and Revolution*, Ithaca, N.Y., Cornell University Press, 1993. A good place to begin with French science is R. Rappaport, "Government Patronage of Science in Eighteenth Century France," *History of Science*, 8 (1969): 119–36. A good general survey of the Austrian Netherlands appeared in 1983: H. Hasquin, ed., *La vie culturelle dans nos provinces au XVIIIe siècle*, Brussels, Credit Communal de Belgique. An indispensable bibliography is W. Baeten *et al.*, eds., *Belgie in de 18de eeuw: Kritische Bibliografie*, Brussels, 1983), published for the *Contact-groep 18de eeuw* and usable in French as well. The western colony of the Austrians receives an intelligent discussion in Franz A. J. Szabo, *Kaunitz and Enlightened Absolutism, 1753–1780*, Cambridge, Cambridge University Press, 1994.

Chapter 8

The whole of French inventiveness in the eighteenth century has now been mapped, with some good work also on the English, by Liliane Hilaire-Pérez, *Inventions et Inventeurs en France et en Angleterre au xviiiè siècle*, 4 vols., University of Lille, Doctorate de l'Université de Paris I Panthéon-Sorbonne-UFR d'Histoire, January 1994. Some of her work first appeared in "Invention and the State in

18th-Century France," *Technology and Culture*, 32 (1991): 911–31. The role of science in the French Revolution and the whole question of radical science can be approached through an old but good work, L. P. Williams, "The Politics of Science in the French Revolution," in M. Clagett, ed., *Critical Problems in the History of Science*, Madison, University of Wisconsin Press, 1959, pp. 291–308; and R. Darnton, *Mesmerism and the End of the Enlightenment in France*, Cambridge, Mass., Harvard University Press, 1968. On French science in general see: Thomas Hankins, *Science and the Enlightenment*, Cambridge, Cambridge University Press, 1985. All of J.R. Harris's work on technology transfer to France is important. Start with J. R. Harris, "Michael Alcock and the Transfer of Birmingham Technology to France before the Revolution," *Journal of European Economic History*, 15 (1986): 7–59. See also the excellent account in Gwynne Lewis, *The Advent of Modern Capitalism in France, 1770–1840. The Contribution of Pierre-François Tubeuf*, Oxford, Clarendon Press, 1993. The relationship between the Revolution and economic change should be approached through Jean-Pierre Hirsch, "Revolutionary France, Cradle of Free Enterprise," *American Historical Review*, 94 (1989): 1281–89. The Revolution and the spurt forward in science education is nicely summarized in Jean G. Dhombres, "French Textbooks in the Sciences 1750–1850," *History of Education*, 13 (1984): 153–161. See also Robert Fox, ed., *Technological Change*, London, Harwood, 1996.

Chapter 9

More basic research is needed on the day-to-day use of technical knowledge in the Industrial Revolution. The book to help with that research remains A. E. Musson and E. Robinson, *Science in the Industrial Revolution* (1969), cited earlier. A case study of one of the new sciences and its relation to industrialization is R. Porter, "The Industrial Revolution and the Rise of the Science of Geology," in M. Teich and R. M. Young, eds., *Changing Perspectives in the History of Science*, London, Heinemann, 1973, pp. 320–43; and see also A. Thackray, "Science and Technology in the Industrial Revolution," *History of Science*, 9 (1970): 76–89. For intelligent thoughts on science and agriculture see Simon Schaffer, "A Social History of Plausibility: Country, City and Calculation in Augustan Britain," in Adrian Wilson, ed., *Rethinking Social History*, Manchester, Manchester University Press, 1993.

Index